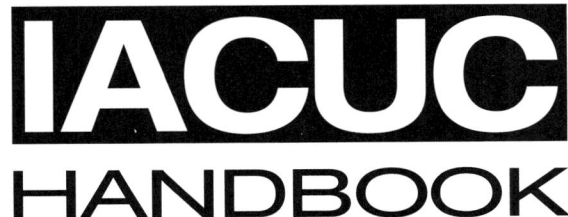

The IACUC HANDBOOK

Edited by
**Jerald Silverman
Mark A. Suckow
Sreekant Murthy**

CRC Press
Boca Raton   London   New York   Washington, D.C.

Library of Congress Cataloging-in-Publication Data

The IACUC handbook / edited by Jerald Silverman, Mark A. Suckow, Sreekant Murthy.
    p. cm.
    Includes bibliographical references and index.
    ISBN 0-8493-1685-5 (alk. paper)
    1. Animal welfare. 2. Laboratory animals. 3. Animals--Treatment. 4. National Institutes of Health (U.S.). Institutional Animal Care and Use Committee. I. Silverman, Jerald. II. Suckow, Mark A. III. Muthy, Sreekant. IV. National Institutes of Health (U.S.). Institutional Animal Care and Use Committee.
    HV4708.123 2000
    179'.4—dc21   99-089619
                                                                                                         CIP

This book contains information obtained from authentic and highly regarded sources. Reprinted material is quoted with permission, and sources are indicated. A wide variety of references are listed. Reasonable efforts have been made to publish reliable data and information, but the author and the publisher cannot assume responsibility for the validity of all materials or for the consequences of their use.

Neither this book nor any part may be reproduced or transmitted in any form or by any means, electronic or mechanical, including photocopying, microfilming, and recording, or by any information storage or retrieval system, without prior permission in writing from the publisher.

All rights reserved. Authorization to photocopy items for internal or personal use, or the personal or internal use of specific clients, may be granted by CRC Press LLC, provided that $.50 per page photocopied is paid directly to Copyright Clearance Center, 222 Rosewood Drive, Danvers, MA 01923 USA. The fee code for users of the Transactional Reporting Service is ISBN 0-8493-1685-5/00/$0.00+$.50. The fee is subject to change without notice. For organizations that have been granted a photocopy license by the CCC, a separate system of payment has been arranged.

The consent of CRC Press LLC does not extend to copying for general distribution, for promotion, for creating new works, or for resale. Specific permission must be obtained in writing from CRC Press LLC for such copying.

Direct all inquiries to CRC Press LLC, 2000 N.W. Corporate Blvd., Boca Raton, Florida 33431.

Trademark Notice: Product or corporate names may be trademarks or registered trademarks, and are used only for identification and explanation, without intent to infringe.

Visit the CRC Press Web site at www.crcpress.com

© 2000 by CRC Press LLC

No claim to original U.S. Government works
International Standard Book Number 0-8493-1685-5
Library of Congress Card Number 99-089619
Printed in the United States of America   3 4 5 6 7 8 9 0
Printed on acid-free paper

Preface

The concept of regulatory oversight as a means to facilitate the welfare of animals used in research has existed for approximately 120 years, beginning with the 1876 Cruelty to Animals Act in England. In the U.S., somewhat disjointed local efforts were undertaken to ensure animal welfare. These efforts met with varied degrees of success.

The Health Research Extension Act of 1985 directed the National Institutes of Health to construct guidelines and recommendations for development of institutional committees that would oversee the use of vertebrate animals in teaching, testing, and research on a local level. The U.S. Department of Agriculture was quick to follow with similar requirements. In this way, the Institutional Animal Care and Use Committee, or IACUC, was born.

The 15 years since the inception of requirements for establishing the IACUC has seen continued evolution of the issues addressed by IACUCs and the means by which review of activities is conducted. For example, the explosion of transgenic technology has presented IACUCs with new challenges in terms of ensuring animal health and welfare. There also has been ongoing development of alternative methods which either replace animal use or significantly reduce animal numbers or distress. This has led investigators to invest deliberate time and effort into the search for alternatives and, whenever possible, to incorporate alternative methods into their research.

Central to the mission of the IACUC is training. The IACUC must ensure that investigators, research technicians, animal care personnel, and others are adequately trained to conduct research with animals in a proper and humane manner. However, it is equally critical that the IACUC itself be adequately trained and knowledgeable with regard to successful interpretation and implementation of regulations and guidelines for proper animal care and treatment. This mission must be accomplished without impeding the scientific enterprise.

The intent of this book is to address those questions and problems often confronting IACUCs and give accurate and succinct answers to them. *The IACUC Handbook* is divided into distinct chapters. For most chapters, the information is provided in "question and answer" format, with the authors offering their personal opinions against a background of regulatory information. We have arbitrarily included, under the regulatory heading, pertinent sections from the Animal Welfare Act and its regulations, the Health Research Extension Act of 1985, the Public Health Service Policy on Humane Care and Use of Laboratory Animals, the 1996 edition of the *Guide for the Care and Use of Laboratory Animals,* and selected publications intended to clarify the Animal Welfare Act regulations and the Public Health Service Policy. Many authors also have provided results from informal surveys as examples of the approaches institutions have taken to solve complex issues. These surveys are not scientific; however, they serve to provide the reader with additional opinions.

This book could not have been written without the chapter authors, chapter contributors, and the many laboratory animal scientists who provided information for the informal surveys. We thank all of you. We would like to give special recognition and thanks to Dr. W. Ron DeHaven and the staff of APHIS/AC, Dr. Nelson Garnett and the staff of NIH/OPRR, and, in particular, Dr. Carol Wigglesworth of NIH/OPRR. APHIS/AC and NIH/OPRR judiciously reviewed every chapter, providing invaluable advice and corrections. Although we have attempted to make the information provided in this book as accurate as possible, we recognize that errors may occur. To that end, we encourage you, the reader, to provide feedback to any of the editors and chapter authors, and to suggest additional questions that might be addressed in future editions of this book.

There can be little question that the role and impact of the Institutional Animal Care and Use Committee will continue to grow in scope and complexity. We hope that this book will provide a foundation — a solid footing — for those attempting to understand the many and varied responsibilities of this committee.

Jerald Silverman, D.V.M.
Mark A. Suckow, D.V.M.
Sreekant Murthy, Ph.D.

Introduction

It has been approximately 13 and 8 years, respectively, since the implementation of the current PHS Policy and Part 2 of the USDA regulations, which first defined the federal requirements for Institutional Animal Care and Use Committees (IACUCs) to oversee certain specific animal care and use program functions. Since then, the IACUC has evolved as the premier instrument of animal welfare oversight within the majority of biomedical research institutions in the U.S. In the interim, the biomedical research community has gained a wealth of experience based, in many cases, on trial and error and on anecdotal information shared through informal communications and other means. Many excellent resources have been developed, but until now, few have provided specific opinion on "best practices" at the facility level.

The format of this book is unique. Answers to a series of practical everyday questions are addressed from the perspective of the applicable regulatory language, from the opinions of knowledgeable and experienced professionals in the field, and in the form of responses to informal surveys on selected institutional policies and practices.

The sections labeled "regulatory" have been reviewed by staff at NIH/OPRR and APHIS/AC for consistency with the Public Health Service Policy on Humane Care and Use of Laboratory Animals (PHS Policy) and the USDA Animal Welfare Act Regulations (AWAR). Every effort has been made to apply correct interpretations in the context of the specific issues being discussed. However, in this highly nuanced field, readers are cautioned not to apply these interpretations out of context or to extend them beyond their intended meaning. Accordingly, such interpretations should not be seen as formulating new federal regulations or policies, and readers are advised to refer to source documents and direct consultations with NIH/OPRR and APHIS/AC when in doubt.

The information provided in the sections labeled "survey" gives an interesting view of how several other institutions may be addressing certain issues, but a note of caution is advised here also. Some responses may appear to indicate deviations from federal policy and regulation. Because variables, such as the type of institution, the species involved, and the type of oversight that is applicable, are not always specified, some answers may reflect differences in applicability of rules or understanding of questions, rather than noncompliance.

For the experienced and sophisticated reader, this book provides a wealth of useful information and taps into the collective experience and wisdom of the various authors. Both NIH/OPRR and APHIS/AC commend the authors and editors for their outstanding efforts and for moving the biomedical research community forward in its formulation of best practices and commonly accepted professional guidance in this complex arena.

Nelson Garnett, D.V.M.
W. Ron DeHaven, D.V.M.

Editors

Jerald Silverman, D.V.M., received degrees in Vertebrate Zoology and Veterinary Medicine from Cornell University and a degree in Nonprofit Organization Management from the New School for Social Research. His background includes private veterinary practice and laboratory animal practice in private and academic research institutions. He is Professor of Pathology and Laboratory Medicine, Asst. Vice President for University Laboratory Animal Resources, and Director of the Master of Laboratory Animal Science program at MCP Hahnemann University in Philadelphia.

Dr. Silverman is a Diplomate of the American College of Laboratory Medicine and past president of the American Society of Laboratory Animal Practitioners. He has published over 40 research papers and book chapters, and has coordinated the Protocol Review column for *Lab Animal* magazine since 1988. His research interest focuses on nutritional means of cancer prevention. He advocates the advancement of the moral stature of research animals and spends his leisure time pursuing nature photography and birdwatching.

Mark A. Suckow, D.V.M., is the Director of the Freimann Life Science Center at the University of Notre Dame. He earned his Veterinary Medicine doctorate from the University of Wisconsin and completed postdoctoral training in laboratory animal medicine at the University of Michigan. He is a Diplomate of the American College of Laboratory Animal Medicine.

Dr. Suckow has published over 40 scientific papers and book chapters. He coauthored *The Laboratory Rabbit,* part of the *Laboratory Animal Pocket Reference Series* for which he serves as Editor-in-Chief. He was honored as the 1996 Young Investigator of the Year by the American Association for Laboratory Animal Science, and in 1998 with the Excellence in Research Award by the American Society of Laboratory Animal Practitioners.

Sreekant Murthy, Ph.D., received his Ph.D. (1973) in Pharmacognosy (Natural Product Chemistry) from Philadelphia College of Pharmacy and Science and his postdoctoral training (1973–1974) at the Clinical Research Center, Temple University Hospital, Philadelphia. Dr. Murthy joined Hahnemann Medical College and Hospital in 1974 and currently holds the title of Associate Vice President for Research Compliance for both Drexel and MCP Hahnemann Universities (formerly known as Allegheny University for the Health Sciences).

Dr. Murthy chaired two IACUCs from 1995 to 1999. He has done both human clinical and basic bench research and his current focus is on inflammatory bowel disease. He has published over 60 articles in peer-reviewed journals, organized both national and international symposiums on inflammatory disorders, and has written a number of book chapters. He currently is responsible for regulatory compliance of Institutional Review Boards, Biosafely Committee, Radiation Safely Committee, and IACUC for both Drexel and MCP Hahnemann Universities.

Authors

David R. Archer, Ph.D. Assistant Professor of Pediatrics, School of Medicine, Emory University, Atlanta, Georgia

Kathryn A. L. Bayne, M.S., Ph.D., D.V.M., Dipl. A.C.L.A.M. Associate Director, AAALAC International, Rockville, Maryland

Stephen K. Curtis, D.V.M., Dipl. A.C.L.A.M. Associate Director, Department of Laboratory Animal Services, Rutgers, The State University of New Jersey, Piscataway, New Jersey

Peggy J. Danneman, V.M.D., M.S., Dipl. A.C.L.A.M. Acting Director of Laboratory Animal Sciences, The Jackson Laboratory, Bar Harbor, Maine

Elizabeth J. Dawe, D.V.M., Dipl. A.C.L.A.M. Director, Surgical Research Services, Wayne State University, School of Medicine, Detroit, Michigan

Ralph B. Dell, M.D. Director, Institute for Laboratory Animal Research, National Research Council, National Academy of Sciences, Washington, D.C.

Bernard J. Doerning, D.V.M., Dipl. A.C.L.A.M. Human and Environmental Safety, The Procter & Gamble Co., Cincinnati, Ohio

Paul Flecknell, M.A., Vet. M.B., Ph.D., D.L.A.S., Dipl. E.C.V.A., M.R.C.V.S. Professor of Laboratory Animal Science, Comparative Biology Centre, Medical School, University of Newcastle, Newcastle, United Kingdom

Diane J. Gaertner, D.V.M., Dipl. A.C.L.A.M. Director, Institute for Animal Studies, Associate Professor of Microbiology and Immunology, Albert Einstein College of Medicine, Bronx, New York

Cynthia Gillett, D.V.M. Director, Research Animal Resources, University of Minnesota, Minneapolis, Minnesota

Edward J. Gracely, Ph.D. Associate Professor of Family, Community, and Preventive Medicine, MCP Hahnemann School of Medicine, MCP Hahnemann University, Philadelphia, Pennsylvania

Michael J. Huerkamp, D.V.M., Dipl. A.C.L.A.M. Assistant Director of Animal Resources, Associate Professor of Pathology and Laboratory Medicine, School of Medicine, Emory University, Atlanta, Georgia

Neil S. Lipman, V.M.D. Director and Associate Professor of Pathology, Weill Medical College of Cornell University, and the Memorial Sloan-Kettering Cancer Center, New York, New York

Kathleen D. Moody, V.M.D., M.S., Dipl. A.C.L.A.M. Consultant in Laboratory Animal Medicine, Assistant Clinical Professor of Comparative Medicine, Yale School of Medicine, Newton, Connecticut

Sreekant Murthy, Ph.D. Prof. of Medicine, Assoc. Vice President for Research Compliance, MCP Hahnemann University, Philadelphia, Pennsylvania

Christian E. Newcomer, V.M.D., M.S., Dipl. A.C.L.A.M. Director, Division of Laboratory Animal Medicine, Research Associate Professor, Department of Pathology and Laboratory Medicine, School of Medicine, The University of North Carolina, Chapel Hill, North Carolina

Gwenn S. F. Oki, M.P.H. Director, Research Subjects Protection, City of Hope National Medical Center, and The Beckman Research Institute, Duarte, California

Scott E. Perkins, V.M.D., M.P.H. Assistant Director of RARC, Assistant Professor of Surgery, Weill Medical College of Cornell University, and Memorial Sloan-Kettering Cancer Center, New York, New York

Ernest D. Prentice, Ph.D. Associate Dean for Research, University of Nebraska Medical Center, Omaha, Nebraska

Harry Rozmiarek, D.V.M., Ph.D. Professor and Chief, Laboratory Animal Medicine, University Veterinarian, University of Pennsylvania, Philadelphia, Pennsylvania

Howard G. Rush, D.V.M. Associate Professor and Assistant Director, Unit for Laboratory Animal Medicine, University of Michigan, Ann Arbor, Michigan

Sallie Thieme Sanford, J.D. Former Assistant Attorney General, State of Washington/University of Washington Division, University of Washington, Seattle, Washington

Jerald Silverman, D.V.M., M.P.S.   Assistant Vice President, University Laboratory Animal Resources; Professor, Pathology and Laboratory Medicine; Director, Master of Laboratory Animal Science Program, MCP Hahnemann University, Philadelphia, Pennsylvania

Jane E. Simpson* Coordinator, Animal Care and Use Committee, University of California, Berkeley, California

Joseph S. Spinelli, D.V.M.   Consultant in Laboratory Animal Medicine, Director Emeritus, Animal Care and Cell Culture Facility, University of California, San Francisco, California

Susan Stein, D.V.M., M.S., Dipl. A.C.L.A.M.   University Laboratory Animal Resources, Michigan State University, East Lansing, Michigan

Harold F. Stills, Jr., D.V.M.   Director, Laboratory Animal Resources, Professor, Microbiology and Immunology, School of Medicine, Wright State University, Dayton, Ohio

Mark A. Suckow, D.V.M., Dipl. A.C.L.A.M.   Director, Freimann Life Science Center, University of Notre Dame, Notre Dame, Indiana

Farol N. Tomson, D.V.M., M.B.A.   Associate Director, Laboratory Animal Resources, University of Florida, Gainesville, Florida

Richard C. Van Sluyters, O.D., Ph.D.   Professor of Optometry and Vision Science; Chair, Animal Care and Use Committee; Faculty Assistant to the Vice Chancellor for Research, University of California, Berkeley, California

Lisa A. Vincler, J.D.   Assistant Attorney General, State of Washington/University of Washington Division, Faculty Associate, Department of Medical History and Ethics, University of Washington School of Medicine, Seattle, Washington

Stefan Wagener, Ph.D., CBSP   Biological Safety Officer, Office of Radiation, Chemical, and Biological Safety, Michigan State University, East Lansing, Michigan

Patricia A. Ward, B.S., M.P.A., L.A.T.G.   Coordinator, Research Animal Standards and Staff Development, Unit for Laboratory Animal Medicine, University of Michigan, Ann Arbor, Michigan

---

* Current Affiliation: Clinical Research Coordinator, Department of Neurology, University of California, San Francisco, California.

Chapter Contributors

Robert J. Ceru   Michigan State University, East Lansing, Michigan

Kristin Erickson   Michigan State University, East Lansing, Michigan

Deborah M. Faryna, B.S.   National Institute of Mental Health, Bethesda, Maryland

Molly Greene.   University of Texas Health Science Center, San Antonio, Texas

Debbie Hampstead, B.S., R.L.A.T.G.   The University of Tennessee, Knoxville, Tennessee

Richard M. Harrison, Ph.D.   Tulane Regional Primate Research Center, Covington, Louisiana

Todd A. Jackson, D.V.M., Dipl. A.C.L.A.M.   University of Cincinnati, Cincinnati, Ohio

Beverly Keniston, B.A., L.A.T.G.   Boston University, Boston, Massachusetts

William W. King, D.V.M., Ph.D., Dipl. A.C.L.A.M.   Hines Veterans Administration Hospital, Hines, Illinois

Eifaang Li, D.V.M., M.P.H.   University of California, Los Angeles, California

Michael Mann, Ph.D.   University of Nebraska Medical Center, Omaha, Nebraska

Joy A. Mench, Ph.D.   University of California, Davis, California

Gregory R. Reinhard, D.V.M., Dipl. A.C.L.A.M.   Schering-Plough Research Institute, Kenilworth, New Jersey

Philip Tillman, D.V.M.   University of California, Davis, California

Julie Watson, M.A., Vet. M.B., M.R.C.V.S., Dipl. A.C.L.A.M.   Baltimore, Maryland

Contents

Abbreviations Used in This Book xix

1. Origins of the IACUC 1
   Harry Rozmiarek

2. Circumstances Requiring an IACUC 11
   Richard C. Van Sluyters, Jane E. Simpson, and Ralph B. Dell

3. Creation of an IACUC 19
   Richard C. Van Sluyters, Jane E. Simpson, and Ralph B. Dell

4. Reporting Lines of the IACUC 31
   Richard C. Van Sluyters, Jane E. Simpson, and Ralph B. Dell

5. General Composition of the IACUC and Specific Roles of the
   IACUC Members 37
   Christian E. Newcomer

6. Frequency and Conduct of Regular IACUC Meetings 59
   Sreekant Murthy

7. General Format of IACUC Protocol Forms 75
   Christian E. Newcomer

8. Submission and Maintenance of IACUC Protocols 87
   Farol N. Tomson

9. General Concepts of Protocol Review 107
   Ernest D. Prentice and Gwenn S. F. Oki

10. Amending IACUC Protocols 133
    Diane J. Gaertner and Kathleen D. Moody

11. Continuing Review of Protocols 141
    Gwenn S. F. Oki

12. Justification for the Use of Animals 149
    Joseph S. Spinelli

13. Justification of the Number of Animals to Be Used 167
    Edward J. Gracely

14. Animal Acquisition and Disposition 179
    Michael J. Huerkamp and David R. Archer

15. Animal Housing and Use Sites 207
    Cynthia Gillett

16. Pain and Distress 221
    Paul Flecknell and Jerald Silverman

17. Euthanasia 251
    Peggy J. Danneman

18. Surgery 277
    Elizabeth J. Dawe

19. Antigens, Antibodies, and Blood Collection 303
    Harold F. Stills, Jr.

20. Occupational Health and Safety 319
    Stefan Wagener and Susan Stein

21. Personnel Training 345
    Howard G. Rush

22. Academic Freedom and Proprietary Information 365
    Sallie Thieme Sanford and Lisa A. Vincler

23. General Concepts of the Facility Inspection and Program
    Review 385
    Stephen K. Curtis

24. Inspection of Animal Housing Areas 401
    Patricia A. Ward

25. Inspection of Individual Laboratories 419
    Neil S. Lipman and Scott E. Perkins

26. Inspection of Surgery Areas 431
    Scott E. Perkins and Neil S. Lipman

27. Assessment of Veterinary Care 447
    Mark A. Suckow and Bernard J. Doerning

28. Laboratory Animal Enrichment 465
    Kathryn A. L. Bayne

29. Animal Mistreatment and Protocol Noncompliance 481
    Jerald Silverman

Appendix A. Animal Welfare Act 505

Appendix B. Animal Welfare Act Regulations 507

Appendix C. Health Research Extension Act of 1985 517

Appendix D. Public Health Service Policy on Humane Care and Use
             of Laboratory Animals 521

Appendix E. U.S. Government Principles for the Utilization and
             Care of Vertebrate Animals Used in Testing, Research,
             and Training 531

Index 533

Abbreviations Used in This Book

AAALAC Association for Assessment and Accreditation of Laboratory Animal Care International. A voluntary accreditation organization that works with institutions and researchers to ensure high standards of animal care and use.

APHIS Animal and Plant Health Inspection Service. Part of the U.S. Department of Agriculture.

APHIS/AC Animal Care division of APHIS. Administers the Animal Welfare Act (AWA) and enforces the AWA through a system of licensing, registration, and inspections.

AV Attending Veterinarian, as defined in the AWAR.

AWA Animal Welfare Act. Public Law 89-544. Title 7 of the U.S. Code (7 USC) §§2131 et. seq. The law that, in part, is intended to ensure that animals used in research facilities (as defined therein) are provided humane care and treatment.

AWAR Animal Welfare Act Regulations. Detailed regulations and standards for implementing the AWA, found in Title 9 of the Code of Federal Regulations (9 CFR), Chapter 1, Subchapter A, Parts 1, 2, and 3.

Chair Chairperson of the IACUC.

Guide Institute of Laboratory Animal Resources, National Research Council, *Guide for the Care and Use of Laboratory Animals,* National Academy Press, Washington, D.C., 1996.

HREA Health Research Extension Act of 1985. Public Law 99-158, November 20, 1985. Section 495. Law that directs the Secretary of the Department of Health and Human Services to establish guidelines for the proper care and use of animals used in PHS-supported biomedical and behavioral research.

IACUC Institutional Animal Care and Use Committee. A committee qualified through the experience and expertise of its members that oversees its institution's animal program, facilities, and procedures.

IO Institutional Official, as defined in the AWAR and PHS Policy.

NIH National Institutes of Health. Part of the U.S. Department of Health and Human Services.

NIH/OPRR Office for Protection from Research Risks of the NIH. Its Division of Animal Welfare has the responsibility for developing, monitoring, and exercising compliance with the PHS Policy.

Opin. Opinion of the chapter's author(s).

PHS U.S. Public Health Service. Part of the Department of Health and Human Services which includes the NIH, Food and Drug Administration, and Centers for Disease Control and Prevention, among others.

PHS Policy   Public Health Service Policy on Humane Care and Use of Laboratory Animals. This policy implements the HREA.
PI   Principal Investigator.
Reg.   Regulatory information offered as guidance to the reader.
SOP   Standard Operating Procedures. A set of written procedures developed by an institution, often used to define how IACUC or animal facility activities will be performed.
Surv.   Survey performed by the chapter's author.
U.S. Government Principles   U.S. Government Principles for the Utilization and Care of Vertebrate Animals Used in Testing, Research, and Training. The nine principles developed by the U.S. Government are incorporated into the PHS Policy.

1
Origins of the IACUC

Harry Rozmiarek

Introduction

The need to provide regulatory oversight to assure animal welfare in the research laboratory was recognized in Victorian England over 120 years ago and led to the 1876 Cruelty to Animals Act. Modifications of this law exist today, and include "Home Office" oversight and registration of individuals who conduct research procedures using animals. In the U.S., the earliest national law addressing animal welfare was the 28-hour law enacted in 1887. This law primarily governs animals being transported for market and does not specifically address animals in the research laboratory.

The New York Anticruelty Bill of 1866 and the formation of Humane Societies in New York (1866), Pennsylvania (1867), Massachusetts (1868), and Washington, D.C. (1870) addressed the use of animals in research, as did the formation of the American Antivivisection Society in Jenkintown, PA in 1883. Nevertheless, none of these actions had national authority or scope. The first official law addressing the care and use of laboratory animals in the U.S. was the Laboratory Animal Welfare Act of 1966. An amendment in 1970 changed the name to the Animal Welfare Act, by which it is known today.

Prior to the 20th century the responsibility for animals used in research in the U.S. was placed directly in the hands of the research investigator. The quality of animal care and animal welfare varied tremendously between research laboratories. Laboratories even within the same school or institution had different animal care policies and standards of care. Animal care was frequently the responsibility of a diener, who provided food, basic care, and much of the animal manipulation. Basic nutrition and sanitation were often inadequate and no environmental or housing standards were available. In many instances, the animal handling staff meant well but was not adequately trained or qualified. In others, animal welfare and even adequate care was of low priority. Rodents were commonly fed cereal grains and dogs and cats received table scraps. Back yard and part-time breeders provided rodents of variable genetic background, and the animals harbored a variety of parasites, pathogenic bacterial agents, and viruses. Long-term studies with rodents were impossible, as animals suffered from malnutrition and clinical disease and accurate research data was difficult to obtain. Dogs and cats were obtained individually, often provided in crates or gunnysacks and obtained from variable sources. Breeders of good-quality, standardized animals for research did not exist.

One of the first records of a major research institution in the U.S. providing central management and care for research animals directed by a highly trained and qualified individual was the University of Chicago in 1945, when Dr. Nathan Brewer was hired to direct a centrally managed program of laboratory animal care. Dr. Brewer had provided animal facility management at the University of Chicago as early as 1930, but had since received formal training as a veterinarian and obtained a Ph.D. in physiology. A number of other institutions in the country were following similar paths as the need for better and more standardized animal care and welfare was recognized. However, it was not until the 1940s that a number of professionals with a major interest in laboratory animal care began meeting in the Chicago area; this was undoubtedly stimulated and encouraged by Dr. Brewer. Among these was Dr. W. T. S. Thorpe, director of the newly organized Laboratory Aids Branch at the NIH, which eventually developed into the National Center for Research Resources. In 1950 these meetings led to the formal establishment of the Animal Care Panel, which in 1967 became the American Association for Laboratory Animal Science (AALAS). Dr. Brewer served as the first national president of this organization and with others stimulated many other activities and organizations instrumental in shaping animal care and use policies and laws in this country.[1]

Officially established and conducting annual conferences on animal care, the Animal Care Panel appointed a committee in 1961, headed by Dr. Bennett Cohen, charged with providing animal care and use guidelines for research facilities. Their product was the publication in 1963 of the first edition of the *Guide*. Future editions of this publication were supported by the NIH and guided by the Institute for Laboratory Animal Research (ILAR). The National Academy Press under the auspices of the National Research Council published the 7th and current edition in 1996.[2] This single document serves as the "bible" for laboratory animal care and use polices and guidelines in the U.S. It is excerpted from and referenced in all major guidelines and regulations, and has been translated into and is being published in at least nine other languages. Over 400,000 copies have been distributed throughout the world.

In 1963, the Animal Care Panel saw a need to evaluate the standard of animal care and use that was developed in their new *Guide* and appointed an Animal Accreditation Committee. This committee visited six major institutions in the U.S. that conducted research with animals and had volunteered to have their policies and programs evaluated by the standards in the new

Guide. This committee soon saw the need to function independently, and in 1965 incorporated in the State of Illinois as the American Association for the Accreditation of Laboratory Animal Care (AAALAC). The first two institutions accredited by AAALAC were the University of Louisville in Kentucky and Howard University in Washington, D.C. This independent accrediting agency changed its name in 1997 to the Association for Assessment and Accreditation of Laboratory Animal Care International. Still using the *Guide* as its primary reference document, AAALAC now has over 600 institutions in the U.S. and 11 foreign countries under its umbrella of accreditation. The AAALAC program has always expected a mechanism of regular review and assurance as part of a good animal care and use program.

Prior to 1960 the U.S. didn't have a national law addressing laboratory animals. Local humane societies were actively promoting protection for pets and sometimes for farm animals as well. Concurrently, the scientific community was improving the quality of animal care and well-being in the research laboratory. The increasing need for research dogs and cats was partially fulfilled by animal dealers who obtained these animals in various ways and sold them to research laboratories. A series of articles and news reports on animal neglect and abuse at the hands of animal dealers culminated in a major article in *Life* magazine in 1965. This was accompanied by pictures and a cover photo showing animal neglect. It suggested a need for regulation and a system of enforcement. Catalyzed in part by this article, The Laboratory Animal Welfare Act was passed by Congress in 1966 (Public Law 89-544), and for the first time in this country there were legal standards for laboratory animal care and use. The USDA was named the responsible agency for implementing and enforcing this new law and it promptly began providing regulations.[3] Although the Act covers all laboratory animals, it was enacted primarily to protect dogs and cats and to counteract the business of pets being stolen by dealers and then sold for research purposes. Research laboratories and dealers were now required to register their facilities or be licensed, and undergo inspection by USDA personnel who made out reports and issued citations for noncompliance. These early inspections did not extend into the research laboratory where animal care and use remained under the direction of the research investigator.

A number of amendments to the Animal Welfare Act have led to regulations that now include animal transportation, marine mammals, and animals in the research laboratory. However, the regulations still exclude common laboratory rats (*Rattus norvegicus*) and mice (*Mus musculus*), birds, and farm animals used in production agriculture. The most probable reason for these exclusions is insufficient funds and staff. As nearly 90% of laboratory animals are rats and mice, their inclusion would have a significant impact on the workload of the USDA inspection staff.

The first PHS Policy on Humane Care and Use of Laboratory Animals went into effect in 1973, was revised in 1979 and again in 1986.[4] All PHS policies on this subject evolved from an NIH policy published in 1971. That policy referenced several NIH and PHS statements on appropriate care and humane treatment of laboratory animals, among them the *Guide*. It introduced the animal care committee as a means of local assurance and included all live vertebrate animals. Each revision placed more specific responsibilities on that committee.

The 1971 NIH Policy stated that institutions or organizations using warm-blooded animals in research or teaching supported by NIH grants, awards, or contracts would "assure the NIH that they will evaluate their animal facilities in regard to the maintenance of acceptable standards for the care, use, and treatment of such animals." The institution could either show that it was accredited by a recognized professional laboratory animal accrediting body or that it had established its own committee to carry out that assurance function. The minimum number of committee members was not stated, but at least one member had to be a doctor of veterinary medicine. Guidelines to be followed include the "NIH Guide," all applicable portions of the AWA, and an appended set of guidelines known as the "Principles for the Use of Laboratory Animals." The committee was required to inspect the animal facilities of the institution at least once a year and report its findings and recommendations to responsible officials of the institution. Records of activities and recommendations were required and were to be available for inspection by NIH representatives. Under this policy, institutions accredited by AAALAC were not required by NIH to have an evaluation committee.

The first PHS Policy replaced the NIH policy on July 1, 1973 and continued to accept AAALAC accreditation in lieu of an institutional committee. The minimum number of committee members was now set at three, and unaccredited or partially accredited institutions were required to have a committee. Only when institutions used a "significant number of animals" was a veterinarian required on the committee. If the number of animals used in activities supported by the Department of Health, Education, and Welfare (DHEW) was not "significant," a veterinarian was not required, but one of the committee members must then be "a scientist with demonstrated expertise in the care and use of laboratory animals." If such a person was not available, a veterinarian on a consultant basis was an acceptable alternative. A "significant number of animals" was not defined, with final determination made by the DHEW from animal inventory information provided as part of the institution's assurance statement. The policy for institutional review of applications and proposals stated, "Grantee and contractor institutions are encouraged to review their applications and proposals in the light of the pertinent provisions of the Animal Welfare Act, the standards set by the Institute of Laboratory Animal Resources (the NIH Guide), and the DHEW Principles for the Use of Laboratory Animals, and to familiarize their staff with these provisions, standards, and principles." However, there was no requirement under this policy that institutional committees perform review of individual proposals or regularly provide to the DHEW summaries or certifications of such committee actions. The policy did not specify who should perform the reviews. Under a different action, there was a requirement to keep records of committee activities for at least 3 years after the budget period. Assurance statements had to list the facilities and components of the program and committee. Activities included periodic facility inspections and reports to responsible officials at least once a year.

The PHS Policy was revised on January 1, 1979 and now required all animal-using grantee institutions to have "a committee to maintain oversight of its animal care program." The policy also required an institution to submit an assurance statement to the NIH/OPRR and have it found to be acceptable before receiving a PHS grant for studies in which animals or animal facilities were used. In addition, to assure compliance with the edition of the *Guide* then in use, AWA, and other applicable laws and regulations, it also required that "such assurance must also indicate that the institution has appointed and will maintain a committee to maintain oversight of its animal care program."

AAALAC accreditation was again recommended as the best means of demonstrating conformance with good animal care and use provisions. No explicit name was given to this committee, but the key words and all references to it were "animal care." Grantee institutions, as in the past, were obliged to keep records of committee activities, including recommendations and determinations, with these records being available for inspection by authorized PHS officials. Review of individual proposals or projects by the institution's committee was encouraged, but not required. Even though the review remained merely a recommendation, the suggestion of the committee as a reviewing body further strengthened its role. A list of committee members was required on the assurance form and information on each member was to include degrees, position, title, and a short description of the member's relevant background. A sample statement on the form was preceded by the following:

> "The samples given below for committee members are not intended to dictate numbers of or qualifications for committee members. However, except in unusual circumstances, the committee should be of at least five members and include at least one veterinarian. Any such unusual circumstances should be explained in a statement accompanying the assurance."

The following was contained in the sample statement:

> "We have appointed and will maintain a committee of at least five members to maintain oversight of our animal care program. The members have appropriate education and experience to perform their duties with respect to the types of animals and species used and the kinds of projects to be undertaken. If the conduct of a specific project is to be reviewed, the quorum will not include any member having an active role in the project. Changes in membership will be reported annually to the Office for Protection from Research Risks (OPRR), National Institutes of Health (NIH), and PHS."

Three options were provided for reporting an institution's degree of compliance: AAALAC accreditation, certification by its own institutional committee, or committee recommendations of improvements needed for compliance.

If the committee recommended any immediate or future improvements, the NIH/OPRR would expect to receive an annual report of the progress of these improvements toward compliance. The policy now required that the committee would review its facilities and procedures at least once a year and that responsible officials would receive and consider all reports from the committee. They also would make an annual review of committee activities for compliance and would keep records of committee meetings and related administrative actions. A new assurance statement was required every 5 years.

After allowing 5 years to study the effectiveness of the assurance system, it was concluded in 1984 that the committees frequently seemed "less than fully assertive" in carrying out their responsibility.[5] A recommendation following this study suggested that "the PHS policy should be further modified to define more precisely the responsibilities of the awardee institutions, particularly the role of the animal care committee. It is imperative that the experience and expertise of the members of such committees be used to conduct full and effective reviews of proposals involving research with animals. The appointment of a nonscientist and an individual unaffiliated with the institution should be given serious consideration."[5] This study resulted in the latest revision of the PHS policy, which further defines and outlines requirements of an animal care committee. This policy now includes provisions of the Health Research Extension Act of 1985 that was enacted on November 20, 1985 as Public Law 99-158. The most significant changes required by this action are that the policy now applies to research that the PHS conducts intramurally, and that the IACUC is appointed by the chief executive officer of the institution.[4] The policy requires that the program description include an explanation of the training or instruction available to scientists, animal technicians, and other personnel involved in animal care, treatment, or use. This training or instruction must include information on the humane practice of animal care and use and the concept, availability, and use of research or testing methods that minimize the number of animals required to obtain valid results and minimize animal distress. The IACUC must now evaluate and prepare reports on all of the institution's programs and facilities (including satellite facilities) for activities involving animals at least twice instead of once each year. The IACUC, through the IO, is responsible for reporting requirements; minority views filed by members of the IACUC must be included in reports filed under this policy.

While the scientific community and the public sector were gradually evolving guidelines and policies to assure animal welfare and good animal care,[6-15] research conducted by the Department of Defense (DOD) had similar concerns. As early as 1961, the Defense Department issued a Policy on Experimental Animals[16-18] which directed that "all aspects of investigative programs involving the use of laboratory animals and sponsored by Department of Defense agencies will be conducted according to the Principles of

Laboratory Animal Care as promulgated by the National Society for Medical Research." While this early policy provided few specifics, a number of revisions followed and included all animals used in DOD laboratories, both in this country and abroad. A joint regulation issued by the Army, Navy, Air Force, Defense Nuclear Agency, and Uniformed Services University on June 1, 1984, entitled "The Use of Animals in DOD Programs," required that "all DOD organizations having animals (other than military working, recreational, and ceremonial) will seek accreditation by AAALAC."[19] It further required that "the local commanders of each DOD organization conducting or sponsoring activities involving animals in RDTE, clinical investigations, diagnostic procedures, or instructional programs will form a committee to oversee the care and use of animals." Such committees were to be appointed by the local commander, include a doctor of veterinary medicine, be made up of at least three members, and to review protocols as well as assure compliance with policies, standards, and regulations. The concept and practice of such committees to review and assure appropriate animal care and use, while not known by the acronym IACUC, were in place at many military installations prior to their being regularly formed at most academic institutions.

Public groups concerned about the acquisition and welfare of animals destined for research use continued their activities, which ranged from local and national humane societies concerned about animal welfare and well-being to antivivisection groups opposed to the use of animals for any reason. The activity of such groups seemed to escalate and become more vocal in the early 1980s. This activity peaked in a series of illegal break-ins and vandalism by a terrorist group, and was brought to the forefront of public opinion soon after two incidents involving "animal cruelty" and insensitivity in two well-known research institutions.[20,21] This climate raised the public concern and visibility of animals in research and served as a catalyst for amendments and clarifications of guidelines and regulations providing for animal welfare and animal well-being.

The 1986 PHS Policy first described the IACUC in its present form and membership, and only provided general guidance on how such a committee should be formed and operated. The Scientists Center for Animal Welfare (SCAW) was instrumental in providing early guidance to institutions on IACUC functions and organization through regional conferences and workshops, culminating in a special AALAS publication entitled, *Effective Animal Care and Use Committees*.[22] Since 1983, training and regular conferences of this type are provided by annual animal care and use conferences sponsored by the Public Responsibility for Medicine and Research (PRIM&R), regional workshops supported by the NIH/OPRR, and numerous similar activities. In an amendment to the Animal Welfare Act, effective October 30, 1989, the USDA for the first time required that each registered research institution appoint an IACUC of not less than three members which "serves as the agent of the research facility that ensures that the facility is in full compliance with the Act." In fulfilling this responsibility, the committee is to "prepare evaluation reports of reviews and inspections and submit them to a designated institutional officer."[3] While originally borrowed from the human Institutional Review Board structure, the concept of local animal care committees to review and assure animal welfare and well-being is now common practice in the animal research community. The goal of each committee is compliance with guidelines and regulations while allowing for the flexibility of individual institutional tailoring to best meet unique needs of the institution. Active participation by research scientists allows for the unique needs of research investigators to be voiced, participation by unaffiliated members protects the public conscience, and veterinarians assure appropriate medical care and provisions. There is a continuing need for education to assure that this concept works as well as possible.

References

1. Rozmiarek, H., Current and future policies regarding animal welfare. *Invest. Radiol.*, 22, 175, 1986.
2. Committee to Revise the Guide for the Care and Use of Laboratory Animals, *Guide for the Care and Use of Laboratory Animals*, National Academy Press, Washington, D.C., 1996.
3. Office of the Federal Register, Code of Federal Regulations, Title 9, Animals and Animal Products, subchapter A, parts 1, 2, and 3, Animal Welfare, Washington, D.C., 1985.
4. U.S. Department of Health and Human Services, Public Health Service, National Institutes of Health, Institutional Animal Care and Use Committee Guidebook, NIH Publ. No. 92-3415, Washington, D.C., 1992.
5. Whitney, R.A., Jr., Animal care and use committees: history and current national policies in the United States, *Lab. Anim. Sci.*, Special issue:18, 1987.
6. *Guide for the Care and Use of Agricultural Animals in Agricultural Research and Teaching*, Consortium for Developing a Guide for the Care and Use of Agricultural Animals in Agricultural Research and Teaching, Champaign, IL, 1988.
7. Committee on Animal Research and Experimentation of the Board of Scientific Affairs, *Guidelines for Ethical Conduct in the Care and Use of Animals*, American Psychological Association, Washington, D.C., 1991.
8. U.S. Department of Health and Human Services, National Institutes of Health, Institutional Administrators' Manual for Laboratory Animal Care and Use, NIH Publication No. 88-2959, Washington, D.C., 1988.
9. ILAR Committee on the Use of Laboratory Animals in Biomedical and Behavioral Research, Use of Laboratory Animals in Biomedical and Behavioral Research, National Research Council, National Academy Press, Washington, D.C., 1988.
10. ILAR Committee on the Use of Animals in Precollege Education, Principles and Guidelines for the Use of Animals in Precollege Education, Institute for Laboratory Animal Resources, Washington, D.C., 1989.

11. Joint AAMC-AAU Committee, *Recommendations for Governance and Management of Institutional Animal Resources*, Association of American Medical Colleges and the Association of American Universities, Washington, D.C., 1985.
12. U.S. Department of Health and Human Services, National Institutes of Health, Preparation and Maintenance of Higher Mammals during Neuroscience Experiments, NIH Publication No. 91-3207, Washington, D.C., 1991.
13. Universities Federation for Animal Welfare, *Guidelines on the Care of Laboratory Animals and Their Use for Scientific Purposes*, Universities Federation for Animal Welfare, South Mimms, Potters Bar, U.K., 1989.
14. International Air Transportation Association, *Live Animal Regulations*, Montreal, 1986.
15. National Association for Biomedical Research, *State Laws Concerning the Use of Animals in Research*, National Association for Biomedical Research, Washington, D.C., 1987.
16. Department of Defense Directive 5129.1, Department of Defense Instruction, Washington, D.C., 1959.
17. Department of Defense Directive 51225.5, Department of Defense Instruction, Washington, D.C., 1961.
18. Policy on Experimental Animals in Department of Defense Research, Department of Defense Instruction, Washington, D.C., September 1, 1961.
19. Army Regulation 70-18, SECNAVINST 3900.388, AFR 100-2, DARPAINST 18, DNAINST 3216.1B, USUHSINST 3203. The Use of Animals in DOD Programs, United States Army, Washington, D.C., June 1, 1984.
20. Fraser, C., The raid at Silver Spring, *The New Yorker*, 66, April 19, 1993.
21. McCabe, K., Who will live, who will die, *The Washingtonian*, 21, 11, August 1986.
22. Orlans, F.B., Simmonds, R.C., and Dodds, W.J., Effective animal care committees, *Lab. Anim. Sci.* Special Issue, January 1987 (published in collaboration with the Scientists Center for Animal Welfare).

2
Circumstances Requiring an IACUC

Richard C. Van Sluyters, Jane E. Simpson, and Ralph B. Dell

Introduction

All of the human components of an animal care and use program — the IO, the AV, the animal care staff, and the IACUC — have important roles to play. Additionally, federal law, regulation, and policy also place the IACUC in a pivotal position for ensuring animal welfare. This chapter reviews the various circumstances in which institutions are required to appoint an IACUC. The fact that separate mandates to appoint an IACUC come from the USDA and the PHS can be a major source of confusion since the circumstances under which these two agencies require IACUCs differ. In an effort to alleviate this confusion, the reader is directed to the specific sections of the AWA, AWAR, and PHS Policy that describe the regulatory requirements for the IACUC. These regulatory requirements are then summarized. The remainder of the chapter provides the answers to a series of questions designed to reveal whether IACUCs are always required at research institutions, and whether they are ever required at grade schools, secondary schools, science fairs, zoos, aquaria, animal shelters, or humane societies.

2:1 What regulatory agencies require the appointment of an IACUC?

Reg. There are two: the USDA and the PHS (see 2:2). The USDA, through APHIS/AC, requires an IACUC at any institution that uses animals in research, provided the species is among those listed under the definition of "animal" in §1.1 of the AWAR (see 12:1). The PHS, through NIH/OPRR, requires an IACUC at any institution that conducts PHS-supported activities involving any live vertebrate animal.

Opin. Within APHIS/AC, IACUC activities are monitored by yearly (or more frequent) unannounced inspections by Veterinary Medical

Officers of APHIS/AC. Within the PHS, they are monitored by occasional random site visits and for-cause site visits by an *ad hoc* team assembled by the NIH/OPRR.

2:2 What specific documents describe the regulatory requirements for the IACUC?

Reg. The AWA is found in Title 7 of the U.S. Code (7 USC) §2131 et. seq. The 1985 amendments to the AWA require the appointment of an IACUC, mandate its minimum composition, and require it to perform at least semiannual inspections of animal facilities and study areas. Detailed regulations and standards for implementing the AWA are set forth by the USDA in Title 9 of the Code of Federal Regulations (9 CFR), Chapter 1, Subchapter A.

Opin. Although most of the regulations that specifically govern the IACUC are found in §2.31, references to the IACUC are scattered throughout the AWAR as follows:

Part 2 — Regulations

Subpart C — Research Facilities

- §2.31   Institutional Animal Care and Use Committee
- §2.33,a,3   Attending Veterinarian and Adequate Veterinary Care
- §2.35,a–§2.35,f   Recordkeeping Requirements
- §2.36,b,3   Annual Report
- §2.37   Federal Research Facilities
- §2.38,f,2,ii; §2.38,k,1   Miscellaneous

Part 3 — Standards

Subpart A — Specifications for the Humane Handling, Care, Treatment, and Transportation of Dogs and Cats

- §3.6,d   Primary Enclosures
- §3.8,b,1; §3.8,c; §3.8,d,2   Exercise for Dogs

Subpart B — Specifications for the Humane Handling, Care, Treatment, and Transportation of Guinea Pigs and Hamsters

- §3.28,c,3   Primary Enclosures

Subpart C — Specifications for the Humane Handling, Care, Treatment, and Transportation of Rabbits

- §3.53,c,3   Primary Enclosures

Subpart D — Specifications for the Humane Handling, Care, Treatment, and Transportation of Nonhuman Primates
- §3.80,b,2,iii; §2.38,c   Primary Enclosures
- §3.81,c,3; §3.81,d; §3.81,e,ii   Environment Enhancement to Promote Psychological Well-Being
- §3.83   Watering

The PHS Policy sets forth detailed requirements governing IACUCs. References to IACUCs are found primarily in the following sections:

- IV,A,3   Institutional Animal Care and Use Committee
- IV,B   Functions of the IACUC
- IV,C   Review of PHS-Conducted or -Supported Research Projects
- IV,E   Recordkeeping Requirements
- IV,F   Reporting Requirements

The PHS Policy requires institutions to use the *Guide* as a basis for developing and implementing an institutional program for activities involving animals. The regulations that govern the appointment, composition, and responsibilities of the IACUC are summarized in the *Guide* (pages 9 and 10).

2:3 What is the difference between the Animal Welfare Act and the Animal Welfare Act regulations in relation to the need for an IACUC?

Opin. Both the AWA and the AWAR require research facilities to appoint an IACUC (see 2:1; 2:2). However, the AWAR provides directives for the implementation of the AWA, and sets forth a more extensive list of duties for the IACUC. The AWAR were written so as to harmonize the USDA requirements for IACUCs with those required by PHS Policy.

2:4 What are the regulatory charges of the IACUC?

Reg. The regulatory charges of the IACUC are described in the AWA (§13,b), the AWAR (§2.31,c), PHS Policy (IV), and the *Guide* (pages 9 and 10).

Opin. In summary, the above regulations require the IACUC to:

- Review, at least once every 6 months, the institution's program for animal care and use, using the AWAR and the *Guide* as a basis for evaluation.
- Inspect, at least once every 6 months, all of the institution's animal facilities, including animal study areas and satellite facilities, using the AWAR and the *Guide* as a basis for evaluation.

- Submit to the IO reports of the above evaluations which distinguish significant deficiencies from minor deficiencies, contain a reasonable and specific plan and schedule for correcting each deficiency, describe and justify any departures from the *Guide* and PHS Policy, which are signed by a majority of the committee and include any minority views.
- Review and investigate concerns involving the care and use of animals at the institution resulting from public complaints or from reports of noncompliance received from personnel at the institution.
- Make recommendations to the IO regarding any aspect of the institution's animal program, facilities, or personnel training.
- Review and approve, require modifications in (to secure approval), or withhold approval of those components of proposed activities related to the care and use of animals.
- Review and approve, require modifications in (to secure approval), or withhold approval of proposed significant changes regarding the care and use of animals in ongoing activities.
- Suspend an activity involving animals if it does not comply with the AWAR, the *Guide*, PHS Policy, or the institution's Animal Welfare Assurance (Assurance) approved by NIH/OPRR.

2:5 Is an IACUC always required where animals are used for research, teaching, or product safety evaluation?

Reg. No. An IACUC is required only if one of the following applies:

- The species used for research, teaching, or testing is covered by the definition of animal given in §1.1 of the AWAR. The AWAR definition of animal specifically excludes birds, rats of the genus *Rattus*, and mice of the genus *Mus* bred for use in research, and farm animals used in agricultural research. See 12:1.
- The research, teaching, or product safety evaluation is supported by the PHS, or the activity will be performed at an institution with an Animal Welfare Assurance that commits the institution to comply with the PHS Policy, regardless of the source of funding.
- The institution is receiving support for animal research, teaching, or testing from an agency that requires compliance with PHS Policy (e.g., the National Science Foundation).

Opin. If an institution is or desires to be accredited by AAALAC, it must have an IACUC. (See 8:5, 8:6.)

2:6 Do elementary schools need to have an IACUC to keep pet animals?

Opin. No. A grade school that keeps pet animals is not considered an exhibitor under the definitions in §1.1 of the AWAR because it does not purchase animals and exhibit them to the public for compensation. Furthermore, both the AWAR (§1.1, Research facility) and the AWA (§2,e) specifically exclude elementary and secondary schools from being designated as research facilities. Accordingly, the regulations do not apply and an IACUC is not required. An IACUC also is not required under PHS Policy, since it applies only to institutions that receive support for vertebrate animal activities from an agency of the PHS or from an agency that requires compliance with PHS Policy. (See 2:5 and Reference 1.)

2:7 Is it necessary to have IACUC oversight of elementary or secondary school science fair projects or other educational activities?

Opin. No. Elementary and secondary schools are specifically excluded from the definition of research facility in §1.1 of the AWAR and the AWA (2,e). Accordingly, the regulations do not apply and an IACUC is not required. An IACUC also would not be required under PHS Policy, since it applies only to institutions that receive funds for vertebrate animal activities from an agency of the PHS or from an agency that requires compliance with PHS Policy.

2:8 What guidelines exist for use of animals in elementary or secondary school educational activities?

Opin. Although elementary or secondary schools are not covered by federal regulations or policies governing the use of animals, they can use the guidelines contained in two National Research Council documents: the *Guide* and the *Principles and Guidelines for the Use of Animals in Precollege Education*.[1] The latter document states:

> *"A plan for conducting an experiment with living animals must be prepared in writing and approved prior to initiating the experiment or to obtaining the animals. Proper experimental design of projects and concern for animal welfare are important learning experiences and contribute to respect for and appropriate care of animals. The plan shall be reviewed by a committee composed of individuals who have the knowledge to understand and evaluate it and who have the authority to approve or disapprove it."*

The above guidelines also describe what the plan should include. Another set of guidelines that may be useful are the "Principles for the Ethical Care and Use of Animals" developed by the National Aeronautics and Space Administration.[2]

In addition to these national guidelines, individual state governments or even local school districts may develop standards that are either guidelines or requirements. Also, in at least 30 states there are nonprofit biomedical research societies that schools can consult for information on guidelines. In general, these societies recommend the formation of a review committee that is composed of three people: a teacher (preferably a biologist), a parent, and a veterinarian.

2:9 Do colleges or universities with live mascots need to have the care and use of these animals approved by the IACUC?

Reg. Technically, institutions using AWAR regulated species as mascots fall under the definition of exhibitor (§1.1 of the AWAR) and, accordingly, should be licensed as such. Although the technical definition requires mascots fall under the definition of exhibitor, common sense says it would not effectuate the purposes of the AWA to license most universities for this purpose. To the extent the mascot is housed in university facilities, it should be cared for and housed in a manner consistent with the AWAR, and the housing and care would be subject to APHIS/AC inspection.[3]

Opin. The question of whether the care and use of a mascot must be approved by the IACUC depends upon an institution's policies since it is not required under the AWAR. For example, it is the policy at some institutions that all uses of live vertebrate animals must be approved by an IACUC, regardless of whether the animals are an APHIS/AC-regulated species or are used in PHS-supported activities. Other institutions state in their NIH/OPRR Assurance that all uses of animals at the institution will be conducted in accord with the *Guide*. At such institutions, both NIH/OPRR and AAALAC expect the care and use of mascots to be covered by an IACUC-approved protocol.

2:10 Do zoological gardens and aquariums fall under the jurisdiction of the AWAR or PHS Policy?

Reg. Yes. Zoos and some aquariums (e.g., those that maintain AWAR-regulated species, such as marine mammals) are defined as exhibitors under §1.1 of the AWAR and must comply with all regulations governing exhibitors.

Opin. Strictly speaking, only zoos and aquariums that receive support for research using animals from an agency of the PHS, such as the NIH, are required by NIH/OPRR to comply with its policy. However, other federal and private funding agencies (e.g., the National Science

Foundation) may require adherence to the PHS Policy as a condition of receiving their support. (See 2:5.)

2:11 Under the AWAR, are zoological gardens and aquariums expected to adhere to the same research criteria as biomedical research institutions?

Reg. Zoological gardens need to be registered as a research facility if they are conducting invasive procedures in a research context, i.e., to gather and disseminate data and information. Typical behavioral and observational studies do not include invasive procedures and, therefore, do not require registration as a research facility.[3] In early 1999 APHIS/AC initiated a process to define research for AWA purposes as it applies to zoos, wildlife studies, and euthanasia, but at the time of this publication the policy has not been finalized.[3]

Opin. Behavioral training or studies designed to improve the management or care of animals in a zoo or aquarium setting are not considered regulated research. However, if a nonobservational procedure is performed on an animal for the purpose of obtaining data not directly related to the management or care of the animal or its species in a zoo or aquarium setting, the activity is considered regulated research. In such cases, the zoo, aquarium, or one of their subunits must register as a research facility with the USDA (in addition to being licensed as an exhibitor), and must meet the requirements stipulated by the AWAR for research facilities. (For further information on this topic, see Reference 4.)

2:12 Is research conducted at zoological gardens and aquariums required to come under the purview of an IACUC?

Reg. (See 2:11.)

Opin. If the activity involves an APHIS/AC-regulated species and is considered regulated research by APHIS/AC (see 2:11), it comes under the purview of an IACUC. Similarly, if the activity is supported by the PHS or other agencies that require compliance with PHS Policy, it must be reviewed and overseen by an IACUC (see 2:5).

2:13 Do private and municipal animal shelters and humane societies fall under the jurisdiction of the AWA and its regulations? Of the PHS Policy?

Opin. Generally, no. Animal shelters and humane societies do not fall under the jurisdiction of the AWAR unless they sell dogs or cats to a research facility, licensed dealer, exhibitor, or sell animals wholesale as pets. If a *private* shelter provides animals to a research facility, it must be licensed with APHIS/AC as a dealer and must comply with all AWA regulations governing dealers. A *municipal* animal shelter or humane society that provides animals to a research facility is not defined as "a person" by APHIS and, consequently, does not meet the definition of dealer under §1.1 of the AWAR. Accordingly, municipal shelters and humane societies are exempt from the regulations governing dealers. However, any shelter (private or municipal) that provides animals to a research facility must comply with animal holding and recordkeeping requirements stipulated under of the AWA (§28) and the AWAR (§2.101). This section requires the shelter to hold any dog or cat for at least 5 days before selling it to a research facility, and to provide the research facility with a certification that:

- The person who gave the dog or cat to the shelter was notified that the animal may be used for research or education.
- The shelter satisfied the 5-day holding requirement.

The original certification must accompany the dog or cat to the research facility, and the shelter is required to retain a copy of the certification for at least 1 year. APHIS/AC is empowered to review these records at shelters and may assess criminal or civil penalties against any shelter that does not comply with these requirements.

Animal shelters and humane societies fall under the jurisdiction of PHS Policy only if they receive support from the NIH or some other agency that requires compliance with PHS Policy.

References

1. Principles and Guidelines for the Use of Animals in Precollege Education, Institute for Laboratory Animal Resources, Washington, D.C., 1989. Available on the World Wide Web at: *http://www2.nas.edu/ilarhome/23da.html*
2. Principles for the Ethical Care and Use of Animals, National Aeronautics and Space Administration, NPD 8910, Washington, D.C., 1996. Available on the World Wide Web at: *http://lifesci.arc.nasa.gov/PECA.html*
3. DeHaven, W.R., personal communication, 1999.
4. Kohn, B. and Monfort, S.L., Research at zoos and aquariums: regulations and reality, *J. Zoo Wildlife Med.*, 28, 241, 1997.

3
Creation of an IACUC

Richard C. Van Sluyters, Jane E. Simpson, and Ralph B. Dell

Introduction

This chapter provides information on how institutions can form an IACUC. It describes whose interests the IACUC should serve and whose responsibility it is to appoint the members. Since most IACUCs conduct their meetings in private, the nonaffiliated member of the committee can be the general public's only link to the federally mandated oversight process for animal welfare. Accordingly, information is provided on the critical role of the nonaffiliated member and how institutions can go about finding an effective one. For an IACUC to fulfill its regulatory requirements properly, its members must be informed of their responsibilities under federal law, regulation, and policy. In addition to these federally mandated obligations, IACUCs may be charged with additional responsibilities by their state, city, or institution. Given the complexity of this environment, institutions have a responsibility to provide their IACUC members with specialized training and this chapter describes some methods by which this can be done. The latter part of the chapter suggests ways that larger institutions can utilize multiple IACUCs to reduce the workload associated with reviewing hundreds of animal use protocols and inspecting dozens of animal facilities. Finally, there is a discussion of the recordkeeping requirements for IACUCs and the use of computerized databases to store and retrieve these records.

3:1 In general, what is expected of the IACUC and whose interests does it represent?

Reg. The AWA (§13,b,1) charges the members of the IACUC with representing, "… society's concerns regarding the welfare of animal subjects. …" In addition, the nonaffiliated (outside) member is specifically charged with representing general community interests in the proper care and treatment of animals (AWA §13,b,1,B,iii; AWAR §2.31,b,3,ii; *Guide,* page 9). The PHS Policy (IV,B) references the IACUC as "an agent of the institution" that will "oversee the institution's animal program, facilities, and procedures" (PHS Policy IV,A,3,a).

Opin. The IACUC is expected to oversee and evaluate the institution's animal care and use program to ensure that they are consistent with the recommendations of the *Guide,* the AWAR, and PHS Policy. It represents multiple interests, including those of the institution and the community. It serves as the local oversight arm of federal agencies and accrediting bodies, such as APHIS/AC, NIH/OPRR, and AAALAC.

3:2 Who is responsible for appointing the IACUC at an academic institution?

Reg. The chief executive officer (CEO) at an academic institution is charged with appointing an IACUC by the AWA (§13,b,1), the AWAR (§2.31,a), and PHS Policy (IV,A,3,a.). The CEO is defined in PHS Policy (IV,A,3,a, Footnote 5) as the highest operating official of the organization, such as the president of a university. PHS Policy allows the CEO to delegate the authority to appoint the IACUC, but requires that such delegation be specific and in writing.

The *Guide* requires institutions to comply with the AWAR and PHS Policy, which both require the CEO to appoint the IACUC. However, the *Guide* (page 9) refers to the individual who is responsible for appointing the IACUC as the "responsible administrative official."

Opin. In larger institutions it is not unusual for the CEO to delegate authority for appointing the IACUC to a senior administrator who is more directly responsible for the institution's research program, such as a Vice Chancellor, Dean, or Vice President for Research. This individual also frequently serves as the organization's IO (PHS Policy III,G) and signs its Animal Welfare Assurance. It is important to note that the CEO and the IO do not have to be the same person.

3:3 Who is responsible for appointing the IACUC at an industrial setting?

Opin. The AWAR and PHS Policy do not discriminate between academic and industrial settings in describing who is responsible for appointing an IACUC. In both settings, it is the CEO or the CEO's designee. (See 3:2.)

3:4 What is the definition of the nonaffiliated (outside) member of the IACUC?

Reg. The nonaffiliated member is defined as an individual who represents general community interests in the proper care and use of animals, is not a laboratory animal user, is not affiliated with the institution, and is not an immediate family member of a person who is affiliated with the institution (AWA §13,b,1,B; AWAR §2.31,b,3,ii; PHS Policy IV,A,3,b,4; *Guide*, page 9; APHIS/AC Policy #15).[1]

3:5 How might IACUCs find individuals to serve as nonaffiliated members?

Opin. Nonaffiliated members can be found in a number of ways. Perhaps the most common source is through personal contacts of the CEO, members of the IACUC, AV, or other personnel at the institution. Other useful sources for recruiting nonaffiliated members may include professional societies (e.g., lawyers, ethicists, clergy, teachers, librarians, healthcare professionals), local humane societies, and nonprofit service organizations. Some institutions run classified advertisements in local newspapers seeking interested individuals.

3:6 What should be the duration of the IACUC membership for the Chair and individual committee members?

Opin. Serving as an effective member of an IACUC is a difficult task. To perform their duties competently, the members of an IACUC must become familiar with a large body of regulations, policies, and guidelines. They also must learn about the history, special problems, and idiosyncrasies of their institution. Experience has shown that it takes a new IACUC member 6 months to a year to gain the experience required to function as a fully effective member of an IACUC. In light of this, most institutions ask members to serve for at least 2 years, and many for 3 years. Some institutions also ask capable and willing members to serve a second 2- or 3-year term.

Serving as an effective IACUC chair requires all of the knowledge and expertise needed to become an effective committee member, plus the additional administrative skills required to manage the committee's staff and interact with the institution's administration. For these reasons, the person selected to Chair the IACUC has usually served at least one term as a committee member. Many institutions ask the IACUC Chair to serve multiple terms in order to retain someone with a high level of regulatory and institutional knowledge in this position. (See 5:10.)

3:7 How can institutions train IACUC members?

Reg. Recognizing the complexity of the responsibilities that new IACUC members are asked to fulfill, the *Guide* (page 9) states that, "It is the institution's responsibility to provide suitable orientation, background materials, access to appropriate resources, and, if necessary, specific training to assist IACUC members in understanding and evaluating issues brought before the committee." Accordingly, AAALAC-accredited institutions are expected both to provide adequate training for IACUC members and to document that such training has been provided.

Opin. There are many ways that institutions can train IACUC members. A common method is to prepare an IACUC member's handbook that contains copies of relevant regulations, policies, and guidelines, as well as copies of the institution's forms and SOPs. New members typically undergo one or more orientation sessions under the guidance of the IACUC staff or Chair, or the AV. New committee members also may be assigned a mentor who is a more experienced member of the IACUC, and they may be invited to attend one or two IACUC meetings as observers prior to the start of their official term. In addition, members may attend an ever-growing number of meetings and workshops that address animal care and use issues of importance to IACUCs. Other useful information for training new IACUC members includes a tutorial on the PHS Policy and an online version of the OPRR–ARENA publication "Institutional Animal Care and Use Committee Guidebook."[2]

The American Association for Laboratory Animal Science (AALAS) sponsors a World Wide Web page called *The IACUC Training and Learning Consortium*[3] that is an excellent source of training materials for IACUC members, and the Scientists Center for Animal Welfare (SCAW) hosts *IACUC Talk*,[4] a forum for IACUC members to voice their opinions, questions, and concerns.

Several Institute for Laboratory Animal Research reports and training manuals also contain information of value to IACUC members.[5-9] Each of these documents was written by a committee of experts and carefully reviewed. They all are available from the National Academy Press.[10]

3:8 Can an institution have more than one IACUC?

Opin. Yes. There is no regulatory language that precludes multiple IACUCs at one institution. In fact, the AWA (§13,b,1) states, "The Secretary shall require that each research facility establish at least one committee," clearly leaving the door open for appointing more than one committee at an institution. Interestingly, the AWAR (§2.31,a), which implement the AWA, read, "The Chief Executive Officer of the research facility shall appoint an Institutional Animal Care and Use Committee (IACUC) ...," language that is echoed in PHS Policy (IV,A,3,a) and the *Guide* (page 9). In practice, a number of institutions have opted to appoint more than one IACUC, and their systems have been approved by both NIH/OPRR and APHIS/AC. Multiple IACUCs are found most commonly at large institutions that have diverse scientific missions, many animal facilities, and large numbers of investigators and protocols.

3:9 Under what circumstances would multiple IACUCs be advantageous?

Opin. Multiple IACUCs are advantageous at large institutions where the sheer volume of research activity exceeds the capacity of a single IACUC to review the associated protocols and program in an effective and timely manner. Dividing the work between a number of IACUCs, each appointed to oversee a discrete portion of the research program, can be the only way to handle the workload. An added advantage of dividing the workload is that it allows committees to gain expertise in reviewing protocols in particular research areas, since they are usually appointed within academic units or to cover specific research disciplines. More localized IACUCs also make it easier for IACUC members to maintain a personal touch with investigators and a sense of ownership and pride in the portion of the program they oversee.

Another situation where more than one IACUC may be advantageous is in institutions that have bipartite research programs. In some cases, this occurs when an organization engages in two very distinctly different types of research. In others, it occurs when the research program is physically divided between two locations. In both situations, the organization may find it effective to appoint individual IACUCs to review protocols from and oversee the separate components of its overall program.

3:10 What might be the disadvantages of multiple IACUCs?

Opin. A major challenge for a multiple IACUC system is to develop mechanisms for maintaining consistency and quality in program and protocol review across the institution. If there is not an over-arching coordinating group or committee, the institution might become fragmented and inconsistent, or even conflicting practices might arise. Another disadvantage is the need for increased IACUC staffing and funding to sustain the necessarily redundant components of multiple committees. The cost of maintaining an IACUC is considerable and, in the absence of centralized support, it can place a large burden on smaller administrative units such as colleges or departments.

Another possible problem with the multiple IACUC system is that each committee must meet all the membership requirements for an individual IACUC (i.e., veterinarian, nonaffiliated member, nonscientist, etc.), including the requirement that no more than three members from the same administrative unit may serve on the IACUC (AWAR §2.31,b,4). It can be difficult to appoint a properly constituted committee from within a small administrative unit. Moreover, difficulties can arise with locally constituted IACUCs because no member may participate in the IACUC review or approval of a research project in which the member has a conflicting interest, nor may a member who has a conflicting interest contribute to the constitution of a quorum (PHS Policy IV,C,2; AWAR §2.31,d,2). Even when there is no obvious conflict of interest, it is important to recognize that there can appear to be a conflict when IACUC members drawn from a small administrative unit review each other's protocols. Finally, when IACUC members are drawn from a limited pool, individuals can find themselves having an excess time commitment to the IACUC.

3:11 Should multiple IACUCs all report to a central IACUC, advisory committee, or other authority?

Opin. Yes. Experience at institutions with multiple IACUCs has shown that it is essential to have the committees report either to the same IO or to an institution-wide IACUC that then reports to the IO as per PHS Policy (IV,B,3; IV,B,5). An institution could designate a central policy or advisory body, but all IACUCs must have a direct reporting channel to the IO. This organizational structure is needed to establish uniform policies and procedures, and to maintain consistent performance standards across the institution. (See 3:12; 4:1.)

3:12 What are some ways in which multiple IACUCs can effectively interact?

Opin. In institutions where multiple IACUCs report to a single IO, the IO can appoint an IACUC advisory committee to address general issues that affect the entire animal care and use program. The committee should consist, at a minimum, of the IO or the IO's designee, the Chairs of the individual IACUCs, and the AV. Issues could be brought before this committee by any of its members.

Another way institutions with multiple IACUCs can ensure effective interaction is for the IO to appoint an over-arching institutional IACUC to which the individual IACUCs report. That is, the individual IACUCs function more as a pre-review and advisory group, providing recommendations to the investigator and the institutional IACUC. In this model, the composition of this institution-wide committee meets the regulatory requirements for and serves as the institution's official IACUC. The products of the individual IACUCs (e.g., pre-reviewed protocols, facility inspection reports, etc.) flow to the institution-wide IACUC for final review and approval as needed. A simple strategy for helping ensure effective interaction between multiple IACUCs is to have some members serve on more than one committee.

In these situations, the goal of the institution-wide IACUC is to make certain that policies, procedures, and performance standards are consistent throughout the institution's animal use program. Examples of the types of activities in which it might engage include development or modification of standardized forms and guidelines (e.g., animal use protocol form, SOPs, animal record forms, etc.), compilation or preparation of semiannual reports for submission to the IO and annual reports for submission to APHIS/AC and NIH/OPRR, coordination of AAALAC site visits, development and maintenance of centralized information resources, evaluation of facility improvement requests, and so forth.

3:13 What strategies can be used to ensure that protocol review is consistent throughout an institution with multiple IACUCs?

Opin. The most important step in ensuring consistent protocol review is to develop a single animal use protocol form that can be used throughout the institution. The form should be designed to allow sufficient flexibility in describing proposed procedures, so as to encompass the full range of activities at the institution (e.g., antibody production, survival surgery, product safety testing, field research, breeding, etc.), while at the same time requiring all investigators to provide the information that is required for any proposed use of animals (e.g., rationale for animal use, justification for animal species and numbers, alternatives to painful procedures, humane endpoints, etc.).

It is also important for the IO to make certain that the membership of each IACUC not only meets the minimum regulatory requirements, but includes individuals with sufficient expertise to competently evaluate the protocols they receive. Each committee needs to have scientists familiar with the kinds of animal use they will be asked to review and veterinarians knowledgeable in the health and husbandry of the proposed animal species. Depending on the nature of the activities to be reviewed, sufficient technical expertise must be available for each IACUC to evaluate potential risks associated with protocols (e.g., biosafety, radiation safety, occupational health, etc.) and to ensure that appropriate safeguards are proposed to mitigate these risks. This can be accomplished by having a health and safety specialist on every committee or by providing a centralized health and safety resource for use by all IACUCs.

It is important to note that absolute consistency is unobtainable, even in programs with a single IACUC. Inevitably, criteria change over time and mistakes are made. It is important to maintain a consistent approach to big issues that have the potential to impact the entire program.

3:14 What mechanism might be used to assure consistent IACUC semiannual inspection and review of animal care facilities and programs at institutions having multiple IACUCs?

Opin. One way for institutions with multiple IACUCs to ensure that semiannual inspections and reviews are consistent is to utilize an institution-wide committee either to perform these functions or to review the product of the individual IACUCs that perform them. The development of institution-wide forms to be used in conducting inspections and preparing reviews also can help assure consistency. Another strategy is for personnel from a central unit (e.g., AV, institutional compliance officer, etc.) to participate in the facility inspections conducted by each IACUC. Finally, institutions that require all of their components to be accredited by AAALAC have an independent method for ensuring uniform compliance with applicable regulations, policies, and guidelines.

3:15 What is the least common denominator (e.g., school, college, department, individual investigator) around which an IACUC can form?

Opin. Institutions with multiple IACUCs may form them around:

- Discrete units within the institution's administrative structure (e.g., medical school, veterinary school, graduate school, college of agriculture).
- The nature of the proposed research (e.g., agricultural, biomedical, field research, product testing).
- The animal species being used (e.g., farm animals, laboratory animals, nonhuman primates, exotic species).
- Geographically separate facilities.

When selecting the size of a unit upon which to base an IACUC, consideration needs to be given to:

- The potential workload of the IACUC (e.g., number of protocols per year; number, size, and location of animal facilities).
- The distance of the proposed IACUC from the IO in the reporting lines of the institution.
- Whether the unit has the resources to support an IACUC.

- Whether the unit has sufficient breadth of animal use to avoid frequent potential conflicts of interest on the IACUC.
- Whether the pool of animal users from which IACUC members can be drawn is sufficiently large.
- The requirement in the AWAR (§2.31,b,4) that no more than three members of the IACUC can be from the same administrative unit.

3:16 How might records for IACUCs be effectively maintained?

Opin. Institutions are obliged to maintain various records related to the animal care and use program. The AWAR (§2.35) and PHS Policy (IV,E) require each institution to maintain:

- Minutes of IACUC meetings, including records of attendance, activities of the committee and committee deliberations.
- Records of applications, proposals, and proposed significant changes in the care and use of animals, and whether IACUC approval was given or withheld.
- Records of semiannual IACUC reports and recommendations, including minority views, that have been forwarded to the IO.

In addition, PHS Policy (IV,E) requires each institution to maintain:

- A copy of its Assurance.
- Records of accrediting body determinations.

The AWAR (§2.35,f) and PHS Policy (IV,E) stipulate that all records must be maintained for at least 3 years and that records relating to applications, proposals, and proposed significant changes must be maintained for 3 years after completion of the activity. All records must be accessible for inspection and copying by authorized representatives of APHIS/AC, NIH/OPRR, other PHS representatives, and (for accredited institutions) AAALAC site visitors.

The IACUC should either maintain these records or ensure that they are maintained in a secure and accessible location at the institution. Additional records that should be kept include:

- IACUC correspondence, including records of electronic mail.
- Institutional policies, guidelines, and SOPs for animal care and use.
- Reports of IACUC investigations of concerns involving animals.

- Records of the training provided to animal care personnel, research personnel, and IACUC members.

Methods of maintaining records will vary depending upon the organization of the institution and the size and complexity of its animal use program. However, any method that is used to maintain records must be efficient, reliable, and secure.

In addition to traditional paper records, many institutions use computerized databases to maintain records of approved animal use protocols. Such databases may contain a minimum of information — such as PI name, protocol number and expiration date, approved animal species and numbers — or they may contain a complete summary of the protocol, including a list of all procedures, drugs, and personnel. Computerized databases have many advantages, including the ability to sort and retrieve information quickly by various characteristics (e.g., species used, type of procedure, protocol expiration date, location of procedure/housing, etc.). The database also can be shared with the unit that orders animals to ensure that only approved animal species and numbers are ordered for a particular investigator. Information stored in the database also can facilitate preparation of the institution's annual report to APHIS/AC and the "Animal Use Summary" that is required as part of the application for AAALAC accreditation or reaccreditation.

References

1. USDA-APHIS Animal Care Policy Manual, available on the World Wide Web at: *http://www.aphis.usda.gov/ac/polman.html*
2. U.S. Department of Health and Human Services, Public Health Service, National Institutes of Health, Institutional Animal Care and Use Committee Guidebook, NIH publication No. 92-3415, 1992. Available on the World Wide Web at: *http://www.nih.gov:80/grants/oprr/library_animal.htm*
3. American Association for Laboratory Animal Science, *The IACUC Training and Learning Consortium*. Available on the World Wide Web at: *http://www.iacuc.org*
4. Scientists Center for Animal Welfare, *IACUC Talk*, Scientists Center for Animal Welfare, Greenbelt, MD. Available on the World Wide Web at: *http://www.scaw.com/forum.html*
5. Committee on Educational Programs in Laboratory Animal Science, Education and Training in the Care and Use of Laboratory Animals: A Guide for Developing Institutional Programs, National Research Council, National Academy Press, Washington, D.C., 1991.
6. Committee on Dogs, *Laboratory Animal Management: Dogs,* National Research Council, National Academy Press, Washington, D.C., 1994.

7. Committee on Rodents, *Laboratory Animal Management: Rodents*, National Research Council, National Academy Press, Washington, D.C., 1996.
8. Committee on Infectious Diseases of Mice and Rats, *Infectious Diseases of Mice and Rats*, National Research Council, National Academy Press, Washington, D.C., 1991.
9. Committee on Occupational Safety and Health in Research Animal Facilities, *Occupational Health and Safety in the Care and Use of Research Animals*, National Research Council, National Academy Press, Washington, D.C., 1997.
10. Publications of the National Academy Press are available through the World Wide Web at: *http://www.nap.edu*

4
Reporting Lines of the IACUC

Richard C. Van Sluyters, Jane E. Simpson, and Ralph B. Dell

Introduction

Federal law, regulation, and policy stipulate that the IACUC must report to the individual who has been designated to serve as the institution's IO. This chapter clarifies what it means to be the IO, by reviewing this individual's many important responsibilities under the PHS Policy and the AWAR. Examples of common job titles for IOs in various types of institutions are used to provide the reader with a sense of the level of power or authority that this person should have within an institution's overall administrative organization. The chapter concludes with brief discussions of useful methods for educating the personnel within an institution about the functions of its IACUC, and the circumstances in which it may be desirable for an institution's internal policies with respect to animal care and use to go beyond merely adhering to the federal requirements.

4:1 To what person or organization does the IACUC report?

Reg. The IACUC reports to the institution's IO and, in some circumstances, through the IO to APHIS/AC, NIH/OPRR, and other federal funding agencies. According to PHS Policy (IV,B,3; IV,B,5) and the AWAR (§2.31,c,3), the IACUC is required to submit reports of its semiannual program evaluations and facility inspections to the IO, and to make recommendations to the IO regarding any aspect of the institution's program, facilities, or personnel training. PHS Policy (IV,F,1; IV,F,2) requires that the IACUC, through the IO, submit an annual report to NIH/OPRR. This report must describe any changes in the program, facilities, or IACUC membership, and list the dates the IACUC conducted its semiannual evaluations of the institution's program and facilities and submitted the evaluations to the IO. In addition, PHS Policy (IV,F,3; IV,C7) requires the IACUC, through the IO, to promptly provide NIH/OPRR with a full explanation of the circumstances and actions taken with respect to any serious or continuing noncompliance with PHS Policy, any serious deviation from the *Guide*, or any IACUC suspension of an activity.

The AWAR (§2.31,c,3) require the IACUC to report through the IO to APHIS/AC and any federal funding agency in writing within 15 days, whenever the institution fails to adhere to a reasonable plan and schedule for correcting a significant deficiency in the institution's program or facilities (see 25:13; 26:16). In addition, if the IACUC suspends an activity involving animals, PHS Policy (IV,C,7) and the AWAR (§2.31,d,7) require the IO, in consultation with the IACUC, to review the reasons for suspension, take appropriate corrective action, and report that action with a full explanation to NIH/OPRR, APHIS/AC, and any federal funding agency.

4:2 What is meant by the "Institutional Official?"

Reg. The AWAR (§1.1, Institutional Official) state that the IO is the individual at a research facility who is authorized to legally commit on behalf of the research facility that it will meet the requirements of the AWAR. Similarly, PHS Policy (III,G) defines the IO as the individual who signs and has the authority to sign the institution's Assurance, which commits the institution to meet the requirements of PHS Policy. The *Guide* (page 9) refers to the "responsible administrative official" as the individual who must appoint the IACUC.

Opin. If the Chief Executive Officer (CEO) does not designate an IO, it is the authors' opinion that the CEO becomes the IO by default and must be identified as such on the NIH/OPRR Assurance and any other pertinent documents. NIH/OPRR will not approve an assurance without the signature of an individual identified as the IO. (See 3:2; 3:3.)

4:3 What are the responsibilities of the IO?

Reg. The IO is the individual who is responsible for ensuring that an institution complies with all applicable animal welfare laws, regulations, and policies. The IO signs forms, reports, and letters on behalf of the institution, and interacts with the IACUC in overseeing the institution's animal care and use program. According to PHS Policy, the IO:

- Signs and has the authority to sign the Assurance (III,G).
- Commits the institution to meet the requirements of PHS Policy (III,G).

- Receives inspection reports and recommendations (IV,B,3; IV,B,5).
- In consultation with the IACUC, determines whether deficiencies are significant or minor (IV,B,3).
- Receives notification of the IACUC's decision to approve or withhold its approval of animal activities.
- Receives and transmits annual reports to NIH/OPRR (IV,F).
- Consults with the IACUC regarding suspensions and corrective actions; reports to regulatory and funding agencies (IV,C,7).
- May subject protocols that have been approved by the IACUC to further review and approval, but may not approve an activity that has not been approved by the IACUC (IV,C,8).
- Ensures that the institution maintains required records for the specified period of time (IV,E).

According to the AWAR, the IO:

- Legally commits the institution to meet the requirements of the AWAR (§1.1).
- Signs and submits the registration form and is responsible for notifying APHIS/AC of any changes in the institution's registration (§2.30,a–§2.30,c).
- If not done by the CEO, signs and submits the annual report to APHIS (§2.36,a).
- Receives inspection reports and recommendations from the IACUC (§2.31,c,3; §2.31,c,5).
- In consultation with the IACUC, determines whether deficiencies are significant or minor (§2.31,c,3).
- Forwards IACUC reports of uncorrected significant deficiencies to APHIS/AC and any federal agency funding that research (§2.31,c,3).
- Receives notification of the IACUC's decision to approve or withhold its approval of animal activities (§2.31,d,4).
- Consults with the IACUC regarding suspensions and corrective actions; reports to regulatory and funding agencies (§2.31,d,7).
- May subject protocols that have been approved by the IACUC to further review and approval, but may not approve an activity that has not been approved by the IACUC (§2.31,d,8).

- Ensures that all personnel involved in animal care, treatment, and use are qualified to perform their duties and that training and instruction in specific areas are provided to those personnel (§2.32,a; §2.32,c).
- Ensures that training and instruction are made available and that the qualifications of personnel are reviewed with sufficient frequency to fulfill the research facility's responsibilities (§2.32,b).
- Ensures that the institution has an attending veterinarian who provides adequate veterinary care to its animals in compliance with the AWAR (§2.33).
- Ensures that any part-time attending veterinarians are employed under a formal written program of veterinary care (§2.33,a,1).
- Ensures that the institution maintains required records for the specified period of time (§2.35).
- If applicable, certifies to APHIS that any outside facilities holding animals for the institution are recognized animal sites under the institution's research facility registration (§2.38,i,3).

4:4 Who can serve as the IO if that role is delegated by the Chief Executive Officer?

Reg. (See 4:2.)

Opin. The AWAR and PHS Policy describe no restrictions nor do they stipulate any minimum qualifications for individuals who serve as IOs. However, to be effective, an IO must have the level of authority described in 4:6. (See 3:2; 3:3; 4:5.)

4:5 Who typically serves the role of IO in academia and in industry?

Opin. Some common titles of individuals who serve as IOs in academia and industry include Chancellor; President; Provost; Vice Provost; Vice Chancellor for Health Affairs; Vice Chancellor, Administration; Vice Chancellor for Academic Affairs; Vice Chancellor for Research; Executive Director, Research Resources; Director, Research Administration; Director of Sponsored Research; Dean, School of Medicine; Associate Dean, Research and Sponsored Programs; Chief Executive Officer; Executive Vice President for Health Affairs; Senior Vice President and Vice Provost for Health Affairs; Vice President for Health Affairs; Vice President for Research; Vice President Research and

Graduate Studies; Vice President for Clinical Affairs; Vice President, Worldwide Toxicology. (See 3:2; 3:3.)

4:6 What institutional power or authority should the IO have?

Reg. (See 4:2.)

Opin. The IO must be authorized to legally commit on behalf of the institution that the requirements of the AWAR and PHS Policy will be met. To be effective, the IO must have sufficient administrative authority to promulgate, implement, and enforce policies across departmental lines. In addition, the IO must have sufficient fiscal authority to approve and fund a level of staffing that is adequate to meet the needs of the program, as well as any needed program improvements, facility repairs, and renovations. (See 4:3.)

4:7 How involved does the IO need to be in the general activities of the IACUC?

Opin. The IO's level of involvement in the activities of the IACUC varies widely from institution to institution. Some factors that can influence the IO's involvement include the size and complexity of the institution's research program, whether the IO has a background in biomedical research, and the management styles of the organization and the IO. In some institutions, the IO or the IO's direct representative actually serves as a member of the IACUC and is directly involved in the general activities of the committee. (NIH/OPRR recommends against the IO serving on the IACUC, since the IACUC reports to the IO.[1]) At a minimum, the IO must understand the functions of the IACUC as they are defined by the AWAR and PHS Policy. Furthermore, IOs that are organizationally distant from the IACUC must ensure that there is a mechanism by which they are promptly informed of any potential threats to animal welfare, or any violations of the AWAR, PHS Policy, the *Guide*, or the institution's PHS Assurance.

4:8 What means might be useful for educating personnel with respect to the functions of the IACUC and related policies of the institution?

Opin. Personnel who should be educated about the functions of the IACUC include research and teaching personnel who use animals, animal care staff, and administrative personnel in departments or units that use animals or administer research grants or projects. Some useful means for educating these personnel include seminars, the development and promulgation of institutional policies and guidelines, training handbooks, Web pages, video tutorials, and posters or brochures. (See Chapter 21.)

4:9 Under what circumstances might institutional policy go beyond federal regulations or policy?

Opin. Many institutions choose to adopt uniform policies that apply to all live vertebrate animals used in research and teaching, regardless of whether the species are covered by the AWAR or the activities are funded by federal agencies. This approach has a number of advantages. It avoids the development of nonuniform standards for the care and use of animals. It promotes the development of standardized procedures, forms, and records, thereby making it easier to monitor for compliance and simplifying recordkeeping. Finally, it helps assure the public that the institution adheres to the highest standards of animal care and use, regardless of the species of animal being used or the funding source.

Reference

1. Garnett, N., personal communication, 1999.

5

General Composition of the IACUC and Specific Roles of the IACUC Members

Christian E. Newcomer

Introduction

The general composition requirements for IACUCs established by federal regulations since 1985 give institutions considerable latitude in fashioning their IACUCs to reflect the scientific expertise and meet the needs of their animal care and use programs. Although the specific IACUC composition requirements vary according to the regulatory oversight agency involved, both the AWAR and the PHS Policy require the IACUC to have a diversified membership, including a veterinarian and a member unaffiliated with the institution. The PHS Policy further extends the diversification by requiring the institution to include a nonscientist on the IACUC. Membership diversification on the IACUC is intended to broaden the perspective and add depth to the important IACUC review processes.

Most institutions recognize that developing and sustaining effective IACUCs takes planning and an ongoing commitment. The selection of individuals to meet specific membership requirements, new IACUC member orientation or education activities, and ensuring that new members are successfully integrated into the IACUC as contributing members takes considerable effort. Even in fully functional and effective IACUCs, self-assessment and improvement efforts are necessary for continued high-quality service to the institution and to maintain an appreciation for and understanding of the emerging trends in animal care and use programs.

This chapter is intended to examine the overall composition of the IACUC and provide information on the roles, responsibilities, and issues involving the IACUC and its various member categories to assist in IACUC development and self-review efforts. Some of the trend information provided was acquired through the author's review of approximately 100 institutions during the conduct of AAALAC International site visits over the course of a decade, and from the author's personal experience as an IACUC member at eight institutions during this same period.

5:1 How many members must an IACUC have?

Reg. The number of IACUC members required depends upon whether the institution is a recipient of funding from the PHS or other cooperating federal agencies, such as the National Science Foundation or the Department of Defense. In these cases, the institution must have at least five IACUC members (PHS Policy IV,A,3,b). Institutions that use animal species covered by the AWAR and funded through internal or private sources are required to have at least three IACUC members (AWAR §2.31,b,2). The *Guide* (page 9) states, "The size of the institution and the nature and extent of the research, testing and educational programs will determine the number of members of the committee."

Opin. The organizations visited by this author usually exceeded the minimal membership requirements set by the federal regulations and, as shown in the survey below, the majority of organizations recognize that some benefits accrue from expanding IACUC membership. Although a minimum of five IACUC members may be adequate for a small organization with a focused research program and a limited spectrum of animal use, generally seven or eight members are required at a minimum in larger, more complex programs. This facilitates adequate representation of the various areas of research expertise in the organization and fulfills the IACUC's protocol and program review responsibilities.

Surv. How many members are there on your IACUC? (Survey conducted by Sreekant Murthy, author of Chapter 6)

- Minimum number as required by regulation   9/63
- 6 to 10   27/63
- 10 to 15   19/63
- More than 15   8/63

5:2 What specific members are required for the IACUC?

Reg. The AWAR (§2.31,b) state that the committee shall be composed of a Chairman and at least two additional members. The AWAR does not describe the qualification of the Chairman, but it states that the other two members of the committee will include at least one veterinarian and at least one shall not be affiliated in any way with the facility and shall not be a member of the immediate family of a person who is affiliated with the facility.

The AWAR.(§2.31,b,4) state that if the committee consists of more than three members, not more than three members shall be from the same administrative unit of the facility. The administrative unit is defined as the organizational or management unit at the departmental level of a research facility which, for example, may include the Office of Research Administration and the University (Institutional) Laboratory Animal Resources.

The *Guide* (page 9) states that the IACUC should have a veterinarian "who is certified by the American College of Laboratory and Animal Medicine (ACLAM)" or one who has the experience "in the use of the species in question." In addition, there should be "at least one practicing scientist experienced in research involving animals." Finally, there should be "at least one public member to represent general community interests in the proper care and use of animals. Public members should not be laboratory animal users, be affiliated with the institution, or be members of the immediate family of a person who is affiliated with the institution."

PHS Policy (IV,A,3,b), on the other hand, states that membership requirements consist of not less than *five* members. The PHS Policy (IV,A,3,b,1–IV,A,3,b,4; IV,A,3,c) states that the committee shall include at least one veterinarian, one practicing scientist experienced in research involving animals, one member whose primary concerns are in a nonscientific area (for example, ethicist, lawyer, member of the clergy), and one individual who is not affiliated with the institution in any way other than a member of the IACUC, and is not a member of the immediate family of a person who is affiliated with the institution. One individual may fulfill more than one requirement. However, no IACUC constituted under the PHS Policy may consist of less than five members. Also note that PHS Policy II requires institutions to comply with the AWAR as applicable and to follow the *Guide*.

Opin. Organizations involved in agricultural research and teaching should be aware that there are guidelines for the composition of animal care and use committees operating in this setting.[1] These guidelines have no regulatory empowerment and closely parallel the PHS Policy differing only in that they specify two types of scientific individuals on the committee. One should be a scientist from the institution who has experience in agricultural research or teaching involving agricultural animals and the other should be an animal, dairy, or poultry scientist who has training and experience in the management of agricultural animals. This document further recommends that a separate committee not be established for this purpose, but that the composition of the IACUC be modified according to the recommendations to provide for the centralized and uniform oversight of the institution's animal care and use program.

5:3 Can one person fill more than one position on the IACUC?

Reg. The PHS Policy (IV,A,3,c) provides a written assent to dual representation as long as the IACUC totals five members. The AWAR offer no direct discussion of this matter but do not preclude this practice.

Opin. It is permissible for an individual to fill more than one of the membership categories on an IACUC under both the AWAR and the PHS Policy. While it is permissible, APHIS/AC strongly discourages one person filling more than one role.[2] Very few institutions have exploited dual category representation as a long-term strategy for their IACUC and, when it occurs, it frequently involves the unaffiliated member serving in the dual capacity as a nonscientist. The more common use of this strategy occurs when committee membership is in transition following the departure or resignation of a member.

5:4 What problems can occur if the number of IACUC members is too large or too small?

Opin. There can be significant problems related to the size of the IACUC. IACUCs with a small membership often have a narrower base of expertise and, depending upon the size of program they service, they may encounter an onerous workload. Also, small IACUCs may often have a more difficult time making a quorum and, even in the presence of a legitimate quorum, member absences can have marked effect on decision making.

The challenges for IACUCs with large memberships are very different. In large IACUCs the principal problem is that members may become disengaged from the activities of the IACUC. This might be due to impediments to open communication in the IACUC or because the scope and size of the program and its challenges are daunting. Thus, members have no sense that their efforts are having any impact on program quality or progress. In the depths of ennui, these members contribute to the quorum in number only. The challenges for the IACUC Chair are to ensure that all members have the opportunity for real contribution, to help members feel a sense of accomplishment and, when necessary, to inform the IO that the IACUC may be too large for the tasks at hand.

5:5 What institutional constituencies should be represented on the IACUC?

Opin. The majority of organizations go beyond the federally mandated composition of the IACUC to diversify the IACUC membership (see

5:2). The first priority in this effort generally is to enlarge the complement of scientific members on the IACUC to match the predominant areas of scientific expertise in the program. In addition to the potential political benefits this has for the IACUC, it brings tangible benefits to the IACUC's deliberations on protocol matters. It also may improve the quality of correspondence exchanged between the IACUC and PIs.

Other institutional constituencies that this author finds to be favored (in approximate order of frequency) are

- Health and safety personnel (including occupational health and safety).
- Senior animal management or supervisory personnel.
- Research laboratory technicians (with extensive animal involvement).
- Legal counsel.
- Public relations personnel.
- Grants and financial personnel.
- Student body representatives.

The merits of having these people on the IACUC should be evaluated in the context of the program for which they are intended. In the author's opinion, health and safety personnel, legal counsel, and public relations personnel should be given first consideration. Health and safety personnel are comfortable working at the interface of science and society and can make a significant contribution to the IACUC by keeping it apprised of the status of health and safety regulations, and recommendations relating to the animal care and use environment.[3] Legal counsel and public relations personnel also can be of great assistance in the consideration of issues relating to science, policy, perception, and the public.

5:6 Should the selection of IACUC members depend upon the institution's research goals and expertise?

Opin. The two most important functions of the IACUC, protocol review and programmatic review, should be important driving forces for the IACUC to select members with a share in the institution's research goals and expertise. Protocol review should be insightful, thorough, and objective. Members with scientific expertise in the areas under review can help focus the discussion of relevant issues improving the quality of the review. IACUC programmatic review is a mandated responsibility of the IACUC (PHS Policy IV,B,1–IV,B,5; AWAR §2.31,c,1–§2.31,c,5), but it really is best performed as a bidirectional process involving the community served. This implies that the IACUC should foster conduits for the flow of information from all participants in the animal care and use program. The appointment of IACUC members representing specialty interests or areas of expertise can prove advantageous to this process. Most institutions make a concerted effort to compose their IACUCs with members reflecting the research expertise of the program.

5:7 What is the typical IACUC composition in a large vs. a small institution and are the members' roles different in these settings?

Opin. There are different trends in the composition of the IACUC and in how the IACUC uses its members in large vs. small institutions. In small *commercial* institutions or independent research organizations based around a particular area of scientific inquiry, the IACUCs tend to be smaller (5 to 7 members) because the number and diversity of issues requiring addressment in the IACUC is limited. Small *academic* institutions tend to have slightly larger IACUCs (7 to 10 members). Although the research portfolio may be limited compared to a large academic institution, the relative diversity of the research activities can be quite high even though each department may only have a few active faculty researchers. The intimacy of the small institutional environment sometimes imposes a higher level of expectation that the IACUC members will be able to speak authoritatively about the research and programmatic needs of their colleagues. IACUC members in small institutions also are more likely to be active participants in developing and implementing programmatic components (e.g., biosafety review and IACUC educational efforts) that may fall under the IACUC's purview.

Large institutions sometimes have as many as 15 to 20 or more IACUC members simply to provide ample representation of the animal user groups and investigators (e.g., 150 to 200) in the program. In this author's experience, these IACUCs tend to be more formal, impersonal, and insular. Frequently, institutionally provided administrative support obviates the need for the personal involvement of the IACUC members in ushering new initiatives forward. As noted in 5:4, a significant challenge for IACUC members in this position is to combat these tendencies and convince the faculty that the IACUC is a receptive, helpful, and compassionate body that works for the institution in support of the scientific mission by encouraging improvement of the animal care and use program. Another issue germane to extremely large IACUCs is the number of veterinarians appointed to that committee. Although federal law has stipulated that an IACUC need only have one veterinary member, most institutions with large IACUCs (greater than 15 persons) have increased the number of veterinarians appointed. This action helps ensure that the veterinary perspective will be represented consistently and affords the veterinarians an opportunity to divide the workload and concentrate on specialty areas of interest.

5:8 Should the IACUC Chair, individual IACUC members, or their departments receive compensation for their efforts?

Opin. In the academic setting, the compensation of the IACUC Chair has become more common and large institutions seem to be more receptive to this approach. Approximately one quarter to one third of academic institutions visited by this author have implemented some method of reimbursement for the Chair's time commitment. In most instances, salary support is provided to the academic department of the Chair. The author is not aware of any cases where the other IACUC members have been compensated for their efforts. (See 5:28.)

5:9 How can the performance of the IACUC be evaluated?

Opin. Several mechanisms are available and should be utilized to evaluate the performance of the IACUC on an ongoing basis. The IO should meet periodically with the Chair of the IACUC to discuss the content, style, and timeliness of IACUC reports and correspondence with members of the institution's research community. This is the most common method reported by the institutions known to this author, although few institutions couch such meetings in terms of an evaluation. Also, without the inducement of a particular incident, the IO should consider soliciting the comments of the faculty or staff who use laboratory animals as well as from the members of the IACUC. The assessment of the IACUC as a working group seems particularly important because of the profound impact group dynamics can have on IACUC morale, attention to detail, and, ultimately, decision making.

Participation in the voluntary, peer review process of the AAALAC entails a substantial review of IACUC function through interviews and the review of written materials. Generally, AAALAC site visits are conducted at triennial intervals and should be supplemented by the institution's internal evaluation mechanisms. In some instances, expert consultants also have been used by institutions.

5:10 Who can remove the IACUC Chairperson or other IACUC members?

Reg. The PHS Policy requires the Chief Executive Officer (CEO) to appoint the IACUC, and the AWAR specifically directs the CEO to appoint

Opin. the Chair. Thus, although once established an IACUC possesses autonomous review and reporting functions, service on the IACUC is at the discretion of the CEO. Under the PHS Policy (IV,A,3,a, footnote 5, 1996 reprint), the CEO may delegate the authority to appoint the IACUC if the delegation is specific and in writing.

Opin. It is very fortunate that institutions are rarely, if ever, confronted with this quandary because very few institutions have clearly written bylaws for their IACUCs that stipulate who has jurisdiction or what the process is in this matter, and no clarification is offered in the PHS Policy or the AWAR. For the most part, informal discussions originating from the Chair, in the case of other IACUC members, or from the CEO, in the case of the Chair, apparently have proven adequate for the silent and amicable departure of IACUC members in the vast majority of cases.

In the absence of bylaws clearly delineating who has authority in this matter, the CEO retains that authority. In a related matter, without bylaws it is also questionable to what degree and in what situations parliamentary procedure should be applied in IACUC meetings and deliberations or to its ability to self regulate through the use of censure or expulsion procedures.[4]

5:11 What criteria can be applied for removing the IACUC Chairperson or other IACUC members?

Opin. Most organizations agree that inappropriate personal conduct, poor attendance or participation at IACUC activities, or repeated inadequate preparation for assigned IACUC duties might constitute sufficient reason to seek the removal of a member. Ideally, these broad areas should be included in the bylaws developed for the IACUC, and the IACUC members would be informed during their initial IACUC training of the general performance criteria for committee members. The IACUC Chair or IO should inquire if there are any mitigating circumstances for those members who cannot regularly attend two thirds of the IACUC meetings held annually and consider the replacement of these individuals if better attendance is not forthcoming.

5:12 What is the procedure for selecting and appointing the IACUC Chair?

Opin. Most IACUCs attempt to prepare for a transition in the position of Chair through the appointment of a Vice Chair or an ad hoc Chair-in-training position (see 3:6). In this regard, the IACUC usually has the opportunity to serve in an advisory capacity to the CEO who ultimately has the responsibility of appointing the Chair (AWAR §2.31,a;

§2.31,b). (See 3:2; 3:3.) Other groups that can prove useful in identifying and supporting candidates for the position of IACUC Chair include the faculty senate and faculty research advisory committees.

5:13 What traits or qualifications are desirable for the IACUC Chair?

Opin. In most institutions, a concerted effort is made to recruit a Chair who is a respected scientist and who has had significant experience using laboratory animals in research. Regardless of whether or not the Chair has a scientific background, the Chair should be regarded as a good colleague in the context of the institution's scientific mission.

The Chair should be patient, tolerant, diplomatic, tactful, and efficient in the handling of the IACUC's business and sensitive issues. He or she should be able to dedicate the time necessary with sufficient flexibility to plan, oversee, and/or participate in all critical IACUC functions and activities. In addition, it is likely that the Chair will encounter occasional urgent matters that require immediate attention.

5:14 What is the Chair's role in the oversight of IACUC activities?

Opin. The Chair should play an active role in the oversight of all IACUC activities regardless of his or her role as an actual participant. The Chair serves five important constituent groups:

- The senior administration (embodied in the CEO and IO).
- The scientific community.
- Other members of the IACUC.
- The federal government.
- The public.

The immense commitment within the institution of time, resources, and interest to the IACUC's activities should compel the Chair's enthusiastic involvement in IACUC oversight. (See 3:1.)

5:15 Can the AV serve as the IACUC Chair?

Opin. While there is no prohibition in the PHS or AWAR against the AV serving as the IACUC Chair, this is not recommended.[2] Only a few institutions among the approximately 100 known to this author have taken this approach. NIH/OPRR has issued an explanation that strongly suggests that having either the AV or IO serve as the IACUC Chair may be inappropriate due to real or perceived conflicts of interest and the disruption of the necessary checks and balances intended in the institutional reporting structure and IACUC review processes.[6] The programmatic review activities of the IACUC encompass laboratory animal management and veterinary care, both of which are often under the direct purview of the AV. Thus, the appointment of the AV as the Chair affords the AV an opportunity to influence the IACUC openly or subliminally.

Similar efforts to dissuade institutions from continuing the use of this practice have been advanced routinely by AAALAC during site visits. AAALAC has made the argument that the responsibilities of the AV are already sufficiently demanding and should not be compounded by adding the complex job of the IACUC Chair. Moreover, AAALAC has recognized some instances were the AV does not carry the title of IACUC Chair but is functioning de facto as the Chair and carrying the workload of the IACUC.[6] Institutions should be wary of this devolutionary development.

5:16 Should the manager of the institution's laboratory animal resource unit serve as the IACUC Chair?

Opin. The appointment of the animal facility manager as the IACUC Chair is much more common than the appointment of the AV in this capacity. Facility managers have a broad understanding of the various aspects of the animal care and use program but generally have a narrower scope of responsibilities than the AV. This lessens the conflict of interest issue discussed in 5:15. The conflict of interest issue is not entirely eliminated, however, since the review of facilities and the diverse provisions for laboratory animal management are critical elements of the IACUC's programmatic review that generally fall within the domain of the facility manager. The involvement of the facility manager in the financial management of the resource may be yet another deterrent to the appointment of this individual as the Chair in some institutions.

5:17 In an academic setting, what should be the faculty rank of the IACUC Chair? Is a person with academic tenure preferred?

Opin. In many large academic institutions, the appointment of a full professor with tenure (and with an active research program or prior history of laboratory animal use) is deemed desirable. Academic rank, per se, is a less critical factor than an individual's scientific stature, collegiality, and reputation for fairness. Institutions should recognize that the faculty often directs its antagonism toward the Chair for adverse decisions made by the IACUC. Hence, the appointment of a nontenured faculty member (in a tenure track position) as the IACUC Chair conceivably might enhance that individual's vulnerability during tenure review decisions.

5:18 Can an IACUC be composed entirely of persons not employed by the university (or other institutions)?

Reg. Both the AWAR and the PHS Policy (see 5:2) require that at least one member of the IACUC not be affiliated with the institution in any way other than as a member of the IACUC. Other members are not required to be affiliated *except* that the veterinarian must have direct or delegated program authority (PHS Policy IV,A,3,b,2) and responsibility (AWAR §2.31,b,3,i) for activities involving animals at the institution. Also, pursuant to PHS Policy (IV,B), the IACUC functions as an "agent of the institution."

Opin. Neither the PHS Policy nor the AWAR stipulates that some (or any) of those appointed must be institutional employees. The pivotal issue in the matter of IACUC appointment is not whether the Chief Executive Officer (CEO) holds any financial sway over the committee members through employment, but rather, whether the CEO is capable of disbanding the IACUC if it proves to be incompetent, inattentive to animal welfare issues, or fails to discharge its federally mandated responsibilities with due diligence. It is conceivable that one organization participating in a cooperating research partnership with another organization may elect to co-appoint a knowledgeable IACUC which serves the research missions of both organizations, irrespective of the employment associations of the membership. This type of arrangement is rare and, in the experience of the author, is likely to be carefully examined by federal oversight and accreditation bodies.

5:19 Can or should an outside consultant serve as the IACUC Chair?

Opin. No institution known to this author uses an outside consultant as the Chair of the IACUC. This practice is not prohibited by the PHS Policy or the AWAR. The difficulty in appointing an outside consultant as the Chair is that such an individual lacks an appreciation for the scientific landscape and key players of the institution and will less effectively respond to issues in a timely and situationally appropriate manner. These difficulties might be overcome by an outside consultant with a long history of involvement with the institution.

5:20 What are the specific duties of the IACUC Chair?

Opin. The IACUC Chair has the responsibility for overseeing the coordination and implementation of effective, efficient systems for protocol review and program review by the IACUC in compliance with the PHS Policy and the AWAR. These review activities can only be performed at a properly convened meeting of the IACUC (see Chapter 6). Thus, the Chair should:

- Ensure that a quorum of the IACUC is present.
- Declare the loss of a quorum resulting in the end of official business if a sufficient number of members depart.
- Prepare or oversee the preparation of meeting minutes and reports and submit these documents to the IO in accordance PHS Policy (IV, E; IV,F) and the AWAR (§2.31,c,3; §2.31,d,2; §2.31,d,4; §2.35,a).
- Report to the IO any activities that have been suspended by the IACUC for noncompliance with PHS Policy (see Chapter 29).
- Establish a sound system of written communication for the IACUC with investigators concerning the approval status of protocols and the steps necessary to secure approval.

Beyond these relegated duties that the Chair performs on behalf of the Committee, most institutions expect the Chair to keep abreast of new regulatory trends and interpretations, and evaluate and champion policy and practice initiatives (e.g., new training and educational programs) to improve the animal care and use program. This involves the Chair's regular interaction with other areas of expertise within the organization, ranging from other institutional committees and occupational health to safety to human resources and the physical plant.

5:21 Must the institutional AV also serve as the veterinary member of the IACUC?

Reg. The PHS Policy (IV,A,3,b,1) and the AWAR (§2.31,b,3,i) state that the veterinarian on the IACUC shall have appropriate training or experience and have direct or delegated program responsibilities for activities involving animals at the institution.

Opin. The AV is not *required* to serve on the IACUC, although this is the strategy exercised in most institutions. Some institutions have deliberately avoided placing the AV or the veterinary program director on the IACUC. This is done to dissociate the overall program of veterinary care and its associated scientific interactions with investigators, from the monitoring and policing functions inherent in IACUC activities. Other institutions have used this allowance to rotate different veterinary members onto the IACUC periodically. In this author's view, these approaches have considerable merit.

5:22 What training or experience is useful for the veterinary member of the IACUC?

Reg. The PHS Policy (IV,A,3,b,1) and AWAR (§2.31,b,3,i) stipulate that the veterinary member of the IACUC should have training or experience in laboratory animal medicine and science. (See 27:2.) The AWAR (§1.1) further clarifies under its definition of the "Attending veterinarian" that this individual either should have graduated from a veterinary school accredited by the American Veterinary Medical Association Council on Education, have acquired a certificate issued by the American Veterinary Medical Association's Education Commission for Foreign Veterinary Graduates, or have received equivalent formal education as deemed appropriate by the APHIS administrator.

Opin. The intent of the PHS Policy and AWAR is to help ensure that the IACUC can rely upon the veterinary member not only for competent clinical insights, but also for information on ancillary areas such as unusual animal models, zoonoses and other occupational health and safety concerns, hazard containment, genetics, and unique nutritional and husbandry requirements of laboratory animal species. Proficiency in these diverse subject areas can be demonstrated formally by board certification in the American College of Laboratory Animal Medicine (ACLAM), and it is advisable for one or more of the veterinarians responsible for a large and complex program of research animal use to be board certified. However, the needs of research animal care and use programs that are smaller or of a more limited scope may be met by veterinarians who are not ACLAM board certified. Regardless of their ACLAM certification status, all veterinary personnel involved in the oversight of research animal use should be aware and stay abreast of the central research, clinical, and regulatory concerns pertaining to the species under their care. This can be accomplished by reviewing the literature and attending the national meetings sponsored by various organizations including: the American Association for Laboratory Animal Science, ACLAM, American Society of Laboratory Practitioners, Institute for Laboratory Animal Research, Association of Primate Veterinarians, American Veterinary Medical Association, AAALAC, and others.

5:23 What is the role of the veterinarian on the IACUC?

Reg. The role of the veterinarian within the context of the IACUC is defined under the AWAR (§2.31,d,iv,B; §2.31,d,vi; §2.31,d,vii; §2.31d,ix; §2.31,d,x,B) and by the *Guide* (pages 12 and 13) which serves as an extension and amplification of the PHS Policy. In summary, the AWAR include provisions for:

- Veterinary consultation on the recognition and palliation of pain.
- Direction of animal care and use (this is a specific role of the AV).
- Medical care.
- Aseptic surgery and post-operative care.
- Oversight of multiple major survival surgery resulting from a veterinary condition in an animal that also had experimental surgery.

Specific provisions within the *Guide* include:

- Advising the IACUC on new procedures or procedures with the potential to cause pain and distress that cannot be reliably controlled.
- Ensuring that veterinary care is available to mitigate the illnesses, lesions, or behavioral abnormalities associated with animal restraint.

These sources also expand on the responsibilities of the AV to the institutional animal care and use program that are unrelated to IACUC membership per se. (See Chapter 27.)

Opin. Veterinarians play a unique role on IACUCs, due to their expertise in laboratory animal medicine and science and broader understanding of the physiological, behavioral, nutritional, and husbandry needs of laboratory animal species. Although by no means a lone voice on the IACUC on matters of animal health and welfare, the veterinarian should make the concerted effort to address these areas with precision and authority. Special areas of emphasis for the IACUC veterinarian should include the discussion of the use of proper anesthesia and analgesia in laboratory animals in the relief of pain and distress, the analysis of possible iatrogenic complications to the procedures used or the disease model proposed, and a review of the plans for appropriate and timely medical intervention. (See 27:6.)

5:24 Under what conditions can the veterinarian delegate some of his responsibilities to other IACUC members?

Reg. According to the AWAR (§2.31,d,1,iv,B), the AV may delegate his or her responsibility for playing a role in the planning and consultation on procedures that may cause more than momentary or slight pain or distress.

Opin. Depending on the background, training, and experience of the other IACUC members, and the size, complexity, and maturity of the animal care and use program, many of the veterinarian's responsibilities can be delegated to other members for defined periods (e.g., a particular IACUC meeting or semiannual review). Very few of the institutions known to this author would proceed with an IACUC meeting in the absence of the veterinarian, but none is prepared to neglect significant issues due to insufficient veterinary input. The strategy most often used when the IACUC veterinarian is absent is to invite another veterinarian, knowledgeable in laboratory animal medicine, to participate as a guest in an advisory capacity. Secondarily, IACUCs can table issues requiring the veterinarian's special analysis and comment. (See 6:7, 27:8.)

5:25 What is meant by the "nonaffiliated (outside) member" of the committee and what is the intent of including such an individual on the committee?

Reg. The intent of including a nonaffiliated member on the IACUC is to ensure that someone who is not affiliated with (or beholden to) the institution in any manner is involved in the review of the institution's animal care and use activities. This individual also may not be a member of the immediate family of a person affiliated with the institution (AWAR §2.31,b,3,ii; PHS Policy IV,A,3,b,4; *Guide,* page 9**).** The *Guide* also stipulates that public members should not be laboratory animal users. According to the AWAR (§2.31,b,3,ii), this individual is intended to represent the "general community interests in the proper care and treatment of animals." (See 3:4.)

Opin. Subsequent clarification on this matter by NIH/OPRR has indicated that it is inappropriate for an institution to have an individual who currently uses laboratory animals serving in this capacity.[7] It does not rule out individuals who have used laboratory animals in the past. Ostensibly, because the nonaffiliated member is not expected to benefit personally from any of the activities proposed and does not depend on the institution for her livelihood, she enhances the public's confidence in the unfettered objectivity of the IACUC review processes.

5:26 What are the backgrounds of individuals typically serving as nonaffiliated members of IACUCs?

Opin. There is no typical background for these individuals. Institutions have chosen a wide variety of individuals to serve in this capacity. Without implying that the following categories are mutually exclusive of one another, the range of individuals includes broadly educated, erudite humanists; business persons; public servants; educators; persons involved in other types of animal care and use;

physicians, veterinarians, scientists, and technicians who do not use laboratory animals; and concerned citizens seeking to make a public contribution.

5:27 Can the nonaffiliated member be a scientist, veterinarian, ethicist, or biostatistician who is not affiliated with the institution?

Reg. Yes.

Opin. Any of the above would qualify as long as they are not involved in the care and use of laboratory animals, either directly or through consulting and similar tangential service. (See 5:25.)

5:28 Can or should the nonaffiliated member be compensated for service on the IACUC?

Opin. Neither the PHS Policy nor the AWAR prohibits the compensation of the outside member. NIH/OPRR has stated that nominal compensation is permissible without jeopardizing a member's nonaffiliated status if it is only in conjunction with service on the IACUC, and is not so substantial as to be considered an important source of income or to influence voting on the IACUC.[8] Most institutions offer no more than payment of incidental expenses and a modest meal during IACUC activities. (See 5:8.)

Several institutions have an annual dinner or luncheon honoring the IACUC with the CEO and other senior administrators in attendance. Some institutions have taken the next step in compensation by offering the nonaffiliated member a modest honorarium or by making a contribution in her name to a charitable organization of her choosing. In the rarest situation, organizations compensate their nonaffiliated members at the rate of the member's employer or at a rate which they will not disclose. They argue that this does not bias the individuals because it is not their primary source of income and that integrity cannot be bought. Most readers and most of the lay public are likely to agree that this latter scenario produces the perception of a conflict of interest and cannot be recommended.

5:29 Can the nonaffiliated member serve on more than one IACUC?

Opin. There is no proscription against nonaffiliated members serving on more than one IACUC if they have an interest in that level of involvement. Skeptics may speculate that the excessive involvement of an individual as a nonaffiliated member on several IACUCs is an indication of a shift in the individual's neutrality and objectivity, and that this might be a reason for institutions to avoid the "overzealous" contributor. Also, scientific staff may be concerned that the individual is serving as a source for the leak of scientific ideas to other institutions.

5:30 What is the role of the nonaffiliated member on the IACUC?

Opin. The role of the nonaffiliated member is not really defined other than as described in 5:25. Most institutions are hopeful that this individual will play an active role in all IACUC activities and will be comfortable with and learn to become adept at making persistent, straightforward, and disarming inquiries about matters that are undetected by the institutional members on the IACUC.

5:31 What are the background and qualifications of individuals typically filling the role of the nonscientist?

Reg. The PHS Policy (IV,A,3,b,3) indicates that the "primary concerns" of individuals serving in this capacity should be "in a nonscientific area," and the specific examples cited (e.g., ethicist, lawyer, member of the clergy) have no obvious connections to any area of science. The AWAR do not identify the need to appoint an IACUC member in this particular category.

Opin. In many institutions, individuals have been chosen who have some scientific training and perhaps even some responsibilities in a scientific area, but who clearly do not qualify as a "practicing scientist with experience in research involving animals" as noted in PHS Policy (IV,A,3,b,2). The types of individuals seen in this role among the institutions surveyed include lawyers, clergy, health and safety personnel, business and human resources personnel, public relations personnel, quality assurance/control personnel, and technicians not involved with animal care or use.

5:32 Is a biostatistician considered a nonscientist?

Opin. Surely, statisticians are scientists because statistics is a branch of mathematics, a pure science. However, as noted above, most institutions would classify a biostatistician as a "nonscientist" even if his work may entail the analysis of animal studies. Although the biostatistician might be involved in the application of a mathematical science to the analysis of animal studies, few people would be seduced by the proposition that a biostatistician would be tempted to encourage animal studies simply to perpetuate the data needed for the practice of his science. (See 5:37.)

5:33 What interests does the nonscientist represent on the IACUC?

Opin. This individual serves to further diversify the IACUC membership adding to the balance of foils for the scientific members who may be regarded as having a vested interest in the promotion of animal studies.

5:34 What is the role of the nonscientist on the IACUC?

Opin. As with the nonaffiliated member, the nonscientist should participate in and contribute to all of the IACUC's mandated activities. With regard to protocol review, this member can be especially valuable by working to ensure that the approach and justification for the proposed animal studies is understandable to nonscientific persons and that humane care and study endpoints have been included in the study design.

5:35 Should the scientist on the IACUC have animal research experience?

Reg. PHS Policy (IV,A,3,b,2) and the *Guide* (page 9), but not the AWAR, indicate that an IACUC should include at least one practicing scientist with laboratory animal experience in its membership.

Opin. This is a universal practice among the institutions visited by this author. In large institutions, typically a number of scientists are appointed to the IACUC to reflect the interests of different user groups in the organization. The appointment of scientists with laboratory animal experience aids the IACUC's discussion of relevant issues during protocol and program review. It helps the IACUC better understand the selection, use, and limitations of animal models and certain aspects of experimental design.

5:36 Should the scientific member of the IACUC be a senior-level scientist or is a junior-level scientist acceptable?

Opin. Scientists at any level are appropriate, but many institutions feel that senior scientists bring more authority and credibility to the IACUC. However, most academic organizations have found the recruitment of senior scientists to the IACUC to be difficult because these individuals usually have already made significant committee contributions to the institution in other areas and have other significant obligations within the institution.

5:37 What is the role of the scientist on the IACUC?

Opin. The principal role of the scientist on the IACUC is to ensure that the interests of scientific colleagues are being fairly represented in the review process and to aid in the IACUC's assessment of the relevance, validity, and technical aspects of the studies proposed. Of course, the scientist recognizes the confluence of animal health, welfare, and scientific interests in research animal studies and plays an important role as a proponent for the prudent, ethical, and humane use of animals. In broader issues involving program development and implementation, the scientist can bring to the IACUC perspectives on how to best launch new initiatives to engender the support of the scientific community and others involved in the care and use of laboratory animals.

5:38 Should the IACUC include an ethicist as a member?

Opin. Some institutions have been successful in recruiting one or more individuals knowledgeable in ethics with an education and work focus in the arts and humanities or in business, but few have been able to identify a bona fide ethicist. Clearly, issues in ethics may be important in the IACUC's deliberations and there may be occasions when an ethicist can be helpful in leading this process.

5:39 Should the IACUC include a biostatistician as a member?

Opin. Most IACUCs have not perceived the need to include a biostatistician to ensure that investigators are using "the minimum number (of animals) to obtain valid results" (PHS Policy IV,D,1,a–IV,D,1,b; *Guide* (page 10) and Appendix D; U.S. Government Principle III; AWAR §2.31,e,1; §2.31,e,2) because they expect the scientist and veterinarian on the IACUC to evaluate this area. Indeed, in some cases the scientific, veterinary, or other IACUC members may be capable of performing this function if they can commit the effort and are given sufficient information about the assumptions on the underlying database anticipated in the experimental study. However, to confirm that the investigator is proposing to use the minimal number of animals necessary can involve an extensive and potentially redundant effort on behalf of the IACUC. If an IACUC routinely encounters egregious explanations for the numbers of animals requested in protocols, the addition of a biostatistician to the IACUC may be helpful.

5:40 Should animal care technicians be considered for IACUC membership?

Opin. In this author's experience, several institutions have appointed animal care technicians to their IACUCs in an effort to develop a better sense of how the basic animal care program functions. An added benefit is that it may promote a sense of empowerment and inclusion in the IACUC review processes for the entire animal care staff.

5:41 Should individuals with specialty expertise be included as voting, nonvoting, or ad hoc IACUC members?

Reg. PHS Policy (IV,C,2; IV,C,3) and the AWAR (§2.31,d,2; §2.31,d,3) clearly indicate that voting privileges are reserved for full IACUC members who are appointed by the chief executive officer. Nevertheless, the PHS Policy and AWAR do not preclude the use of nonvoting consultants.

Opin. Individuals with specialty expertise might include properly appointed ad hoc members who are willing to participate in the full range of IACUC activities. Individuals with specialty expertise who are not interested in assuming all of the responsibilities of an IACUC member but who would be helpful to the IACUC on an episodic basis, can be invited to participate in an advisory capacity as a consultant without voting privileges. The types of ad hoc appointees by function that might be considered include: librarians for alternative searches, environmental or occupational health and safety personnel for hazardous studies, ichthyologists, herpetologists, or ornithologists for studies involving species in their respective disciplines. These contingencies should be addressed in the bylaws developed by the institution for the IACUC.

5:42 Should the IACUC seek the advice of consultants for issues that may require special expertise, such as pain and distress concerns?

Opin. IACUCs have the freedom to exercise their ability to use consultants whenever necessary. However, in most institutions, their use is reserved for truly extraordinary issues as opposed to any areas in which the IACUC feels it might improve its decision making. In this author's opinion, many IACUCs would potentially benefit from the more liberal use of consultants.

5:43 Should consultants be from inside or outside of the institution and should they be anonymous?

Opin. Consultants can be selected from either inside or outside the institution depending upon the availability of relevant expertise and the resolution of concerns about objectivity. Most organizations have not found it practical, necessary, or desirable to maintain the anonymity of the consultant. However, if the need for a consultant is likely to involve a highly contentious and disputed, high-stakes issue, anonymity may be elected by the IACUC or imposed by the consultant as a condition of participation. Regardless of the circumstances, the IACUC has the obligation to foster an environment conducive for the investigator to respond to the consultant's critique.

5:44 Should consultants to the IACUC be compensated?

Opin. External consultants frequently are compensated for their efforts, and it seems reasonable that these individuals should be compensated according to the institutional policies established for other types of external academic review.

5:45 Should there be a confidentiality agreement between the consultant and the institution?

Opin. It is advisable for institutions to have a confidentiality agreement with consultants to prevent the disclosure of important, and potentially patentable, scientific or technical information. In addition, confidentiality agreements are important because they establish or reaffirm a precedent for the manner in which all materials relevant to the IACUC's deliberations are handled. This becomes particularly important from the legal standpoint if the institution is faced with the prospect of retrieving documents that may have been obtained illegally by adversarial parties.

5:46 What support staff is useful for the IACUC?

Opin. Many institutions have developed a support staff to assist in the administrative, monitoring, and training activities of the IACUC. The composition of this staff depends upon the size and scope of the animal care and use program and the degree to which the IACUC members are willing and capable of becoming personally involved in the fine details of committee function. The type of positions that have proven to be very useful on the IACUC support staff in different settings include a director or senior assistant to the IACUC Chair, training, monitoring, and compliance personnel, computer/database specialists, biostatisticians, and clerical staff. In small institutions with limited programs, a support staff of 0.25 full-time equivalents (FTEs) or less may be sufficient, but in large, complex programs it is not uncommon for the IACUC support staff to include 3 to 5 FTEs.

5:47 Who typically provides support staff to the IACUC?

Opin. The support staff is generally recognized and provided as an administrative cost that is provided under the authority of the CEO or IO of the organization.

5:48 What are some of the typical responsibilities of the support staff?

Opin. Support staff is often involved in preparing the correspondence related to semiannual reviews, protocol matters, policies and procedures of the IACUC, and maintaining the records thereof. The staff also may be responsible for making the arrangements for all IACUC activities. If staff members have appropriate training and technical expertise, they may be involved in conducting institutional training seminars and visiting laboratories and animal use areas to provide instruction and to monitor ongoing activities. This staff may maintain databases of IACUC protocol approval and of the appropriate ancillary approvals in other areas such as use of biohazards, and the enrollment in occupational health and safety activities. In a few institutions, support staff has been available to conduct literature searches for alternatives to animal use and to provide a statistical analysis to confirm that the investigator has properly justified the number of animals requested.

References

1. Committee to Revise the Guide for the Care and Use of Agricultural Animals in Agricultural Research and Teaching, *Guide for the Care and Use of Agricultural Animals in Agricultural Research and Teaching*, 1st revised ed., Federation of Animal Science Societies, Savoy, IL, 1999.
2. U.S. Department of Agriculture, Animal and Plant Health Inspection Service, Policy #15, IACUC Membership, April 14, 1997. Available on the World Wide Web at: *http://www.aphis.usda.gov/ac/policy15.html*
3. National Research Council, Institute of Laboratory Animal Resources, Occupational Health and Safety in the Care and Use of Research Animals, National Academy Press, Washington, D.C., 1997.
4. Robert, H.M., *The Scott Foresman Robert's Rules of Order Newly Revised*, Robert, S.C., Robert, H.M. III., and Evans, W. J., Eds., Scott, Foresman and Company, Glenview, IL, 1990.
5. Division of Animal Welfare, Office for Protection from Research Risks, National Institutes of Health, Frequently asked questions about the Public Health Service Policy on Humane Care and Use of Laboratory Animals, *ILAR News*, 35 (3–4), 47, 1993.
6. Newcomer, C.E., Behold! The animal care and use magician! *AAALAC Int. Connect.*, Fall, 1997.
7. Division of Animal Welfare, Office for Protection from Research Risks, National Institutes of Health, Maintenance of properly constituted IACUCs, *OPRR Reports*, No. 97-03, Animal Welfare, June 2, 1997. Available on the World Wide Web at: *http://www.nih.gov/grants/oprr/dc97-3.htm*
8. Division of Animal Welfare, Office for Protection from Research Risks, National Institutes of Health, Frequently asked questions about the Public Health Service Policy on Humane Care and Use of Laboratory Animals, *ILAR News*, 33 (4), 68, 1991.

6

Frequency and Conduct of Regular IACUC Meetings

Sreekant Murthy

Introduction

A comparison of the general issues for IACUC composition and functions shows that the PHS endorses the U.S. Government Principles I to XII for the Care of Vertebrate Animals Used in Testing, Research, and Training. Principle I states: "The transportation, care, and use of animals should be in accordance with the Animal Welfare Act (7 U.S.C. 2131 et. seq.) and other applicable federal laws, guidelines, and policies." The guidelines include those in the *Guide*. The AWA requires that proposed activities using animals be conducted in accordance with the AWAR. Based on these acts, regulations, policies, and principles, IACUCs are constituted to protect the health and well-being of the animals used in testing, research, education, and training. The AWAR (§2.31,a) state that "nothing in this part shall be deemed to permit the Committee or IACUC to prescribe methods or set standards for the design, performance, or conduct of actual research or experimentation by a research facility." This rule, and Principle II of the U.S. Government Policy, ensure animal care and use without compromising research for the good of the society. Therefore, a carefully composed IACUC that oversees humane use and well-being of the animals and facilitates animal research is fundamental to a research facility's teaching and research objectives and ultimately to the good of society. This chapter addresses critical issues that are pertinent to the construction and function of the IACUC.

The surveys presented in this chapter originate from questionnaires sent to largely academic institutions.

6:1 How is the frequency of meetings of the IACUC decided?

Reg. IACUCs are required by the AWAR (§2.31,c,1; §2.31,c,2) and PHS Policy (IV,B,1–IV,B,3) to conduct a review of its animal care and use program every 6 months. Thus, the AWAR and PHS Policy require at least one IACUC meeting every 6 months to review animal care and use policies. APHIS/AC requires that a majority of the IACUC review and sign the ensuing report (§2.31,c,3), although for this purpose the AWAR does not require a quorum to be present. NIH/OPRR has stated that "… final reports must be reviewed and endorsed by a convened quorum. …"[1] However, unlike the AWAR, there is no requirement for signatures.

There is a mandated not-less-than annual review of all IACUC-approved protocols (AWAR §2.31,d,5), and not less than every 3 years under the PHS Policy (IV,C,5). These reviews *do* require a quorum of the IACUC to be present.

The *Guide* (page 9) states that "the IACUC must meet as often as necessary to fulfill institutional responsibilities, but it should meet at least once every 6 months." Thus, the absolute number of convened meetings can vary from two to as many as needed to meet regulatory and institutional requirements.

The PHS Policy (IV,B,2) uses the *Guide* as the basis for evaluation; therefore, it endorses the requirements of the *Guide* for the committee to meet as often as necessary to fulfill its responsibility.

Opin. Each institution must evaluate its responsibilities to the investigator and funding organizations. Based on the number of IACUC applications, the institution should conduct as many meetings as necessary to review protocols. Many institutions decide on the frequency based on NIH or other sponsor deadlines that may include industry-related and internal applications. It is this author's opinion that animal welfare is best served by having the semiannual review of the program of animal care and use (including facility inspections) at a convened meeting of the IACUC with a quorum present.

6:2 Who determines the frequency of IACUC meetings?

Opin. There is no regulation or policy that authorizes specific individuals to determine the frequency of meetings. Either the Chair, or the Chair after consulting members of the IACUC, the IO, or the Office of Research Administration can determine the frequency of meetings. In some instances, the institution's internal policy (including its bylaws) can authorize the Chair or Office of Research Administration to make such decisions. The scheduling of a meeting is recommended to be done by the IACUC after consultation with the Chair.

Surv. At your institution, who determines the frequency of IACUC meetings?

- There is no policy   19/63
- IACUC Chair   13/63
- IACUC Chair after consulting IACUC members   29/63
- Institutional Official   2/63

6:3 What parliamentary rules are used in conducting IACUC meetings?

Reg. The PHS Policy (e.g., IV,C,2; IV,C,6) and AWAR (e.g., §2.31,d,2; §2.31,d,6) address some aspects of parliamentary procedure, such as conducting full reviews of protocols at a convened meeting of a quorum of the IACUC, and voting in case of a suspension. NIH/OPRR has noted in a Dear Colleague letter[2] that the validity of IACUC activities is always predicated on the existence of a properly constituted IACUC, although there is no requirement that all of the members be present at all meetings. Institutions may not allow Robert's Rules of Order or other parliamentary rules to supplant those procedures which are specified in PHS Policy or the AWAR.

Opin. None of the regulatory agencies or the *Guide* prescribe a rule for conducting the IACUC parliamentary procedures. Nevertheless, many institutions use either Robert's Rules of Order or their institution's own parliamentary rules as prescribed in the bylaws of that institution.

Surv. At your institution, what parliamentary rules are used to conduct IACUC meetings?

- Robert's Rules of Order   26/63
- Institutional internal rules   12/63
- Robert's Rules of Order and internal institutional rules   25/63

6:4 What information should be provided to IACUC members prior to a meeting?

Reg. The AWAR (§2.31,d,2) and PHS Policy (IV,C,2) state: "prior to IACUC review, each member of the committee shall be provided with a list of proposed activities to be reviewed. Written descriptions of all proposed activities that involve the care and use of animals shall be available to all IACUC members, and any member of the IACUC may obtain, upon request, full committee review of those activities."

Opin. It is customary for many IACUCs to provide all correspondence concerning a review of a protocol. This includes correspondence concerning allegations of protocol violations or animal mistreatment, results of subcommittee investigations (particularly those that may lead to suspensions that require review and approval by the full committee). Also, reports of any task forces that the IACUC has requested, updates on institutional policies that affect the conduct of an IACUC meeting, membership changes, updates on APHIS/AC and NIH/OPRR regulations, copies of all pertinent publications, and meeting notices that may enhance the review process.

6:5 What additional verbal or written information might be given to IACUC members at the time of a regular meeting, but before research protocols are discussed?

Opin. Verbal information that can be presented at a convened meeting may include reports from the Chief Executive Officer, IO, Director of Office of Research, Chair of the IACUC, AV, and Supervisor of the animal care facilities. Other materials may include opinions by external consultants and any other reports that the Chair or the institution deems necessary for the proper conduct of IACUC functions.

Written information may include minutes of a previous meeting for approval, agenda for the convened meeting, summaries of protocol reviews, external consultant's reports, and any other handouts that are pertinent for the conduct of IACUC functions.

Some IACUCs conduct separate quarterly, semiannual, or annual business meetings to discuss various issues that have been the general concern to the IACUC. These meetings are held for the purpose of discussing IACUCs administrative procedures. Such meetings may include:

- Evaluation of surveys conducted by the IACUC.
- Recruitment of new members.
- Review of attendance records.
- Training new members.
- Updating IACUC forms.
- Educational matters.
- Methods to train and certify animal users.
- Creation and dissemination of newsletters for animal users.
- Security issues.
- Ethical issues for the appropriate use of animals in research and education.

An additional purpose is to reserve the IACUC's routine meetings to focus its undivided attention on the review of the institution's animal care and use program and protocols for animal use.

6:6 What specific business should be conducted at regular IACUC meetings? Is a formal vote required for any business conducted?

Reg. The AWAR (§2.31,c,1–§2.31,c,8; §2.31,d,1), PHS Policy (IV,B,1–IV,B,8; IV,C,1–IV,C,8) and the *Guide* (pages 9 and 10) require the following specific business to be conducted at the meetings:

- Review and approval of reports of an institution's program for the humane care and use of animals and inspections of animal facilities, including animal study areas (laboratories), before submitting them to the IO.
- Discuss and distinguish significant deficiencies from minor deficiencies included in the report above report.
- Discuss and provide a reasonable and specific plan and schedule (with dates) for correcting each deficiency.
- Review, and if review warrants, investigate concerns involving animal care and use complaints from all sources.
- Make recommendations to the IO on all aspects of institutional animal program and training of personnel who handle animals.
- Review and approve or require modifications or clarifications or withhold approval of proposed and ongoing activities with significant changes related to animal use.
- Suspend an activity involving animals that compromise the health and well-being of an animal used in research (PHS Policy IV,C,6).
- Any business that the Committee considers threatening to, or has already affected, the health and well-being of the animals used in their research facility.

Opin. Many of the above regulations are pertinent for the semiannual review of institution's animal care and use program. Any institutional program that is reviewed at this meeting, whether the program does or does not have deficiencies (minor or significant) and a plan of action recommended by the committee must be formally approved by a majority of the quorum present, and all minority opinions also must be recorded. These items are distinctly different compared to those that are discussed in a routine meeting. Routine meetings are usually reserved for reviewing research protocols. However, there is no reason for not conducting both aspects of IACUC business in a single convened meeting.

Many IACUCs have agendas ranging from discussion of the institutional program of animal care and use, review of protocols, and other general business of the IACUC. All business matters that pertain to institutional program evaluations, protocol reviews, suspensions, punitive actions (if the institution has authorized the IACUC to do so), and any other matters that are part of the AWAR and PHS Policy (or institutional policy) must be approved by the majority of quorum present. All other business matters should be discussed and, if the committee is developing a policy, such policies must be approved by a majority of members.

Surv. What specific business should be conducted at the IACUC meetings? Is a formal vote required for any business conducted?

- All IACUC business matters 55/63
- Review and approval of protocols only 2/63
- A separate business meeting which excludes protocol review 6/63
- A separate business meeting which includes protocol review pending from a previous meeting 0/63

6:7 Which IACUC members must be in attendance to legally conduct a meeting?

Reg. The AWAR (§2.31,b,1–§2.31,b,4) and the *Guide* (page 9) require that the IACUC consist of at least three members, including a veterinarian and a person not affiliated with the institution. PHS Policy (IV,A,3,b; IV,A,3,c) requires a minimum of five members, including a veterinarian, a scientist, a nonscientist, and a nonaffiliated member. For a quorum, the minimum attendance is three for the PHS Policy; thus, any of the two members required could be absent, yet the meeting could still be legally conducted. (See 6:8 for a definition of quorum.)

The AWAR (§2.31,d,1,iv,A; §2.31,d,1,iv,B) require that for procedures involving more than momentary or slight pain or distress, either the veterinarian or his or her designee must be consulted in the planning of the research project.

Opin. Ideally, the nonaffiliated and nonscientist (under PHS Policy) members should be present at meetings to address ethical issues and appropriate use of animals from the noninstitutional and nonscientific perspectives. However, it is not legally required for any specific member to be present at all meetings, as long as the committee membership is properly constituted and a quorum is present.[2]

When research projects involve pain or distress, the veterinary consultation on analgesia or anesthesia is normally provided prior to IACUC review by the AV. Another approach to this requirement is to have the AV identify one or more designees who can be involved in planning and consultation. Another veterinarian with knowledge in

	this area is an appropriate choice for such designation. Nonveterinarians also may be involved in the consultation at the discretion of the AV. (See 5:24; 27:8.)
Surv.	What is your IACUC's policy in regard to attendance of specific IACUC members at a meeting?

- There is no policy — 29/63
- All members as required by regulation or policy must be present — 9/63
- Attending veterinarian must be present — 9/63
- At least some members as required by regulation or policy must be present — 16/63

6:8 What constitutes the quorum needed to conduct an IACUC meeting?

Reg.   A quorum is an assembly of a majority (more than 50%) of the voting members of the IACUC. For AWAR (§2.31,b), the minimum number is two and for PHS Policy (IV,A,3,b), the minimum number is three (provided the IACUC has only three and five members, respectively). Otherwise, it requires more than 50% of members to be present in a convened meeting to constitute a quorum.

No member may participate in the IACUC review of an activity in which that member has a conflicting interest (e.g., is personally involved in the activity), except to provide information requested by the IACUC; nor may a member who has conflicting interest contribute to the constitution of a quorum at which a proposed activity relevant to the conflict is being considered for approval (AWAR §2.31,d,2; PHS Policy IV,C,2). The IACUC may invite consultants to assist in the review, but they may not approve or withhold approval of an IACUC activity and may not vote unless they are members of the IACUC.

NIH/OPRR has accepted the practice of designated alternates for IACUC members, if the alternates are formally appointed by the CEO and identified in the Animal Welfare Assurance. In these instances, the individual(s) serving as alternate(s) do not add to the total number of committee members.

Opin.   There is often a problem convening a quorum of the IACUC. Some institutions apply many methods to attain or maintain quorum at a convened meeting. These methods can include calling members of the committee ahead of the meeting date to alert them about an upcoming meeting or arranging a meeting on a day that is most suitable for all or a majority of the members. Other ideas include designating alternative members or providing food at the meetings so that the members can devote their lunch, breakfast, or dinner time for the meeting. All of these methods are workable and there are no AWAR or PHS Policy limitations for applying these methods.

Generally it is advisable to have the IACUC membership at more than the minimum number mandated by either the AWAR or PHS Policy. For example, it is possible that one or more members may leave the meeting at the most inopportune time due to prior commitments or unavoidable circumstances. This might result in the loss of a quorum if only the minimum number of persons are on the IACUC and present at the meeting. Conflicting interest and abstentions are difficult to resolve if an IACUC membership is small.

The problem of a member leaving a convened meeting can often be avoided if protocols reviewed by that individual are placed on the top of the agenda. Another method to avoid problems with maintaining a quorum is by a full committee review via mail and subsequent polling of votes. (See 8:20.) *The reader is cautioned, however, that this opinion is not consistent with the NIH/OPRR position,*[1] *which states:*

> *The simple polling of IACUC members does not, however, satisfy the definition of a meeting of a convened quorum and should not be used for conducting IACUC business that requires the vote of a convened quorum of the committee. For example, polling should not be considered a valid method of voting under the "full committee review" method of protocol review and is not an acceptable substitute for having a vote of a convened quorum on the suspension of a previously approved activity involving animals.*

The author's opinion also is not consistent with APHIS/AC policy[3] *which is that:*

> *E-mail, polling, conference calls, etc., are typically not acceptable methods for full committee review. This requires a face-to-face meeting (see 6:11). It is acceptable to poll members to see if any want a full committee review.*

Surv. What constitutes the quorum needed to conduct an IACUC meeting?

- Above 50% (a quorum) of the membership   63/63

At 15% of the institutions surveyed, the IACUC of those institutions required all five members mandated by PHS Policy to be present at each meeting. The AWAR require only three members to constitute an IACUC. However, the majority of institutions that submit federal grants must adhere to the PHS Policy, which requires the institution to constitute an IACUC consisting of at least five members. Thus, this question was posed requiring at least five members as voting members on the IACUC.

6:9 If a quorum is in place at the beginning of an IACUC meeting, but the quorum is lost during the meeting, is there any way that the meeting can continue?

Opin. There are no regulations that stipulate whether a meeting should be suspended if the quorum is lost in the middle of a meeting. However, the AWAR and PHS Policy (see 6:8) state that a quorum of members must be present to approve, withhold approval, or perform other IACUC actions; therefore, all IACUC activities that relate to voting by a majority of the quorum will become suspended. In such exceptional circumstances, the Chair may decide to review a protocol via the designated member method (see 11:8; 9:19–9:24). Alternatively, to avoid issues related to the maintenance of a quorum, the institution may amend its assurance to conduct a review by electronic means; however, even such electronic reviews require stringent policies that are described in 6:11 and 6:12.

6:10 Should any or selected investigators be invited to the IACUC meeting?

Opin. It is customary for some IACUCs to invite investigators to respond to some questions in a face-to-face meeting. Since the committee has the dual responsibilities of thoroughly reviewing protocols and executing decisions, the PI can answer specific questions raised by the IACUC. This often saves time for the IACUC and the PI. The PI should leave the meeting before a vote is taken. If the protocol involves a committee member that has a conflicting interest, the member is asked to leave the meeting while the committee is discussing that protocol (although the person may give factual information to the committee if requested to do so). The member with a conflicting interest should leave the meeting when the committee votes.

6:11 Can electronic methods be used for conducting IACUC meetings?

Opin. Recent advances in electronic communication methods have raised questions about the acceptability of teleconferences, audio-visual conferences, fax transmissions, electronic mail, and postal mail as alternate methods to face-to-face IACUC meetings. The NIH/OPRR does not fully endorse the use of these alternative methods to a traditional meeting since IACUCs conduct diverse activities and some of these activities require direct interactions, voting, and good recordkeeping of minutes. Be that as it may, NIH/OPRR considers that there may be "exceptional circumstances" in which one or more of the foregoing options could be used.[1] This requires the inclusion of a section on "alternate methods to conduct a meeting" in the institution's NIH/OPRR Assurance, with subsequent approval by that agency. The requirements are

6:12 *Frequency and Conduct of Regular IACUC Meetings*

- The Animal Welfare Assurance file with the NIH/OPRR contains a description of the procedures that the IACUC will follow to fulfill the requirements of the PHS Policy, which may include description of *an alternate method to conduct a meeting under exceptional circumstances.*
- NIH/OPRR expects that all approved Assurances are complete and reflect the use of the nontraditional procedures used to conduct IACUC business, if such procedures are used.
- Conference calls, audio-video conferences, and possibly some forms of highly interactive online computer discussion groups may qualify in exceptional circumstances. The exact definition of "exceptional circumstances" is not provided.
- The conveyance, by fax or electronic mail, of information such as the institutional Assurance, animal study proposals, agendas and minutes of meetings, institutional policies and standard operating procedures, reports, announcements or correspondence from oversight or regulatory agencies, and other matters related to the institutional animal care and use program for consideration and review by IACUC members, would be regarded as appropriate. (See 6:8; 6:12; 8:20 and 8:21.)

Surv. Does your IACUC have face-to-face (in person) meetings or are other means used?

• Face-to-face meetings	57/63
• Conference call using the voice mail system (these IACUCs used conference calls in addition to face-to-face meetings)	6/63
• Video conferencing	0/63
• Electronic mail or letters	0/63

6:12 What are the requirements for conducting other than face-to-face IACUC meetings ("exceptional circumstances" meetings)?

Opin. In exceptional circumstances, telephone and audio-visual conferencing may be appropriate alternatives to face-to-face meetings for full committee review of protocols and other IACUC matters (see 6:11). The requirements for exceptional meetings are[1]

- All members must be given ample prior notice to participate.
- At least a quorum of the voting members (>50%) must be convened on the same conferencing line whether it is telephone or an audio-visual line.

- The quorum of IACUC members must be in direct communication with each other and be given full opportunity to participate for the duration of the meeting.
- The minutes of the meetings must be compiled and maintained on file as required by oversight and regulatory agencies.

6:13   How can a Designated Member Review (see 11:8; 9:19–9:24) of protocols be accomplished using the electronic media meeting format?

Reg.   (See 9:19.)

Opin.   A designated member review means there is an IACUC Chair-appointed reviewer(s) to review an IACUC protocol. Designated member review also can be conducted using electronic media to facilitate the review process as long as the requirements in the PHS Policy (IV,C,2) are met. (See 6:12; 9:20.)

6:14   Can IACUC reports be endorsed by electronic methods and can signatures of members be obtained by using electronic methods?

Opin.   Since most reports, including semiannual program review and inspection reports, require IACUC approval (see 6:1), alternate methods that exactly follow those used at a legally convened face-to-face meeting of the IACUC are appropriate. (See 6:5.)

IACUC activities generally require documenting votes or verifying committee approval. They also may require signatures to be legally binding. Electronic methods of voting are conducted quite rapidly, but signatures are difficult to obtain by this method. Therefore, in an "electronic meeting" the Chair calls for a vote, the IACUC coordinator records the vote, and then prepares and distributes the final document by mail to solicit original signatures for the permanent record.

6:15   Must the minutes of IACUC meetings be recorded? If so, what precautions should be taken to maintain confidentiality?

Reg.   PHS Policy (IV,E,1,b) and the AWAR (§2.35,a,1) require records of IACUC meetings to be kept, whether the meeting is face-to-face or held alternatively using electronic communications (see 6:11; 6:12). They must be compiled and maintained on file for 3 years (AWAR §2.35,f; PHS Policy IV,E,2). Minutes must include records of attendance, activities of the Committee, and Committee deliberations (PHS Policy IV,E,1,b).

Opin.   While recording meeting minutes satisfies regulatory needs, special care must be taken to ensure proper usage of words since meeting minutes can, under certain circumstances, be obtained through a state's Freedom of Information Act (FOIA) or any state open record laws (the federal FOIA is at present only applicable to documents in the possession of a federal agency). (See Chapter 22.)

It is imperative that each IACUC take into account potential security problems of any privileged and trade-secret confidential information which may be protected under the AWA (§27). Since minutes are often generated using computers, special care should be taken to prevent unauthorized access to computer records. The mandatory enforcement of passwords, controlled access and encryption should be practiced to prevent unauthorized access. IACUC members should be advised to return copies of the minutes to the IACUC secretary once the minutes are ratified by the committee. Unwanted and used copies of the minutes should be destroyed using appropriate methods.

Many IACUCs record the meeting on an audiotape. This raises a question as to whether these tapes should also be saved as part of recordkeeping. This is internal policy unique to each IACUC — whether to save or erase the tape. In general, these tapes also contain irrelevant discussions; therefore, it is customary for the IACUC to transcribe portions of the discussion that are relevant and erase the tape for future use. There are no regulations that explicitly deal with audiotaping and archieving of such tapes. (See 8:19; 8:22; Chapter 22.)

Surv. What information does your IACUC record in its meeting minutes?

- Only discussions that are pertinent to IACUC business   61/63
- All items that are discussed at the meeting   2/63

6:16 Should meeting minutes include voting results and specific names of individuals and how they voted?

Reg.   AWAR (§2.35,a,1) and PHS Policy (IV,E,1,b) state that the minutes of IACUC meetings include records of attendance and activities of the committee. The AWAR (§2.31,d,2) and PHS Policy (IV,C,2) do not permit any member who has a conflicting interest to review, approve, or contribute to the constitution of quorum. The AWAR and the PHS Policy do not state whether the names of individuals who voted, approved, opposed, or abstained need to be recorded.

Opin.   It is customary to record the names of individuals attending a meeting and to record the actual number of individuals approving, withholding approval, or abstaining in a vote.

Surv.   What voting information is recorded by your IACUC?

- All votes are recorded with names   3/63
- All votes are recorded without names   60/63

6:17 Should each IACUC protocol be voted on individually?

Opin. In general, each protocol is unique and it is best to approve them individually as they are discussed. This method inherently avoids problems that arise from maintaining a quorum. Exceptions for collective approval arise if two similar (or dissimilar protocols) from the investigator are discussed, and the investigator is called to answer specific questions pertaining to one or more of the protocols.

6:18 Who records IACUC meeting minutes?

Opin. Persons assigned to this task should be knowledgeable of federal and other requirements, and the institution's animal care and use program. The institution should assume responsibility for selecting a person to record meeting minutes, based on input from the IO, the Office of Research, and the Chair of the IACUC.

Surv. Who records the minutes of your IACUC meetings?

- Research administration staff   33/63
- A recording secretary other than the research administration staff   19/63
- One of the voting members on the committee   11/63
- Any member on a rotating basis   0/63

6:19 What should be included in the meeting minutes?

Reg. (See 6:15.)
Opin. Since the AWAR (§2.35,a) and PHS Policy (IV,E,b) state that the minutes require the records of attendance, activities of the committee, and committee deliberations, it is reasonable that the following items be included in IACUC meeting minutes:

- Inspection reports distinguishing significant and minor deficiencies with plans and schedules for correcting each deficiency.
- Investigation of complaints and issues of noncompliance received from the public or research facility personnel.
- Approval, conditions such as modifications and clarifications, withheld approvals, and suspensions of protocols.
- Approval or withheld approval of proposed significant changes to protocols.
- Training and certification guidelines approved by the committee.
- Consultants' reports on protocols.

The minutes should include sufficient detail of the discussions and language that is generally understood by the nonscientific and lay members of the committee.

6:20 How are IACUC minutes used?

Reg. The IACUC minutes must be maintained for the duration of the project plus at least 3 years and be accessible for inspection and copying by authorized NIH/OPRR or other PHS representatives at reasonable times and in a reasonable manner (PHS Policy IV,E,2). The AWAR (§2.35,f) have similar wording for APHIS/AC or federal funding agency representatives.

Opin. There are no guidelines for the way the recorded minutes of the meetings are to be used. In general, minutes are maintained to:

- Meet regulatory requirements.
- Resolve any questions that may arise from a program or protocols review.
- Resolve an ethical issue that was previously discussed.
- Reference a previous discussion.

Surv. A How does your IACUC use the minutes of its meetings? (More than one answer is possible.)

• As an official record of the meeting	57/63
• For APHIS/AC inspection purposes	19/63
• To record humane issues that were discussed in the meeting	23/63
• To record issues pertinent to pain and distress which required consensus of the IACUC	15/63

Surv. B Do your IACUC's minutes contain all discussions held at the IACUC meeting?

• Do not include discussions unrelated to IACUC activities	62/63
• Do include all discussions that occurred at the IACUC meeting	1/63

6:21 Does the Freedom of Information Act (FOIA) have an impact on meeting minutes?

Reg. The federal FOIA, at the present time, only applies to documents in the possession of a federal agency.

Opin. The *Institutional Animal Care and Use Committee Guidebook*[4] warns that the person assigned to prepare reports should be aware of the FOIA and any state open record laws. Many of the reports written may be accessible under such laws, and particular care must be taken to avoid using language that may be misconstrued by the lay public. Furthermore, institutions which use computers to prepare and store meeting minutes and reports must be aware of potential security problems that could lead to unauthorized retrieval, tampering of reports, or obtaining information on trade secrets protected by law (AWA §27). Thus, the general advice is that meeting minutes should be written carefully with appropriate usage of words so that the lay public will not be alarmed by reports obtained through the FOIA or pertinent state laws. (See Chapter 22.)

Surv. Does the FOIA have an impact on your IACUC minutes?

- No effect   44/63
- It has severe effect   0/63
- It depends on the nature of minutes recorded   19/63

Some responders said that the FOIA or some states' open record laws have an impact on the meeting minutes. One respondent stated that meeting minutes may become the subject of the FOIA only if the APHIS/AC inspector copies them and it becomes a federal government document. The remaining respondents said that conscious or carefully chosen words and the elimination of minute details of discussions (ethical or confidential) may help avoid unnecessary problems linked to the FOIA.

6:22 IACUCs often review many protocols at a meeting. The concern is that those protocols considered at the beginning of the meeting may be reviewed more thoroughly than those at the end of the meeting. How can IACUCs avoid this problem?

Opin. The AWAR (§2.31,c,4–§2.31,c,7) and PHS Policy (IV,C) provide appropriate guidelines for reviewing research projects, but the guidelines do not provide a time schedule for discussing reports and research projects in a meeting. The *Guide* (page 9) suggests that each IACUC should meet as often as necessary to fulfill its responsibility and provides topics (page 10) that should be considered in the review of animal care and use protocols.

Surv. How does your IACUC assure that all protocols undergo a thorough review?

- Limit the number of protocols that can be reviewed in a meeting   6/63

- Time-limited review of protocols 2/63
- Lengthen the meeting to accommodate all protocols without losing quorum 36/63
- Postpone to discuss it in the next meeting if funding agency deadline allows 13/63
- This is not an issue 6/63

References

1. Garnett, N. and Potkay, S., Use of electronic communications for IACUC functions, *ILAR J.*, 37(4), 190, 1995.
2. Ellis, G. and Garnett, N.L., Maintenance of properly constituted IACUCs, *OPRR Reports*, No. 97-03, June 2, 1997. Available on the World Wide Web at: *http://www.nih.gov/grants/oprr/dc97-3.htm*
3. DeHaven, W.R., personal communication, 1999.
4. U.S. Department of Health and Human Services, Public Health Service, National Institutes of Health, Institutional Animal Care and Use Committee Guidebook, NIH Publication 92-3415, 1992, Section E, Washington, D.C.

7
General Format of IACUC Protocol Forms

Christian E. Newcomer

Introduction

Protocol review was mandated by federal law for most institutions using animals in research, teaching, and testing during 1985 and 1986 by the passage of an amendment to the AWA known as the "Improved Standards for Laboratory Animals Act" (Public Law 99-198) and by a component of the legislative reauthorization for the NIH, known as the "Health Research Extension Act of 1985" (Public Law 99-158). In advance of this requirement, and as early as 1980, several academic institutions already had broached the issue of protocol review. The veterinarians in these institutions had made a compelling case for needing some information about the nature of the ongoing animal use to be able to sign their institutional APHIS/AC Annual Report in good conscience. In the context of the contemporary standards for protocol review in most institutions, the forms used were rudimentary. The information retrieved in these pioneering efforts was extremely scant and generally would now be insufficient for the issuance of an IACUC approval. Nevertheless, these tentative and inadequate efforts were important because they introduced the setting, players, and ethos of an evolving drama that eventually would be adapted and embellished at diverse institutions across the country as the federal mandates for protocol review took effect.

Many of the factors that influenced the early approaches to protocol review continue to be important today. They have shaped the character and content of the forms used in the protocol review process according to institutional preference. For example, most institutions want the protocol review process to be objective, consistent, and efficient. This fosters the support of research investigators and IACUC members who must commit precious time to fulfill this institutional requirement. However, the actual protocol review forms used to facilitate this process are markedly different from institution to institution. Institutional philosophy, the types of research conducted, pragmatic considerations related to administrative and regulatory detail, and many other ancillary factors can influence the format of the IACUC protocol review form.

The purpose of this chapter is to examine the various strategies used by institutions in the information collection phase of the protocol review process, and to highlight the strategies that appear to be the most successful and widely adaptable. The chapter also will discuss alternative approaches that have proven to be well-suited to particular institutional environments. Information for the preparation of this chapter was acquired through the review of approximately 100 institutions during the conduct of AAALAC International site visits over the course of a decade, and from the author's personal experience as an IACUC member at eight institutions during this same period.

A well-designed protocol form can assist investigators in their efforts to provide the clear, concise, and comprehensive information necessary for the review process. It is very important to the smooth functioning of the IACUC. Most IACUCs have revised their protocol forms several times since the inception of the protocol review process in response to new or augmented regulatory requirements, the need to improve the quality and detail of the information provided, or to enhance the retrieval and sharing of information contained in these documents. While the protocol review form should be considered a living document and an IACUC should not be reluctant to change the protocol review format for due cause, format changes are likely to be disruptive and burdensome during a transition period. For this reason, format changes should be carefully planned and coupled with investigator outreach and education efforts.

7:1 Should there be a standard protocol form for submission to the IACUC?

Opin. The large majority of institutions require investigators to submit their animal care and use protocols to the IACUC for review on an institutionally developed, standardized protocol review form. This approach is usually regarded as an essential first step in the fulfillment of the IACUC's mission to establish a review process that is perceived as unbiased, consistent, efficient, and, under optimal conditions, user-friendly. Most institutions have PIs that file multiple studies with the IACUC, and the use of a standardized form allows these individuals to develop a familiarity with the type of information required by the IACUC for successful completion of that committee's review process. Most institutions have found that the quality of information provided by PIs using standardized protocol forms improves over time. Nevertheless, when changes are incorporated into existing forms to meet newly defined needs of the IACUC, a period of coaching and re-acclimatization for investigators is necessary. Additionally, standardization is generally regarded as helpful and critical to the IACUC members in meeting their review obligations. The compartmentalization of information into predictable areas enables the IACUC members to focus their review and discussion on sensitive areas with greater alacrity, and also facilitates the retrieval of information pertinent to any subsequent IACUC monitoring and compliance activities.

It is quite unusual for institutions to conduct IACUC protocol review without the use of a standardized form, but there are some institutions that do this. It is used mostly by academic institutions with smaller animal care and use programs and a limited research profile. It usually involves 10 or fewer PIs. In these instances, the IACUC reviews the PI's grant application and a supplemental narrative concerning specific areas deemed necessary for IACUC review. The narratives are responsive to a list of queries provided by the IACUC to ensure that all of the federally mandated aspects of protocol review are covered. While it is conceivable that this approach can occasionally produce a superbly integrated, relevant, and coherent document for the IACUC review process, in practice this rarely happens. In addition to requiring IACUC members to demonstrate a unique flexibility and commitment, this approach to protocol review challenges IACUC members to wade through much extraneous information and maintain a running inventory of the quality and location of the investigator's responses (or nonresponses) to important issues. In order for the information gleaned from the grant application and protocol at the time of the review to remain conveniently accessible to the IACUC for later use, a synopsis of the review becomes necessary. For these reasons, a free-form protocol IACUC review process will likely remain limited to animal care and use programs in small institutions imbued with a sense of courtesy, reciprocity, and the good faith principle.

7:2 What general information is necessary for the IACUC to adequately review a protocol?

Opin. Several authoritative documents reiterate the basic components of an animal care and use activity that should be assessed by the IACUC during the protocol review process. These documents include the *Guide*, the PHS Policy, and the AWAR. The AWAR impose specific review requirements on covered animal species, and many institutions have chosen as a matter of policy to extend these requirements to all vertebrate species. Other institutions have maintained a dichotomy in the review criteria based upon exclusions provided in the definition of the term "animal" in the AWAR (§1.1). The essential broad areas of review that can be derived from these documents include:

- Procedures conducted on animals.
- Justification for using animals.
- Rationale for using the animal species selected and number of animals proposed.
- Consideration and adoption of applicable alternatives to animal use where appropriate.
- Assurance that the studies conducted are not unnecessarily duplicative.
- Efforts to minimize animal pain and distress through the use of anesthetics, sedatives and analgesics, or timely intervention.
- Provisions for aseptic survival surgery.
- Provisions for postsurgical and postprocedural care.
- Appropriate justification for multiple major survival surgery.
- Provision of appropriate living conditions for animals including the unique requirements of unusual laboratory animals.
- Method of animal euthanasia and disposal.
- Provisions for a safe working environment for personnel.
- Training of personnel in the aforementioned topics commensurate with their assigned responsibilities and areas of involvement in the conduct of the protocol.

The reader should refer to other chapters of this book where many of these topics are explored in detail.

7:3 Is there an optimal format for an IACUC protocol form?

Opin. A central question in the protocol review process is how to optimize information retrieval from the areas identified in 7:2, in order for the IACUC protocol review process to be sufficiently rigorous to ensure high-quality animal care, safety, etc., and still improve a PI's prospect for receiving protocol approval on a first attempt. There is no apparent consensus on this matter as evidenced by the many different approaches taken to protocol review by IACUCs, which in turn is mirrored by the variety of forms used by different IACUCs. The forms used by institutions generally fall into the following categories:

- Those that elicit a free-form narrative response and consolidate all relevant information onto a single form.
- Those that use a free-form narrative approach but are divided into modules allowing the form to be tailored to the particulars of the submission.

- Those that emphasize a checklist and fill-in-the-blank approach whenever possible, either in the modular or consolidated format.

Institutions using a free-form narrative for protocol review simply ask questions related to each of the items in 7:2, eliciting the investigator's response. The protocol form includes questions covering all areas, and answers are typically provided only for the segments of the protocol that apply. Thus, in studies involving minimal animal procedures, the majority of the form may not be used. However, in protocols involving complicated studies with extensive animal use and entailing specialized expertise (e.g., surgery, postprocedural care, or the use of biohazardous agents), all information deemed necessary for the IACUC review is provided. Most institutions realize that investigators cannot be relied upon to provide the kind of qualitative and quantitative information necessary for protocol approval without assistance, even though this subject may have been extensively covered in institutional training and education activities. Therefore, many institutions using this format generally make an effort to enhance the PI's appreciation for the "hot button" issues or nuances particular to each area of review. This is accomplished through the distribution of guidelines for the completion of a protocol review form, example (mock) protocols, IACUC position statements, regulatory announcements on issues, or articles emphasizing the IACUC's interest or perspective on a topic. In a variant of this approach, some IACUCs incorporate this type of information into sidebar or parenthetical comments in the protocol form to channel the investigator in the right direction.

Institutions using the modular version of the free-form narrative protocol work towards paperwork reduction by capitalizing on the fact that many studies only involve a limited, and somewhat predictable, spectrum of activities around which the "core" protocol form is designed. This core form is linked to appendices that are designed for specific topics such as surgery and postoperative care, prolonged restraint, multiple major survival surgery, other procedures potentially involving significant pain and distress for laboratory animals and the use of biohazards. The psychological benefits attendant to the reduction in paperwork for investigators and IACUC members are additional reasons given by institutions who use this approach.

The use of checklists and "fill-in-the-blanks" as the predominant modes of information collection in a form are used by some IACUCs in an effort to expedite completion of the forms through a series of discrete, informative inquiries into aspects of the protocol. Some areas of inquiry and some research settings are better suited to this approach. For example, organizations having a narrow research mission where many of the studies follow a predictable format and involve techniques that are well-described and repetitive, often use this format to allow investigators to quickly respond. In particular, pharmaceutical companies and commercial laboratories involved in testing compounds in conformance with federal regulations often have extensive standard operating procedures for animal care and use. Many aspects of their IACUC protocol form are addressed simply by citing these procedures or other documents used in project development if these primary sources are relevant and have been reviewed and approved by the IACUC. Some larger animal care and use programs, with diversified research endeavors, have incorporated this approach into their protocol form to service research activities that can benefit. A significant pitfall to this approach, when it is applied to complicated studies, is that it tends to disperse the information about the different types of procedures being performed on the animals. This can disintegrate the reviewer's appreciation for the temporal sequencing and impact of procedures on the animals. Flow charts illustrating the various uses of animals in an application often can remedy this problem. A second problem is that these forms tend to be lengthy because they are structured to capture numerous small bits of information. If poorly worded, the information gathered may be voluminous but meaningless.

Many institutions have blended the narrative, checklist, and fill-in-the-blank strategies in their protocol review efforts to optimize their protocol review efforts. The experience of the IACUC often leads to identifying portions of the protocol form that investigators repeatedly do not answer correctly. These areas of the form can be transformed into a checklist format. For example, in the segment on the search for alternatives to painful procedures, the form might include a list of commonly used databases, along with prompts for the keywords used and the date of the search. As another example, in the discussion on the provisions for postoperative care, the questions might stimulate responses for each important aspect of care (e.g., physiological monitoring, analgesic therapy, fluid and antibiotic therapy, recordkeeping, etc.). A final example involves procedures such as polyclonal antibody production. Here, the questions could channel the investigator into methods that conform to IACUC guidelines.

An approved IACUC protocol form should be regarded as the investigator's contract with the institution to conduct animal care and use activities in conformance with applicable laws, regulations, and sound clinical and scientific practices. In summary, any approach, or mix of approaches, has merit if it satisfies the investigators and IACUC members and produces a well-defined contract through a thorough and appropriate IACUC review and approval process.

7:4 How specific should the information on the protocol form be to secure an approval? Is it acceptable for an IACUC to approve a generic protocol?

Reg. For PHS Policy purposes, IACUC approvals must be project specific in order to address the required review criteria at PHS Policy IV,C,1. The use of a more generic protocol, with project-specific amendments to cover significant changes, is one way to meet this requirement.

Opin. There are many investigators who mistakenly view the IACUC protocol review and approval function only as an approval of the animal procedures. Thus, if they use the same animal procedures over an extended period, they prefer to work under a generic approval to reduce their "unnecessary" interactions with the IACUC bureaucracy. Nevertheless, most IACUCs have not acquiesced to this preference because protocol approval is not simply an approval of procedures. In general, IACUCs require PIs to submit very specific information on their protocol forms. This allows the IACUC to link the scientific objectives with the justification for a particular animal model, the number of animals requested to complete the studies and the particular procedures that are involved. Accordingly, in the few IACUCs that do use this approach for certain types of protocols, the PIs are required to submit supplementary information for each study conducted under the protocol to define its objectives, assure the validity of the animal model, assure the lack of alternatives, and explain the need for the number of animals requested. Information also must be provided and reviewed pertaining to any new animal welfare or personnel safety concerns that may have been introduced by new variables in the studies proposed.

There are times when the approval of a generic protocol, abiding by the contingencies stated above, may be appropriate and efficient. An example is a protocol with a standard set of procedures that is used repetitively to test compounds and materials in a particular animal model. The only variable in these protocols is the material or compound under investigation.

7:5 Should the protocol form be used as a vehicle for the continued instruction of investigators in aspects of the PHS Policy, AWAR, and ethical considerations in animal research?

Opin. Most IACUCs use protocol forms that refer to the PHS Policy, AWAR, and ethical issues in one or more areas. Some include a statement signed by the investigators assuring that all of the activities described will be conducted in conformance with the aforementioned regulations. These references to PHS Policy and the AWAR serve to remind PIs of the regulatory underpinnings of the protocol review process. They also emphasize the contractual nature of an approved protocol. Similarly, many IACUC protocol review forms attempt to evoke the discussion of ethical considerations in response to a variety of experimental practices such as prolonged restraint, food and water restriction, use of aversive stimuli, surgery, multiple major survival surgery, disease induction, postprocedural care, and the selection of study endpoints. While use of the protocol form in this fashion to give these issues continued visibility with the investigators is done with good intention, it is this author's opinion that using the protocol form as a vehicle for this instruction, per se, has little merit. Most IACUCs would likely agree that investigators focus their attention on filling out protocol forms for the purpose of securing approval, not for the purpose of personal edification or enlightenment.

7:6 Could inclusion of a questionnaire on the form on animal welfare considerations be useful to the IACUC?

Opin. Fewer than half of the IACUCs this author is familiar with have attempted to evaluate animal welfare considerations in a separate section or questionnaire. When this is done, it usually is limited to querying the investigator about the allocation of animals to the various APHIS/AC pain categories. Even at this basic level, it is surprising how often the IACUC does not agree with the PI's assessment or how often the IACUC is not able to reach an internal consensus. On most IACUC protocol forms, this type of information is integrated into the sections involving the discussion of procedures performed.

7:7 Should IACUC protocol forms require the PI to include a statement written in language understandable to the lay public about the purpose and relevance of the proposed studies?

Reg. This inquiry has become standard on most IACUC protocol forms. The basis for making this request is U.S. Government Principle II, which states, "Procedures involving animals should be designed and performed with due consideration for relevance to human or animal health, the advancement of knowledge, or the good of society."

Opin. In the age of highly specialized research, most IACUCs regard this type of statement as essential not only for the lay member of the IACUC and their institutional public relations personnel, but also for scientists on the IACUC who work in areas unrelated to the protocol under consideration. In the author's experience, this inquiry on the IACUC protocol form most often reflects the frequent unpreparedness of scientists to address the public. Some institutions do not provide this type of inquiry on the form because they regard all protocols to be subsumed under an organizational or corporate research mission that assures this matter.

The lay summary provided by the investigator should include a brief and general overview of the intended use of animals. The function of the lay summary is to provide the nonscientific members of the IACUC, and particularly the nonaffiliated member, with an understanding of the scientific question being asked. The lay summary should indicate how the animal model contributes to the hypothesis being tested or to the resolution of an important scientific dispute. The importance of the animal work undertaken can (and should) be amplified by linking it to the larger body of scientific work that it is supporting. In this regard, investigators and their institutions should keep in mind that ultimately science competes for the attention, recognition, and the continuous stream of financial support from a public that is easily distracted by many other of life's endeavors.

7:8 Should IACUC protocols submitted for educational activities have the same basic structure as a research protocol?

Opin. Of the institutions known to this author, very few actually have experience with the approval of protocols involving the use of animals in educational activities, at either the undergraduate, graduate, or professional school level. Most institutions use the same protocol form for all activities involving animal care and use. Several institutions have devised separate protocol form attachments to address areas regarded as unique to the use of animals in this capacity. The areas of special emphasis include a review of:

- Course syllabus.
- All relevant handouts provided to students.
- Student profile information.
- Summary of animal care and use educational discussion or materials provided to the students.
- Backup provisions for students not fulfilling their obligations to laboratory animals.
- Occupational health and safety provisions.
- Summary of students' written critiques of the educational activity from the previous year.
- Opportunities for the course director to develop alternatives to the use of animals for subsequent sessions of the course.

7:9 Is a PI required to provide information on the protocol review form that may be proprietary in nature?

Opin. PHS Policy and the AWAR do not require IACUCs to review proprietary information. Most IACUCs have designed their protocol forms and implemented procedures to prevent the unwitting and unnecessary disclosure of proprietary information by investigators. On the other hand, IACUCs must have sufficient information about all aspects of the study to make an informed decision about the assurances for animal welfare during all phases of animal care and use. Where PHS Policy is concerned, the IACUC must be able to identify compounds and other materials, equipment, practices, and procedures that may have an impact on occupational health and safety considerations for personnel involved in the project (*Guide*, pages 14 to 18). Hence, most IACUCs do not require the disclosure of the name of a proprietary compound, but do require information about the chemical class of the compound, toxicity data in the target animal species or other animals, special handling requirements, and the basis for dosage selection. The development of new techniques and the use of new instrumentation in animals for application in humans poses more of a quandary. For example, if a new surgical technique is an integral component of a proprietary package involving a new device, an IACUC would be remiss to waive its responsibility to assess the impact of the procedure on experimental animals. Also, most IACUCs require sufficient information about the purported advantages of the new equipment to support the claim of potential benefit. (See 22:3 to 22:7.)

7:10 Should information such as the PI's name, title of the application, and research sponsor be included on the IACUC protocol form, or could that information bias the IACUC's decisions?

Reg. PHS Policy (IV,C,1,f) and the AWAR (§2.31,d,1,viii) require the IACUC to verify that personnel conducting procedures are appropriately qualified and trained. This would be difficult, if not impossible, without the identification of the PI.

Opin. While there is some variation in the content of the protocol signalment among the institutions known to this author, all identified the PI by name and included a title for the proposed activities. Most protocol forms also asked the PI to provide information about the funding source, particularly when the IACUC is responsible for sending the letters of protocol approval to funding agencies. However, in some institutions a response to this inquiry is at the discretion of the PI. In institutions that fund their research programs internally, this inquiry is not relevant. Virtually none of the institutions believe that this type of information is likely to compromise their objectivity. This is notwithstanding the admission by several institutions that internally funded projects have to provide evidence of peer scientific review via an internal mechanism or consultants, whereas externally funded projects don't. Other information about the PI and associated laboratory personnel often collected in the protocol form include: position or academic title; office, laboratory, and home (emergency) phone numbers; and office and laboratory location. In many institutions the ability to serve as a PI is reserved for individuals with particular positions or academic titles.

7:11 What are some methods used by IACUCs to identify protocols for the purpose of correspondence, timely periodic reviews, and aiding the oversight of ongoing animal-based research activities?

Opin. Most institutions use an alphanumerical identification system for the protocols that are submitted to and approved by the IACUC. This type of system offers several advantages and there are numerous variations in this approach. It provides a brief, content neutral, reference which links both the animals and the investigator's grant application or award to a set of approved animal care and use procedures. Many institutions indicate the year of protocol submission in their log or accession system. This provides rapid identification of those protocols that are due for annual or triennial renewal. Additional numbers are added to reflect the sequence of protocol submission within the year, and sometimes letter or decimal extensions are used to indicate the renewal year within the triennial period. As an example, 99-38.02 might indicate the 38th protocol submitted to the IACUC in 1999, with a full review of the protocol scheduled in the year 2002.

It is not uncommon for the approval of problematic protocols submitted at or near the end of the calendar year to be delayed until the following year. Computer databases are used routinely to track the actual dates of submission and approval to ensure that the requirements for periodic review or renewal are met in a timely fashion. Also, organizations vary in their requirements for the number of animal species that are permitted in a particular protocol application; some allow multiple species, whereas others allow only one. Institutions that allow only one species per protocol application sometime retain the same number but follow it with a letter designator for species (e.g., R=rat, Rb=rabbit, M=mouse, etc.) to associate the procedures described with a particular grant application or award. The same tracking number may be retained, along with a modifying extension, to indicate that animal care and use procedures are identical to those described in the parent protocol, but an alternate project title or funding source is associated with the protocol. (See 7:12.)

7:12 When protocols are due to expire, should the IACUC retain the original number assigned or assign a new number to the protocol?

Reg. The AWAR (§2.31,d,5) require review of IACUC protocols at least annually, while the PHS Policy (IV,C,5) requires the same at least every 3 years.

Opin. Many institutions apply a modifying extension to the same protocol number when an annual review of animal care and use activities is mandated for a species, such as occurs under the AWAR. However, most institutions that receive PHS support for their research programs prefer to retire protocol numbers when the protocols have reached the limit of their 3-year PHS approval period. It should be noted, though, that at least a general review of each protocol involving AWAR covered species must be performed on an annual basis. This approach is convenient in that it ensures individuals involved in animal care and use (and research oversight) that the review process is current. In a few cases in the author's experience, an institution has elected to retain the original protocol number to indicate the continuity and longevity of a research program, both noting the consistency of the animal model and procedures used as well as emphasizing with distinction an investigator's funding history. This approach is also satisfactory as long as the IACUC has performed a rigorous review at the appropriate review cycle.

8

Submission and Maintenance of IACUC Protocols

Farol N. Tomson

Introduction

Questions often arise regarding which protocols actually require an IACUC review and which ones don't. What is the difference between private agencies and the federal government regarding IACUC approval? What kind of information is required on protocol forms for IACUC review? Can electronic media be used to submit and approve IACUC protocols? These and other questions are addressed in this chapter.

8:1 What activities are specifically required to have an IACUC-approved protocol?

Reg.   IACUC-approved protocols are required if the animals to be used are

- Regulated species and used in research, teaching, or testing in registered research facilities as defined by the AWAR (§1.1 and §2.31). (See 12:1.)
- Used in research, research training, and biological and testing activities conducted or supported by any PHS agency (PHS Policy II).

Opin.   An IACUC-approved protocol is also needed when projects supported by other public or private agencies have a requirement for IACUC approval (see 8:4 to 8:6). Alternately, the institution may require IACUC approval of all animal protocols (see 8:7). The terms "biomedical, behavioral, research, teaching, and testing" are not defined in the regulations. For example, a breeding colony of vertebrate animals (mice) may not be part of any biomedical research, teaching, or testing project and, therefore, not required by any of the above regulations to have an IACUC approval. Yet, the institution may require an IACUC review of this animal project (see 8:7, 12:1).

8:2 What activities are specifically exempt from the requirement for an IACUC-approved protocol?

Opin. The best approach to this question is to state the exclusion requirements from Question 8:1. IACUC-approved protocols are *not* required if the animals to be used are

- Not under the auspices of the AWAR or PHS Policy (see 8:1).
- Not supported by any private or public agency requiring IACUC approval.
- Not required by the home institution to be IACUC reviewed.

(See 2:1, 2:5 to 2:7, 2:9, 2:10.)

For example, institutions have many uses of animals ranging from animal demonstrations and displays for open houses to rodeos, theatrical plays using animals, and animals drawn by art students. People often keep companion animals on site as pets, or in cages or aquariums in their offices or dormitory rooms. Institutions may house residential animals for specific purposes like guard dogs, pet therapy animals, mounted police horses, or preschool classroom animals (rabbits and gerbils). Institutions also deal with pest animals such as wild rodents and certain birds. If these animals are not involved in biomedical research or related activities (or intended to be used in a research project), they are exempt from IACUC approval.

8:3 What information is required on an IACUC protocol?

Reg. The HREA (§495,b,3,A–§495,b,3,C) is very general in its information requirement for the IACUC to review. It states: "Each animal care committee of a research entity shall:

  A. Review the care and treatment of animals in all animal study areas and facilities of the research entity at least semiannually to evaluate compliance with applicable guidelines established under subsection (a) for appropriate animal care and treatment.

  B. Keep appropriate records of reviews conducted under subparagraph (A).

C. For each review conducted under subparagraph (A), file with the Director of NIH at least annually (i) a certification that the review has been conducted, and (ii) reports of any violations of guidelines established under subsection (a) or assurances required under paragraph (1) which were observed in such review and which have continued after notice by the committee to the research entity involved of the violations."

The PHS Policy (IV,C,1,a–IV,C,1,g) is more specific. It states: "In order to approve proposed research projects or proposed significant changes in ongoing research projects, the IACUC shall conduct a review of those components related to the care and use of animals and determine that the proposed research projects are in accordance with this policy. In making this determination, the IACUC shall confirm the research project will be conducted in accordance with the Animal Welfare Act insofar as it applies to the research project, and that the research project is consistent with the *Guide* unless acceptable justification for a departure is presented. Further, the IACUC shall determine that the research project conforms with the institution's Assurance and meets the following requirements:

- Procedures with animals will avoid or minimize discomfort, distress, and pain to the animals, consistent with sound research design.
- Procedures that may cause more than momentary or slight pain or distress to the animals will be performed with appropriate sedation, analgesia, or anesthesia, unless the procedure is justified for scientific reasons in writing by the investigator.
- Animals that would otherwise experience severe or chronic pain or distress that cannot be relieved will be painlessly killed at the end of the procedure or, if appropriate, during the procedure.
- The living conditions of animals will be appropriate for their species and contribute to their health and comfort. The housing, feeding, and nonmedical care of the animals will be directed by a veterinarian or other scientist trained and experienced in the proper care, handling, and use of the species being maintained or studied.
- Medical care for animals will be available and provided as necessary by a qualified veterinarian.
- Personnel conducting procedures on the species being maintained or studied will be appropriately qualified and trained in those procedures.
- Methods of euthanasia used will be consistent with the recommendations of the American Veterinary Medical Association

(AVMA) Panel on Euthanasia,[1] unless a deviation is justified for scientific reasons in writing by the investigator."

In addition, the AWAR state that the following components be reviewed and approved by the IACUC:

- The PI must document consideration of alternatives to painful procedures and provide a written narrative description of the methods and sources used to determine that alternatives were not available (§2.31,d,ii).
- The PI has provided written assurance that the activities do not unnecessarily duplicate previous experiments (§2.31,d,iii).
- Deviations from any AWA standards or regulations must be identified and reviewed and approved by the IACUC. The IACUC must report these exceptions to the APHIS/AC in their annual reports (§2.36,b,3).
- The identification of the species and the approximate number of animals to be used (§2.31,e,1).
- The rationale for animal use, species, and number of animals to be used (§2.32,e,2).
- A complete description of the proposed animal use (§2.31,e,3).
- A description of the procedures and drugs used to provide relief from pain or distress (2.31,e,4).
- A description of any euthanasia method to be used (§2.31,e,5).

Opin. The *Institutional Animal Care and Use Committee Guidebook*[2] summarizes the required information items to be included on IACUC forms. Unfortunately, there is no guidance from this book or any federal agency on how to phrase specific questions or how to phrase certain answers. Therefore, each IACUC must determine for themselves the most effective way of obtaining and evaluating this information using their own protocol form.

It is this author's experience that IACUCs generally want more information than is required. For example, the AWAR (§2.31,d,xi) states only that the method of euthanasia must be in accordance with the AWAR (§1:1) definition of euthanasia. PHS Policy (IV,C,g) states only that the method of euthanasia must be consistent with the recommendations of the AVMA Panel on Euthanasia. Some institutions (this author's included) go beyond these regulations and use the protocol form to ask for specific drug names, doses, and routes of administration for these euthanasia methods. Other institutions may require substantiation of Drug Enforcement Administration registration if controlled substances are being used.

8:4 Do government agencies conducting or funding animal research require IACUC-approved protocols?

Reg. (See 8:1.)

Opin. Some agencies, such as the National Science Foundation, have adopted the PHS Policy requirements. In 1985 an Interagency Research Animal Committee, with representatives from various federal agencies, developed a set of principles (U.S. Government Principles for Utilization and Care of Vertebrate Animals Used in Testing, Research, and Training) which applies to institutions and persons using animals in federally funded programs.[3] These principles do not, however, mention or require IACUC reviews. Requirements for IACUC review and approval are made by the different federal agencies. *The reader is cautioned, however, that it is the opinion of APHIS/AC that all government agencies, including federal, are subject to the AWA and the AWAR, including the requirement for IACUC-approved protocols when regulated species are used.*[4]

8:5 Do government agencies using animals but not doing biomedical research require IACUC approval?

Reg. Some government agencies, such as the USDA, use animals in nonbiomedical research and teaching programs. On August 29, 1990, the USDA's Agricultural Research Service (ARS) published new Policies and Procedures (Directive 635.1 — Humane Animal Care and Use) that required an ARS IACUC review of all ARS animal projects.[5] Another USDA agency, the Cooperative State Research, Education, and Extension Service (CSREES) also requires IACUC approval of their animal projects.[6] The "research exemption" in the AWA (§2,g), which exempts horses not used for research purposes and other farm animals, is for *agricultural* research, not nonbiomedical research.

Opin. The U.S. Fish & Wildlife Service, an agency of the Department of Interior which funds research with animals, does not usually require IACUC approval because they conduct no biomedical research. The main regulatory concern for Fish and Wildlife services is the review and issuing of permits. Permits must be obtained by researchers from state or federal wildlife agencies.[7-10] These agencies review permit applications for their merit and their potential impact on native populations. They issue permits that authorize the taking of specified numbers of animals, the methods allowed, the length of study, and other restrictions which are designed to minimize the likelihood that an investigation will have deleterious effects. There are some exceptions: the AWAR (§1.1, Field Study) does cover certain field studies and, as noted in 8:4, APHIS/AC interprets the AWA as requiring all government agencies, including federal ones, to have IACUC approval for biomedical research using covered species. Therefore, the Fish and Wildlife Service and other similar agencies do not require IACUC approval for their funded projects *unless* the study is covered by the AWAR as a "field study." Otherwise, they rely on the permitting process to review the care and use of animals.

Several sets of guidelines for using wildlife animals have been published.[11-13] They refer to an IACUC review, but only if it is required. Some of these guidelines, however, recommend that an IACUC review be sought for any experiments in which these animals are handled or otherwise manipulated, certainly if invasive procedures are involved.[13]

8:6 Do private agencies funding biomedical research require IACUC-approved protocols?

Reg. (See 8:1 for animals and animal activities requiring IACUC approval.)

Opin. There are no federal or state laws that require private funding agencies to obtain IACUC approval of animal use projects. Therefore, not all private agencies require IACUC approval. Some private agencies, however, do require IACUC approval. The American Heart Association, American Cancer Society, and American Lung Association are examples.[14-16] Before releasing funds for animal research, they require the awardee institution to obtain IACUC approval. Some private associations and societies (including some of their state affiliates) make statements such as the following in their grant application package to institutions and researchers:

> "Research involving animals will conform with the current 'Guide for the Care and Use of Laboratory Animals,' NIH publication, DHHS/USPHS; and with federal laws and regulations; and has been approved by the Institutional Animal Care and Use Committee."[17]

8:7 How can institutions require IACUC approval of projects from agencies that do not require it?

Opin. Each institution needs to determine, for themselves, if they want to have their IACUC approve nonbiomedical (i.e., agricultural) research using animals or research using animals not regulated under the conditions of the AWAR or PHS Policy. In this regard, two options are available. First, the institution can establish and implement its own policy that requires this review. Second, if an institution is operating under an NIH/OPRR Assurance statement, that Assurance probably applies to all vertebrate animal projects, regardless of the funding source. The latter is currently recognized as an "institutional-wide" Assurance. NIH/OPRR offers advice regarding whether or not to include the non-PHS-supported animal projects in the Assurance of Compliance.[18]

8:8 Must the IACUC review and approve protocols using animal carcasses and parts of carcasses?

Reg. Based on the AWA (§2,g), the AWAR (§1.1, Animal) define regulated animals as "live or dead." Nevertheless, the AWAR (§1.1, Research Facility) define a research facility as one using or intending to use live animals. There is no reference to using dead animals in research facilities. It is only in the definition of a dealer (AWAR §1.1, Dealer) that the AWAR refer to alive *or* dead animals.

Only live animals are mentioned in PHS Policy III,A. Therefore, only live animals at research facilities are required to have an IACUC review and approval.

Opin. IACUCs often choose to review the use of carcasses and carcass parts even though there are no federal requirements to do so. At the author's institution the review of carcasses and carcass parts is done primarily to protect and defend PIs from outside critics regarding the taking of these samples, i.e., researchers receiving biopsy material (or carcasses) from endangered species. The IACUC review is an effective way of informing the institution of these projects. IACUC members can discuss and advise the institution on projects which may become controversial and worthy of additional consideration. (See 12:1, 12:5, 14:21–14:23.)

8:9 Must the IACUC review and approve protocols using animal tissues, cells, or biological fluids derived from living animals at your institution?

Reg. (See 8:1.)

Opin. The concern for animal welfare is at the time of collection of tissues, cells, or fluids. If the animal is being handled or restrained for the procedure, the IACUC should review that protocol. If the animal is removed from the collection site or not involved with the collection technique (e.g., feces on the floor, remnants of placentas, preserved tissues) being used for additional *in vitro* tests or additional slide stains of preserved tissues, there is little concern (if any) for animal welfare. Nevertheless, there is still concern for the care and housing of the animals. The author's institution has a mechanism of expediting the IACUC review and approval of protocols using tissues, cells, or fluids. Also expedited are protocols for studies using only samples or materials removed from animals under patient care in a veterinary clinic setting (e.g., hair, nails, teeth, blood, urine, cerebrospinal fluid).

The study of existing documents, records, tissues, pathological specimens, diagnostic specimens, and other specimens sent for teaching or diagnostic purposes are processed in an expedited fashion through the IACUC. (See 13:11, 14:20.)

8:10 Must the IACUC review and approve protocols using tissues, cells, or biological fluids acquired from sources other than your own institution?

Reg. (See 8:1.)

Opin. This issue is not directly addressed in the AWAR, but a similar issue is addressed by the NIH/OPRR as it relates to purchasing antibodies from a commercial source (see 8:11). The outside source (vendor) providing the tissues, cells, or fluids must comply with the conditions of 8:1 if live animals are being used to produce these products. This vendor must adhere to the requirements stated in 8:1 and 8:3. (See 8:16.)

8:11 Does the purchase of antibodies from commercial sources require IACUC approval?

Reg. NIH/OPRR has made the following interpretation of the PHS Policy.[18]

> "In the case that standard reagent antibodies (e.g., mouse-antihuman) are produced by a commercial supplier using their own resources and offering them for general sale, for example, through a catalog, the institution may consider the antibodies to be 'off-the-shelf' reagents and the supplier is not required to file an Assurance with OPRR. If, on the other hand, a supplier or contractor produces custom antibodies using antigen(s) provided by or at the request of a principal investigator, the antibodies are considered 'customized' and the vendor or subcontractor must file an Assurance with OPRR.

> "Usually it is known in advance that someone intends to perform this kind of work under a PHS grant. In such cases, the applicant must mark the PHS Grant Application (PHS Form 398) 'yes' for vertebrate animal involvement and include the appropriate Animal Welfare Assurance number(s), verification of project-specific date of IACUC protocol review, and the identification of all project performance sites. All animal-related activities supported by the PHS must be conducted at Assured institutions and must be reviewed and approved by an IACUC. When both the PHS Grantee and its Contractor hold OPRR-approved Assurances, some latitude is allowed in determining which IACUC (if not both) will review the proposal. However, the institution which subcontracts or subgrants any animal activity retains partial accountability for providing effective oversight mechanisms to ensure compliance with the PHS Policy.

Part of that responsibility includes ensuring that subgranted/subcontracted animal-related activities are conducted only at an Assured institution."

Opin. There are no AWAR requirements for a research facility to review protocols of researchers using commercial companies producing their antibodies. However, as noted in APHIS/AC Policy #10,[19] if that commercial company uses rabbits (or any other regulated species) to produce antibodies, it must be registered as a research facility and have its own IACUC review relevant procedures.

8:12 Does the use of fertilized eggs (birds, reptiles, amphibians) or fetuses (mammalian) require an approved protocol?

Reg. If the eggs and fetuses remain as part of the female (dam), it is the dam that is the object of IACUC interest. If the dam is under the auspices of the AWAR or PHS Policy (see 8:1, 12:1), then IACUC approval is required. If these are chicken eggs, NIH/OPRR has made the following interpretation of the PHS Policy for governmental biomedical research.[20]

> "The PHS Policy is applicable to proposed activities that involve live vertebrate animals. While embryonal stages of avian species develop vertebrae at a stage in their development prior to hatching, OPRR has interpreted 'live vertebrate animal' to apply to avians (e.g., chick embryos) only after hatching."

Opin. If the eggs are allowed to hatch and if the animals come under the considerations described in 8:1, then IACUC approval is required. The AWAR does not regulate birds or cold-blooded (poikilothermic) animals. Therefore, at this time there are no AWAR requirements for the IACUC to review the use of eggs (unhatched or hatched) from these species. (See 14:6.)

Using mammalian fetuses cannot be done without using the mammalian mother (dam). An IACUC approval is needed if the dam is under the auspices of the AWAR or PHS Policy. If, however, the mammalian dam is not under the auspices of the AWAR or PHS policy, then an IACUC approval is not required. (See 13:11, 14:20, 16:24, 16:25.)

8:13 Is an IACUC protocol required for breeding colonies, independent of any experimental procedures performed on the animals?

Reg. If the breeding animals come under the considerations described in 8:1 and 12:1, then IACUC approval is required. All PHS-supported activities involving animals must be reviewed and approved by an IACUC (PHS Policy II).

Opin. In some instances, such as the production of genetically unique rodents, some IACUCs might consider the breeding itself as an animal "research" project and they will require IACUC approval. In some instances, other breeding colonies (rodents, fish, birds, domestic livestock, etc.) not involved with any activities described in 8:1 are not required to have IACUC approval unless the institution requires it (see 8:7, 14:32).

8:14 What information should be included for a protocol describing the maintenance of a breeding colony?

Reg. If IACUC approval is required of a breeding colony (see 8:1, 8:13), then colony managers or responsible investigators need to submit certain information to the IACUC (see 8:3). The *Guide* (pages 47 and 48) notes that it is important to maintain the genetic integrity of the colony whether the animals are inbred or outbred, and that the use of standard nomenclature is important.

Opin. In addition, IACUC members may want to know more about the colony husbandry practices. This information may be requested in special forms or formats, or can be submitted as a separate document for review. Conventional IACUC protocol forms can also be used, but because no research is usually being done, the forms include many unrelated questions.

Breeding animal programs require special attention to certain animal welfare concerns, e.g., male to female ratios, artificial insemination, chemicals synchronizing estrus, time male is left with female, caging type, etc. Additional variables influence breeding results and the IACUC may want very detailed information on file (e.g., light cycles, nutrition, sanitation, bedding material, cage size and type, etc.). Additional information often specified involves the age of the breeders, when to cull old breeders, how many litters (offspring) are permitted per dam, and how birthing difficulties are managed. The information requirement will vary from IACUC to IACUC and from species to species.

Some institutions have developed forms to be used for animal breeding programs. Forms used at the author's institution cover breeding operations ranging from dairy, poultry, beef, sheep, and equine units to mouse and rat colonies. Additional information related to breeding colonies may include some or all of the following (in addition to items mentioned above):

- The name of the veterinarian to be contacted if and when needed.
- The source of the breeding animals.

- The number of offspring per year produced or number of breeding pairs per year maintained.
- The disposition of animals when no longer useful.
- The method of euthanasia when needed.
- The justification for maintaining animals.
- A description of painful or stressful husbandry procedures (if any).
- A list of personnel who will handle animals and their training and experience.

8:15 Do sentinel animals or blood donors need IACUC approval? Since these animals are not used in specific research projects, is it necessary to develop protocols for the review and approval of these projects by the IACUC?

Reg. If the animals in question fall under the considerations described in 8:1 and 12:1, then IACUC approval is required. NIH/OPRR has made the following interpretation of the PHS Policy.[21]

> "PHS Policy applicability is not limited to research. It also includes all activities involving animals including testing and teaching. OPRR has determined that although animals used as sentinels, breeding stock, chronic donors of blood and blood products, or for other similar objectives may not be part of specific research protocols, their use for these purposes contributes significantly to the institutional research program and constitutes activities involving animals. Consequently, the IACUC must receive and approve of protocols and appropriate systems to monitor the use of animals prior to the commencement of such activities, and should then perform reviews at the appropriate intervals (IV.C. of the PHS Policy)."

Opin. Since there are regulations defining what is a biomedical research project or even a research project, there are some animal uses that are obviously not research projects. For example, blood donor animals (dogs, cats, birds, ferrets, etc.) in veterinary clinics (at registered research facilities) that are not part of any research or teaching project are not required by regulations to be reviewed by the IACUC. Many management-related decisions are based on sample trials of products on animals to determine if they are going to work or not. For example, different bedding materials, caging, foods, or enrichment devices are often tested on resident animals with some preliminary data obtained to help management make purchase decisions. These animal projects are not required by regulation to be reviewed by the IACUC. If these projects, however, are planned with an experimental design and the results submitted for publication, this then becomes more of a typical research project that requires IACUC review if it comes under the considerations stated in 8:1 or 8:7. (See 14:32.)

8:16 If an investigator wishes to conduct research using animals at another institution, must the IACUC at the investigator's home institution approve the protocol? Must the other institution's IACUC approve it?

Reg. Investigators often collaborate with scientists at other institutions and often in other countries. Although not addressing this question directly, NIH/OPRR has made the following interpretation of the PHS Policy.[21]

> "All animal activities supported by the PHS must be reviewed by the IACUC of the domestic-assured awardee institution that receives such support. Foreign institutions that serve as performance sites must also have Assurances on file with OPRR. The OPRR considers institutions whose scientists are engaged in such collaborative work accountable for the animal-related activities from which they receive animals or animal parts. When a foreign institution holds a PHS Assurance, it also is expected that the institution will conduct the study in accordance with the applicable host-nation's policies and regulations. In the specific case of sample collection, the review should take into account the species involved, nature of the specimen, and the degree of invasiveness of the procedure, giving appropriate consideration to the use of anesthetics and analgesics. In cases in which samples are obtained directly by citizens of a foreign country for subsequent shipment, recipient PHS-supported investigators should determine the proposed methods of collection and present that information to their IACUC for review. Prior to sample collection and regardless of whether specimens are obtained by an awardee institution's investigator directly or by persons in a foreign country, the OPRR strongly recommends that each awardee institution consult with other agencies of the U.S. government concerning importation requirements. Depending on the species involved and the nature of the specimen, the following may be of assistance: the U.S. Fish and Wildlife Service, Department of the Interior (for compliance with the International Convention on Trade in Endangered Species of Fauna and Flora [CITES]), APHIS, U.S. Department of Agriculture (regarding potential animal pathogens), and the Centers for Disease Control and Prevention (concerning importation of nonhuman primates and potential pathogens of human beings)."

Opin. If biomedical animal research activities are performed at domestic or foreign sites other than the investigator's home institution, then the IACUC at the home institution needs to satisfy itself that the PHS

Policies are being followed. The home institution can conduct the IACUC review or they can rely on the performance site to conduct this review, or both may conduct the review.

The home institution can request documentation from the performance site that its IACUC has reviewed and approved the project. Documents also can be requested regarding its NIH/OPRR Assurance statement and even AAALAC accreditation status. Documents can be requested concerning its USDA registration as a research facility, and recent copies of USDA inspection reports. If none of these are available (for example, a PHS-funded rat project in a facility without an Assurance statement), NIH/OPRR should be contacted for assistance. (See 9:50.)

8:17 What minimum qualifications does a person require to submit a protocol to the IACUC? For example, can students, postdoctoral fellows, medical residents, or visiting scientists submit a protocol?

Reg. Personnel conducting procedures on the species being maintained or studied will be appropriately qualified and trained in those procedures (AWAR §2.31,d,1,viii; PHS Policy IV,C,f).

Opin. There is a difference between the person submitting the protocol and the person actually performing the animal work. There are no federal or state minimum requirements or qualifications for anyone to "submit" a protocol to the IACUC. Requirements like this are left to the discretion of each institution. Some institutions allow only their own scientists or faculty to submit protocols. Others may allow outside scientists to submit protocols. Some find a middle ground and have local faculty "sponsor" any nonfaculty members.

In some instances, the PI who submits a protocol may never see or handle the animals. The research staff does the actual work. Nevertheless, the PI is held responsible for the conduct of the study. It is important for the IACUC to know who is working with the animals and who is ultimately responsible for the conduct of the study.

The author's institution permits scientists from private firms, local government research laboratories, and even high school science students to submit protocols for review. In the information provided to the IACUC, it is clear who the responsible person is, and who will be doing the actual animal work. The IACUC can then accept, modify, or reject this information (AWAR §2.31,c,6; PHS Policy IV,B,6).

8:18 Who typically maintains protocols and for how long should they be kept?

Reg. For the regulated species and PHS-funded animal projects, protocol records must be held for 3 years after completion of a study, as are the

IACUC minutes and records of deliberations. These records must be made available to APHIS/AC or NIH/OPRR when requested (AWAR §2.35,a; §2.35,f; PHS Policy IV,E,1; IV,E,2).

Opin. A separate IACUC office or a research administration office usually maintains the protocol files. Investigators should also keep a copy. These records are also considered institutional documents and subjected to whatever institutional requirements are made (above and beyond the federal regulations).

8:19 Who should have access to IACUC protocols?

Reg. For the regulated species and PHS-funded animal projects, protocol records must be made available to APHIS/AC or NIH/OPRR when requested (AWAR §2.35,f; PHS Policy IV,E,2).

Opin. Private institutions can control this access much better than public institutions. Public institutions (including their IACUCs) may be under "Sunshine" laws (open meetings or open records laws). At the author's institution (a public university), IACUC documents are subject to the state's public records law and, except when exempted under the law, IACUC documents are available to the public. For example, Florida's Statute Section 240.241(2) exempts the following information from being released.[22]

> "[M]aterials that relate to methods of manufacture or production, potential trade secrets, potentially patentable material, actual trade secrets, business transactions, or proprietary information received, generated, ascertained, or discovered during the course of research conducted within the state universities shall be confidential and exempt from the provisions of s.119.07(1), except that a division of sponsored research shall make available upon request the title and description of a research project, the name of the researcher, and the amount and source of funding provided for such project."

In Florida, IACUC meetings are also subject to the state's open meetings law, which means that students, faculty, staff, media, and the public at large can have access to them.

In institutions not subjected to sunshine laws, the access to meetings and the access to records become policy matters for the institution. Researchers, certain administrators as well as the IACUC, veterinary, and husbandry staffs are most involved and affected by this information and, therefore, should be afforded access. It is the opinion of this author that conducting open meetings and distributing nonexempted information to the public is in the best interest of animal research. This is an effective means of public education and it helps to negate the veil of secrecy that often hinders the public image of animal research.

Individual state's freedom of information acts may govern who has access to IACUC records. If questions arise regarding the legal access to this information, one should seek legal counsel from the institution. (See 6:15, 6:21; Chapter 22.)

8:20 May protocols be submitted electronically to the IACUC and can protocols be electronically distributed to IACUC members for review?

Opin. Yes. Neither the AWAR nor PHS Policy prohibit this. Protocols can be submitted electronically to the IACUC and electronically posted for members to review. Technology is available for protocols to be completed and submitted to secured Web sites. Members can then gain access to this site to read or download the information and return e-mail messages with their results. It is emphasized that the final approval of protocols submitted in this fashion must not vary from the regulations and requirements set forth by the AWAR and PHS Policy regarding full committee review and designated member review. The use of electronic communication simply speeds up the review process by avoiding the slower mail delivery systems. (See 6:11 to 6:14.)

Several problems can occur with electronic reviewing. For example, reading from a computer monitor can be difficult when scrolling through pages of text. Signature requirements of the investigator and administrators (if they are required) need to be resolved. The following comments from NIH/OPRR lend perspective to the issue of signature requirements.[23]

> "Many IACUC activities involve the need for documenting votes or verifying committee approval. Letters and reports often require signatures in order to be legally binding. Techniques for providing such legally acceptable 'electronic signatures' are being developed by the computer industry to address this problem. Currently, electronic communication is frequently used to facilitate the rapid conduct of business with signed original documents to follow for the permanent record."

The following comments from NIH/OPRR lend perspective to the issue of polling IACUC members.[23]

> "Polling is defined as sequential, one-on-one communication, either in person or via telephone, e-mail, fax, U.S. mail, or by other similar means. Polling is an appropriate mechanism for providing all committee members with the opportunity to call for full review of a protocol prior to initiating the 'designated reviewer' method of protocol review described below. It may also be appropriate as a mechanism for distributing and reviewing drafts of meeting minutes or reports. The simple polling of

IACUC members does not, however, satisfy the definition of a meeting of a convened quorum and should not be used for conducting IACUC business that requires the vote of a convened quorum of the committee. For example, polling should not be considered a valid method of voting under the 'full committee' review method of protocol review and is not an acceptable substitute for having a vote of a convened quorum on the suspension of a previously approved activity involving animals."

The author's institution requires IACUC members to have access to the World Wide Web and be able to accept electronic mail from the IACUC office. Amendments, notices, and other information are distributed instantaneously to all members. Usually 5 business days are given for all IACUC members to review this information. Any member can then call for a convened meeting or ask questions regarding the proposal. At the end of the 5 days, if no one has called for a convened meeting, the Chair or delegate can approve or request modification of proposals. The use of electronic mail to exchange information prior to and just following a convened meeting of the IACUC also effectively hastens this review and approval process. (See 6:8, 6:11.)

8:21 What are some recommendations for maintaining confidentiality if electronic media are used for submission and review?

Reg. The unauthorized release of confidential IACUC information by members is already prohibited by law (AWA §27a; §27b):

> "It shall be unlawful for any member of an Institutional Animal Committee to release any confidential information of the research facility including information that concerns or relates to (1) the trade secrets, processes, operations, style of work, or apparatus; or (2) the identity, confidential statistical data, amount or source of any income, profits, losses, or expenditures, of the research facility. It shall be unlawful for any member of such Committee (1) to use or attempt to use to his advantages, or (2) to reveal to any other person any information which is entitled to protection as confidential information under subsection (a)."

Opin. Security is always a concern when dealing with animal research records, especially those protocols under discussion and not officially approved. NIH/OPRR provides the following perspective:[23]

> "Institutions utilizing innovative modes of communication must be aware of the potential security problems inherent in the method chosen. Some material considered by IACUCs should

be treated as privileged or confidential, especially prior to final committee action. In the case of trade secrets, such information may be protected by law (7 USC 2157, section 27). Because of widespread reports of unauthorized access to computer records, the use of available computer security measures such as passwords, controlled access, and encryption should be considered."

Using electronic mail is not private. Public employees may be subjected to a state's public records law, such as the following policy at the University of Florida.[24]

"E-mail created or received by University of Florida employees in connection with official business, which perpetuates, communicates, or formalizes knowledge, is subject to the public records law and open for inspection."

(See 6:15; Chapter 22.)

References

1. American Veterinary Medical Association, Report of the AVMA panel on euthanasia, *J. Am. Vet. Med. Assoc.*, 202, 230, 1993.
2. U.S. Department of Health and Human Services, Public Health Service, National Institutes of Health, Institutional Animal Care and Use Committee Guidebook, NIH Publication No. 92-3415, 1992. Also available on the World Wide Web at: *http://www.nih.gov/grants/oprr/iacuc_guidebook/iacuc-guidebook.htm*
3. Office of Science and Technology Policy, Interagency Research Animal Committee, U.S. Government Principles for Utilization and Care of Vertebrate Animals Used in Testing, Research, and Training, Federal Register, Washington, D.C., May 20, 1985. (Agencies represented included: Veterans Administration, Department of Energy, National Aeronautics and Space Administration, Environmental Protection Agency, Department of the Interior, Department of State, Department of Defense, National Science Foundation, U.S. Department of Agriculture, and Consumer Product Safety Commission, Department of Health and Human Services, including the National Institutes of Health, Fogarty International Center, Centers for Disease Control and Prevention, Office of International Health, the Health Research Services Administration, and the Food and Drug Administration.)
4. DeHaven, W.R., personal communication, 1998.
5. U.S. Department of Agriculture, ARS, CSREES, ERS, NASS. Policies and Procedures. Available on the World Wide Web at: *http://www.ars.usda.gov:80/afm2/ppweb/*
(This directive states policy, responsibilities, committee membership, committee procedures, including reporting requirements for IACUCs.)

6. U.S. Department of Agriculture Cooperative State Research Service. October 18, 1991, Administrative Manual for the Evans-Allen Cooperative Agricultural Research Program (Section 1445, Public Law 95-113). Available on the World Wide Web at: *http://www.reeusda.gov/1890/man/1890-toc.htm* (The General Provisions section describes the policy using animals.)
7. U.S. Fish and Wildlife Service, Division of Endangered Species. Available on the World Wide Web at: *http://www.fws.gov/r9endspp/poldocs.html.* (Endangered Species Policy Documents. Including policies intended to complement the current public review processes prescribed by sections 4(b)(4)(6) and 10(a)(2)(B) of the Act and associated regulations in Title 50 of the Code of Federal Regulations.)
8. Florida Wildlife Code, Possession of Wildlife in Captivity: Permits, Title 39, 39-6.0022, F.A.C., 1996.
9. U.S. Fish and Wildlife Service, AIA. 106 STAT. 2224 Public Law 102-440, October 23, 1992. Available on the World Wide Web at:
http://www.fws.gov/r9dia/global/law102.html
(This is an Act to promote the conservation of wild exotic birds, to provide for the Great Lakes Fish and Wildlife Tissue Bank, to reauthorize the Fish and Wildlife Conservation Act of 1980, to reauthorize the African Elephant Conservation Act, and for other purposes.)
10. Convention on International Trade in Endangered Species of Wild Fauna and Flora, Article III.2c, IV.2.c, V.2.c, and VIII.3 and 4. Available on the World Wide Web at: *http://www.wcmc.org.uk/CITES/english/text.htm* (The methods of capture, transport, and maintenance of the species minimizes the risk of injury or damage to health, including inhumane treatment.)
11. The Ornithological Council, *Guidelines to the Use of Wild Birds in Research*, Gaunt, A.S. and Oring, L.S., Eds., Washington, D.C., 1997. Available on the World Wide Web at: *http://www.nmnh.si.edu/BIRDNET/GuideToUse*
12. American Society of Ichthyologists and Herpetologists, The Herpetologists' League, Society for the Study of Amphibians and Reptiles, *Guidelines for Use of Live Amphibians and Reptiles in Field Research*. Available on the World Wide Web at: *http://www.utexas.edu/depts/asih/pubs/herpcoll.html*
13. American Society of Ichthyologists and Herpetologists, American Fisheries Society, American Institute of Fisheries Research Biologists, *Fisheries Guidelines for Use of Fishes in Field Research*, 13(2), 16, 1988. Available on the World Wide Web at: *http://www.utexas.edu/depts/asih/pubs/fishguide.html*
14. American Heart Association policy statement regarding the use of animals. Available on the World Wide Web at:
http://www.americanheart.org/Heart_and_Stroke_A_Z_Guide/animals.html
15. American Cancer Society, Inc., Atlanta, GA. Available on the World Wide Web at: *http://www.cancer.org/research/grants.html* (Provides access to grant applications and instructions relative to IACUCs.)
16. American Lung Association. Available on the World Wide Web at:
http://www.lungusa.org (Provides access to grant applications and instructions relative to IACUCs.)
17. American Heart Association. Available on the World Wide Web at:
http://www.americanheart.org (Provides the Research Program Standards and Policies Manual for the 1998–99 Research Awards.)

18. Potkay, S., Garnett, N.L., Miller, J.G., Pond, C.L., and Doyle, D.J., Frequently asked questions about the Public Health Service Policy on humane care and use of laboratory animals, *Lab Anim.*, 24(9), 24, 1995. Available on the World Wide Web at: *http://www.nih.gov:80/grants/oprr/laba95.htm*
19. U.S. Department of Agriculture, Animal and Plant Health Inspection Service, Policy #10, Licensing and registration of antibodies, sera and/or other animal parts and pregnant mare urine (PMU), April 3, 1997. Available on the World Wide Web at: *http://www.aphis/usda.gov/ac/policy10.html*
20. Division of Animal Welfare, Office for Protection from Research Risks, National Institutes of Health, The Public Health Service responds to commonly asked questions, *ILAR News*, 33(4), 68, 1991. Available on the World Wide Web at: *http://www.nih.gov:80/grants/oprr/ilar91.htm*
21. Potkay, S., Garnett, N.L., Miller, J.G., Pond, C.L., and Doyle, D.J., Frequently asked questions about the Public Health Service Policy on humane care and use of laboratory animals, *Contemp. Topics Lab. Anim. Sci.*, 36(2), 47, 1997. Available on the World Wide Web at:
http://www.nih.gov:80/grants/oprr/faq_labanimals1997.htm
22. Florida Statutes, Section 240.241(2).
23. Garnett, N. and Potkay, S., Issues for IACUCs: use of electronic communications for IACUC functions, *ILAR J.*, 37 (4), 190, 1995. Available on the World Wide Web at: *http://www.nih.gov:80/grants/oprr/ilar95.htm*
24. University of Florida. Policy on e-mail as public records. (June 1995). Available on the World Wide Web at: *http://www.ufcn.ufl.edu/email.htm*

9
General Concepts of Protocol Review

Ernest D. Prentice and Gwenn S. F. Oki*

Introduction

Reviewing protocols involving the use of animals is one of the most important responsibilities of the IACUC. While the format employed for this review varies across institutions, the criteria used in reviewing and approving protocols should be as consistent as possible. Consistency of review is a particularly important consideration with regard to animal welfare issues, such as the use of nonanimal model alternatives and the application of refinement techniques to reduce animal pain, discomfort, and distress.

The purpose of this chapter is to present general concepts of protocol review which will help IACUC administrators, IACUC members, AVs, and other interested individuals gain an appreciation for how IACUCs conduct protocol review. Particular attention is given to issues such as scientific merit review, prereview, use of consultants, expedited vs. full committee review, and IACUC actions. Since the answers to the questions posed in this chapter are designed to be as succinct as possible, the reader is encouraged to consult the PHS Policy, the AWAR, and the *Guide*.

9:1 What guidelines do the AWAR and the PHS Policy provide with respect to IACUC review of protocols?

Reg. PHS Policy (IV,C,1,a–IV,C,1,g; IV,D,1,a–IV,D,1,e) specifies criteria which must be met before an IACUC can approve a proposed research protocol. These requirements address issues such as:

---

* Contributing to this chapter were Molly Greene, Eifaang Li, Michael Mann, Joy A. Mench, and Philip Tillman.

- Avoidance or minimization of discomfort, distress, and pain to the animals.
- Appropriate living conditions for species used in the project.
- Availability of medical care to be provided by a qualified veterinarian.
- Euthanasia consistent with the recommendations of the American Veterinary Medical Association Panel on Euthanasia.[1]
- Qualifications of personnel conducting procedures on the species being studied.

PHS Policy (IV,A,1) requires that assured institutions also base their IACUC review of protocols on the *Guide* and that they comply with the AWAR which apply to APHIS/AC regulated species.

The AWAR (§2.31,d,i–§2.31,d,xi; §2.31,e,1–§2.31,e,5) list a total of 16 requirements which must be met before the IACUC can approve a protocol. Some of the requirements mirror the PHS Policy while others are not found in the PHS Policy. For example, the AWAR require an investigator to consider alternatives to painful procedures and to describe the methods and sources (e.g., electronic literature search) used to determine that alternatives are not available. The AWAR also require the investigator to provide written assurance that the animal-related activities do not unnecessarily duplicate previous experiments. The PHS Policy requires that the PI consider alternatives, but does not require a description of the methods and sources used.

The PHS Policy also includes the nine U.S. Government Principles for the Utilization and Care of Vertebrate Animals which provide additional guidance concerning the concepts of Replacement, Reduction, and Refinement (3 Rs). Of particular note is Principle II which states that "procedures involving animals should be designed and performed with due consideration of their relevance to human or animal health, the advancement of knowledge, or the good of society." Principle II represents the scientific and societal justification for using animals in research.

Opin. It should be noted that neither the PHS Policy nor the AWAR provide specific guidelines in all areas of protocol review. It is, therefore, left to the institution to develop specific guidelines for implementation of federal requirements. Many institutions have, therefore, developed investigator handbooks which include information concerning anesthetics, analgesics, blood sampling, antibody production, and other aspects of animal research.

Finally, it should be mentioned that NIH/OPRR periodically publishes *OPRR Reports* in the form of "Dear Colleague" letters that provide information concerning NIH/OPRR's interpretation of the requirements of the PHS Policy.[2] In a somewhat similar vein,

APHIS/AC issues its *Animal Care Policy Manual* that provides further guidance for IACUCs.

9:2 What are some useful pathways to disseminate protocols for review once they have been submitted to the IACUC?

Reg. PHS Policy (IV,C,2) requires that each IACUC member be provided with a list of proposed projects to be reviewed, and that written descriptions of projects be available to all IACUC members. AWAR (§2.31,d,2) have the same requirement.

Opin. There are many methods used to disseminate protocols for review. Most IACUCs meet in a common location to review protocols, but some committees communicate by fax, electronic mail, phone, or video conferencing (see 6:11). It is common practice for IACUCs to assign each protocol to a primary and secondary reviewer (chosen based on their expertise in the subject matter of the protocol) who are given principal responsibility for in-depth review. With this method, all members of the IACUC receive complete copies of the protocols or a summary for review. Some IACUCs, particularly at institutions that have a large amount of research using animals, utilize a prereview system where protocols are clarified and modified as necessary prior to full committee review (see 9:5, 9:6). Many IACUCs also utilize a designated reviewer system (also known as an expedited review system) which in turn helps to decrease the workload of the full committee (see 9:20). In general, the method used to disseminate protocols for review is dictated by the size and specific needs of the institution.

9:3 What are the IACUC and institutional obligations for reviewing a protocol for an internally funded project?

Reg. If an institution has elected, in its NIH/OPRR Assurance, to extend the requirements of the PHS Policy to all activities (regardless of the source of funding), then the IACUC obligations for reviewing internally funded projects do not differ from those obligations under the PHS Policy.

Opin. Internally funded projects may pose special problems for the review process, particularly with respect to scientific merit review. Even IACUC members who maintain steadfastly that the committee has no responsibility or right to review scientific merit agree that the IACUC has some obligation to do so when the proposal will not be subjected to external peer review. At least a minimal level of merit which justifies animal usage must be assured before such protocols are approved. Needless to say, animal welfare considerations should be the same, and the same rigorous review process with regard to regulatory compliance must apply to both internally and externally funded projects.

9:4 At what point in the review of a protocol are consultants best brought in?

Reg. PHS Policy (IV,C,3) and the AWAR (§2.31,d,3) allow the use of consultants for the review of a protocol. Consultants may not vote with the IACUC and the IACUC remains responsible for its actions and decisions.

Opin. The use of consultants is highly variable among IACUCs. In some institutions, it is common practice for IACUC reviewers to use colleagues within the institution as consultants for clarification of issues raised during prereview of a protocol (see 9:5, 9:6). This prereview clarification can greatly speed the final review process. Outside consultants are seldom used unless there is disagreement among the IACUC members as to the merit of a given proposal or unresolved concerns exist with regard to animal welfare. Consultants also are useful if an investigator appeals the decision of the IACUC (see 29:30).

9:5 What is meant by prereview of a protocol?

Opin. Prereview refers to review of a protocol by one or more individuals prior to formal review by the IACUC. Neither the PHS Policy nor the AWAR make any specific mention of the term or concept of prereview, although many institutions find it useful. The AWAR (§2.31,d,1,iv,B) require veterinary consultation during the planning of a procedure that might cause more than momentary pain or distress. (See 9:6, 16:4.)

9:6 What are the reasons for a prereview of protocols?

Opin. Protocols are sometimes incomplete when they are submitted for IACUC consideration. Information that the IACUC needs in order to effectively review the protocol might be absent or unclear. For example, the investigator may be proposing to use an injectable method of anesthesia without specifying the dose. Clearly, this information must be provided before the IACUC can approve the protocol. Prereview saves the IACUC time, since IACUC members do not have to read protocols that are obviously not ready to be reviewed. Prereview also should be helpful to the investigator in terms of speeding up the review process, since it means that the protocol is less likely to be tabled and carried over to the next IACUC meeting pending further information and clarification.

9:7 Is prereview of protocols required by any regulations or policies?

Opin. Prereview of protocols is not a PHS Policy requirement, but the AWAR (§2.31,d,1,iv,B) require PIs to consult with a veterinarian when potentially painful procedures are planned. Prereview of the protocol by a veterinarian is one way that such a consultation can be provided. (See 9:5.)

9:8 Can the prereview team be an IACUC subcommittee composed of IACUC members, non-IACUC members, or a combination of these?

Opin. Since prereview is not a mandated function of an IACUC, the prereview process can be structured in a way that is most helpful to the IACUC and investigators. Prereview can be carried out by a designated IACUC member or members, dedicated staff, veterinarians, or any other experienced individuals who are able to adequately review the protocol.

9:9 What is a workable mechanism for the prereview of protocols?

Opin. The best mechanism depends on the institutional structure and the composition of the prereview team. For example, an institution with multiple vivaria may choose to have incoming protocols immediately submitted to a designated prereviewer. Each prereviewer handles all of the submitted protocols from particular vivaria on the campus, so the prereviewer becomes very familiar with the investigators, facilities, and types of research conducted in those units. The prereviewer reviews the proposal, notes any points of inadequacy in the document, and communicates with the PI. The prereviewer may be able to suggest minor corrections or suggest adding clarifications to the protocol (for example, changing the route, dose or type of anesthetic agent, or suggest adding a sentence to the protocol indicating the volume of blood to be withdrawn) with the permission of the PI. In the case of more substantial corrections, the prereviewer discusses the problems with the PI and may recommend that the investigator rewrite the protocol and submit a corrected version.

9:10 What should the PI do following prereview?

Opin. Assuming that the prereviewer suggests changes that are in accordance with IACUC policies and concerns, it is in the PI's best interests to make the necessary corrections before the IACUC meeting so that the protocol is suitable for review.

9:11 What happens to a protocol after it is prereviewed?

Opin. After a protocol is prereviewed, it must be processed for formal review by the IACUC. This may be accomplished by either a designated reviewer (see 9:19 to 9:24) or review by the full IACUC at a convened meeting.

9:12 Are the results of the prereview presented to the full IACUC? If so, by whom?

Opin. The results of the prereview can be presented to the IACUC in several different ways. At least one institution with a very high volume of research ensures that the protocol which is finally sent to the IACUC reflects the changes that were made during the prereview process. If all of the changes are minor, they are typed directly onto the originally submitted protocol. Major changes are usually incorporated into a rewritten protocol. In this case, the IACUC is usually not presented with information about all of the specific issues and changes that the prereviewer and the PI have discussed. When rewriting the protocol does not seem warranted, the IACUC is sent the protocol with attachments detailing the prereviewer's questions and the PI's clarifications or added information. Other IACUCs request a "clean" protocol (no attachments) for its final review. Last minute changes from the prereview discussion also may be presented at the IACUC meeting. Since the prereviewers are familiar with each project, they often either verbally present the project at the IACUC meeting or are available at the meeting to answer members' questions about the project.

9:13 Are prereview comments binding on investigators?

Opin. No. Prereview is simply a method for giving investigators advice on their protocol. A possible exception is where the AV determines during prereview that a particular anesthetic is inappropriate (AWAR §2.33,b,4). Investigators may occasionally disagree with the prereviewer's suggestions and choose to send an unmodified or only partly modified protocol to the IACUC for formal review. However, an investigator who insists on presenting an inadequate document to the IACUC is likely to find that the IACUC will not approve the protocol, and that the protocol will be returned for modifications or clarification for the same reasons identified by the prereviewer.

9:14 Should the IACUC review protocols for scientific merit? Do the AWAR or the PHS Policy specifically require (or prohibit) review of scientific merit?

Opin. In our experience, many IACUCs review protocols for "scientific merit" or "scientific relevance." It is not clear, however, that these terms have the same or similar meanings to every IACUC. Whether IACUCs *should* perform such review is open to debate. Prentice et al.[3] reviewed the PHS Policy. That policy does not use the term "scientific merit." Rather, it refers to "relevance" in U.S. Government Principle II, in a manner which strongly suggests that the terms are synonymous. The PHS Policy also uses the terms "sound research design" (IV,C,1,a.) and "scientifically valuable research" (IV,D,1,d). When all of these terms are considered together, it supports the position of NIH/OPRR that an IACUC should at least consider the scientific relevance of a proposal.[4] Indeed, NIH/OPRR has written:[4]

> "Although not intended to conduct peer review of research proposals, the IACUC is expected to include consideration of the 'U.S. Government Principles for the Utilization and Care of Vertebrate Animals in Testing, Research and Training ...' in its proposal review process. Principle II calls for an evaluation of the relevance of a procedure to human or animal health, the advancement of knowledge, or the good of society. Other references (sections IV,C.1 and IV,D.1) include language such as 'consistent with sound research design,' 'rationale for involving animals,' and 'in the conduct of scientifically valuable research,' which presumes that the IACUC will consider in its review the general scientific relevance of the proposal. The presumption is that a study that could not meet these basic tests would be inherently invalid or wasteful and, therefore, not justifiable."

The AWAR appear inconsistent in their reference to scientific merit review. In the Public Comment Section of the regulations, APHIS/AC states, "we added the term 'animal care and use procedure' ... to avoid any misunderstanding or implication that APHIS intends to become involved in the evaluation of the design, outlines, guidelines, and scientific merit of proposed research."[5] On the other hand, the AWAR (§2.31,e,4) like the NIH/OPRR, make reference to "scientifically valuable research." Also, the AWAR (§2.31,a) and the AWA itself (§13,a,6,A,i-ii) state that "except as specifically authorized by law or these regulations, nothing in this part shall be deemed to permit the Committee or IACUC to prescribe methods or set standards for the design, performance, or conduct of actual research or experimentation by a research facility."

It appears that NIH/OPRR and APHIS/AC do not want to commit themselves openly to require IACUC review of scientific relevance. Nevertheless, according to NIH/OPRR,[6] an institution cannot totally defer scientific relevance review to the funding agency. IACUC approval, using criteria stated in the PHS Policy, must precede NIH peer review. Therefore, approval by an IACUC of a proposed activity that is conditional upon successful peer review by the funding agency is not in keeping with the PHS Policy requirements. Such conditional action does not constitute the approval required by the PHS Policy prior to review by the NIH Initial Review Group (IRG) or Study Section. The NIH review of merit should, therefore, be viewed as additional assurance rather than the only assurance that the research has value.[3,6] Based on this, Prentice et al.[3] concluded that the IACUC does have a responsibility to review scientific merit for animal projects which are subject to the requirements of the PHS Policy. There is not, however, general agreement with this conclusion. Black[7] argues that merit and relevance are not the same thing; that IACUC review does not constitute peer review, the consequences of IACUC review are different from those of external review, and IACUC review may constitute a violation of researchers' academic freedom. Prentice et al.[8] offer counter arguments. Clearly there is not general agreement.

9:15 Can the IACUC approve judicious use of animals without consideration of scientific merit?

Opin. See 9:14 for a general discussion. Prentice et al.[3] note that there are two levels of review for scientific merit. They refer to a "fundamental level" of review in which scientists form "basic judgments about the adequacy and appropriateness of experimental design in terms of the ability to test the hypothesis, use of controls, sample size, statistical analysis, and the training and experience of investigators." This first-level merit review is necessary for the IACUC to approve the judicious use of animals. The other level of review, "knowledge-based level," requires "an assessment be made of the scientific importance of the study." An assessment of merit at the fundamental level could be made by any appropriately constituted IACUC, even in the absence of special expertise in the topic of the protocol. On this latter level, the jury is out.

9:16 Is it appropriate for the IACUC to use consultants to review scientific merit?

Reg. (See 9:4.)
Opin. If the IACUC assesses the scientific importance of a study (see 9:14, 9:15), then it is appropriate for the IACUC to use consultants, particularly if the IACUC does not have the necessary expertise to assess scientific merit. Also, outside reviewers would likely be most useful when there is disagreement among the IACUC members as to the scientific merit of a protocol or if an investigator can appeal the decision of the IACUC (see 9:53, 29:30). Indeed, the IACUC should seek

advice from consultants with regard to any aspect of the project that is problematic.

9:17 What types of consultants might be useful for review of scientific merit? Who picks them?

Opin. Consultants can be either experts on the topic of the protocol under review or experts in the use of animals, with respect to a particular procedure proposed in the protocol. An example of the former is an expert who is asked to review a protocol because the IACUC is uncertain whether the use of the ascites method for making monoclonal antibodies is required in a particular project as the PI claims. Or, the consultant can be asked to judge whether any useful information will be forthcoming once the antibody is made. An example of the latter is an expert in primate behavior and care who is asked to review a protocol because the IACUC is uncertain whether a rhesus monkey may reasonably be restrained continuously for more than 24 hours. Or, the expert can be asked whether any useful information could be derived from the experiments that require the monkey to be restrained for that period of time.

Consultants should be selected by the IACUC, but it is reasonable to allow the PI to suggest possible experts. This allows the committee to maintain objectivity in the review process while allowing the PI to feel that her point of view is being presented.

9:18 Should the identity of consultants be kept anonymous to the investigator?

Opin. The same arguments apply to anonymity of consultants and to the anonymity of grant proposal reviewers. It has been argued that reviewers refrain from making justifiable negative comments about a proposal if their identity is known to the investigator. They may fear reprisals. It also is possible to argue that reviewers may make unfair statements about a protocol because their identity is withheld. Whether this will happen probably depends upon who the reviewer and investigators are. In any case, if the reviewers of a protocol remain anonymous, then certainly consultants should as well.

9:19 What is "expedited review" or "designated member review," and what is its intent?

Reg. Both the PHS Policy (IV,C,2) and the AWAR (§2.31,d,2) recognize a method of "designated member review" that is widely implemented by research institutions.[9] Many institutions refer to this review process as "expedited review," but "designated member review" is the appropriate term. Designated member review must conform to the following process: written descriptions of research projects that involve the care and use of animals must be made available to all IACUC members, and any member of the IACUC must have the opportunity to obtain, upon request, full committee review of those research projects. "If full committee review is not requested, at least one member of the IACUC, designated by the chairperson and qualified to conduct the review, shall review those activities, and shall have the authority to approve, require modifications in (to secure approval), or request full committee review of any of those activities." (PHS Policy IV,C,2; AWAR §2.31,d,2).

Opin. Although the exact procedures frequently vary between institutions, an expedited (designated review) process can enable IACUCs to review and approve protocols faster than those presented for full committee review. It's important to mention, however, that designated review in no way implies the quality of review is less stringent than a protocol reviewed by the full committee. A successful designated review program allows institutions, particularly larger institutions that process a large number of protocols, the opportunity to reduce some of the IACUC workload, thereby, allowing members to focus on protocols that may warrant more time and attention. (See 11:9.)

9:20 Under what circumstances might a protocol be assigned to designated member (expedited) review?

Reg. A protocol may be assigned to one or more designated reviewers only after all IACUC members have been provided the opportunity to call for full IACUC review (PHS Policy IV,C,2; AWAR §2.31,d,2). (See 9:19.)

Opin. Since the AWAR and the PHS Policy do not define the criteria for assigning protocols to the expedited process, each institution must develop an internal policy for its own designated review process tailored to the individual institution's needs and idiosyncrasies. Even if an institution develops criteria to determine the kinds of activities or protocols that it will permit to be reviewed utilizing the designated review method, all IACUC members must still be provided with the opportunity to call for full IACUC review of each individual project.

An institution should first decide whether it needs to establish a designated review process. Then, the IACUC should create a list of well-defined criteria to determine the types of protocols qualifying for designated vs. full committee review. For example, some institutions restrict designated review to protocols involving noninvasive or acute procedures that do not cause more than momentary or slight pain or distress to the animals (e.g., the procedures only involve euthanizing the animals in order to harvest tissues).

If an institution elects to adopt a designated review process, the IACUC should document the assignment criteria and the procedures for processing expedited protocols. The IACUC should develop a set of SOPs in order to ensure that the designated review process is not misused. The SOPs should be included in the institution's Animal Welfare Assurance, approved by NIH/OPRR, and included in the institution's IACUC Policy and Procedural Manual.

9:21 Under what circumstances is a designated review inappropriate?

Opin. Although each institution must determine the criteria appropriate for its own program, designated member review may be unacceptable under many circumstances, including the following:

- IACUC members are not provided with sufficient information concerning the proposed research activities.
- IACUC members are not given an opportunity to call for full committee review prior to the approval of the proposed protocol via an designated review process (PHS Policy IV,C,2; AWAR §2.31,d,2; OPRR "Dear Colleague" letter of 5/21/90).[10]
- If any IACUC member requests full committee review (PHS Policy IV,C,2; AWAR §2.31,d,2).
- One or more members of the designated subcommittee authorized to review and approve research activities has a conflict of interest (PHS Policy IV,C,2; AWAR §2.31,d,2).
- The proposed protocol does not meet the requirements for designated review delineated in the institution's NIH/OPRR-approved Animal Welfare Assurance or the institution's IACUC Policy and Procedural Manual. For example, when the invasive nature of the research does not permit the protocol to be reviewed by the expedited method.

9:22 Should the entire IACUC be involved in the designated review process or can a subcommittee conduct designated review?

Reg. (See 9:19, 9:20.)
Opin. The entire committee must have the opportunity to review the information provided on each designated member protocol review list and to request review by the full committee. However, a subcommittee of one or more members who are qualified to review the protocol (e.g., the Chair and the AV) can be designated to review the protocol

in its entirety. Questions and concerns raised by any member of the committee should always be addressed prior to approval.

9:23 What is required for approval of a protocol via designated review?

Reg. The designated reviewer(s) must use the same criteria that are applicable to protocols undergoing full IACUC review.

Opin. Although the intent of the designated review process is to conduct rapid review of protocols, by no means does this process require less information than the full committee review process. The reviewer(s) conducting designated review must ensure that the method is in full compliance with all of the appropriate requirements of the PHS Policy (IV,C,1,a–IV,C,1,g; IV, D,1,a–IV,D,1,e) and the AWAR (§2.31,d,i–§2.31,d,xi; §2.31,e,1–§2.31,e,5). Before approval may be granted via a designated review process each member of the IACUC must have the opportunity to review the protocol and the opportunity to request full committee review.

9:24 What is an effective means for administering a designated member review?

Opin. One significant difference between designated member review and full committee review is that the designated member review process does not require a fully convened meeting of the IACUC. Despite this fact, after a protocol is assigned for designated review, at least one or two IACUC members should review the entire protocol with the same degree of thoroughness that is given to a protocol reviewed by the full committee.

Another difference between the two processes involves the way in which protocol information can be disseminated to the IACUC members. Unlike the full committee protocols, the information pertinent to protocols presented for designated review can be more easily disseminated through the use of electronic means (e.g., fax or e-mail) in order to facilitate the most expeditious type of review. (See 6:11.)

The following is an example of an designated member review process. This particular example goes beyond the requirements of the AWAR and PHS Policy in that it requires a summary to be sent to all members, a quorum of the IACUC members to respond, and documentation of each member's decision with regard to designated member review:

- The IACUC administrator reviews the protocol submitted to the committee and decides whether a protocol qualifies for designated review based upon the IACUC's predetermined criteria.

- A subcommittee, composed of the IACUC Chair and the AV, is authorized to review and approve the protocol. In a case where the Chair has a conflicting interest (e.g., is personally involved in the project), he must be replaced by another member qualified to conduct the review (PHS Policy VI,C,2; AWAR §2.31,d,2). (See 6:8.)
- A summary of the protocol qualifying for designated review (including title of project; species, number of animals requested; type of experimental procedures) is forwarded to members of the IACUC via postal mail, fax, or electronic mail (see 6:11). In some instances, additional information may be included (e.g., supporting grant materials or parts of the original protocol).
- Each IACUC member reviews the summary and, if necessary, requests a copy of the complete protocol to determine whether clarification or changes are needed, or has the opportunity to call for full committee review.
- If no member calls for full committee review, the protocol can be reviewed and approved by one or more qualified reviewers who are designated by the Chair.
- Documentation, including voting sheets, are maintained as evidence of each member's decision with regard to these protocols. The designated review lists and voting sheets are kept in a centralized location in the event the review of a designated member protocol is ever questioned. No matter what process an institution adopts, the entire designated review procedure must be documented from assignment to approval.

9:25 What is "full committee" review?

Reg. Full committee review of an IACUC protocol is one that is conducted by a quorum of the IACUC at a regularly scheduled or specially convened meeting (PHS Policy IV,C,2; AWAR §2.31,d,2).

Opin. Due to the volume of protocols reviewed at many institutions, a good portion of the actual work involved in protocol review may be done before the meeting via a mechanism such as prereview (see 9:5 to 9:13). Thus, at the meeting, the committee members are able to concentrate on the proposed protocol, its merit, and the use of laboratory animals without the necessity of obtaining clarifications related to incomplete information and/or confusing points.

9:26 What is the purpose of a full committee review?

Opin. The purpose of full committee review is to have all IACUC members involved in the review and decision making on the disposition of

9:27 When is full committee review of a protocol appropriate?

Reg. Full committee review is appropriate at any time and required when requested by any member of the IACUC (PHS Policy IV,C,2; AWAR §2.31,d,2). Each member must have the opportunity to review any protocol and the opportunity to request full committee review before approval may be granted.[9] There are no other federal requirements specifying when a protocol should receive full committee review or the type of protocol that should receive full committee review.

Opin. Many IACUCs have developed criteria to determine which protocols will receive full committee or designated review. Items to be considered in developing such criteria can include:

- Invasiveness of procedures.
- Level of pain or distress.
- Species used and number requested.
- Experimental design.
- "Controversial" procedures, e.g., death as an endpoint.
- Procedures that request exceptions to regulations, e.g., multiple survival surgeries.
- Whether the protocol will receive peer review by the funding agency or other group prior to funding.

9:28 What is an effective full committee review process?

Opin. Each IACUC should establish SOPs regarding conduct of full committee review of protocols. One effective method of full committee review could include the following:

- Prereview (see 9:5 to 9:13).
- Review by IACUC staff, or designated member of the committee, for completeness of the application and compliance with information requirements, as indicated in PHS Policy (IV,C,1,a–IV,C,1,g; IV,D,1,a–IV,D,1,e; AWAR §2.31,d,1,i–§2.31,d, 1,xi; §2.31,e,1–§2.31,e,5).
- Review by the AV in addition to the consultation required during design of the study (AWAR §2.31,d,1,iv,B).

- Review by outside consultants if the IACUC members do not possess relevant expertise.
- Review by committee members assigned as primary and secondary reviewers.
- Primary and secondary reviewers communicate with the investigator prior to the meeting, during which time the reviewers may recommend modifications in the protocol (PHS Policy IV,C,2; AWAR §2.31,d,2).
- Presentation of the protocol to either a protocol review subcommittee or the full committee by the primary and secondary reviewers followed by a general discussion of the protocol.
- If reviewed by a subcommittee, it is recommended that a synopsis of that group's deliberations be presented to the full committee to allow for any additional comments.
- Every member of the committee should be provided with a copy of the protocol, or, at a minimum, a summary of the protocol to be reviewed at the meeting.
- Attendance by the investigator to present the protocol or answer questions.
- Following discussion at a convened quorum of the committee, a vote is taken to determine final disposition of the protocol.

9:29 What are the PHS Policy and AWAR requirements for approval of a protocol by full committee review?

Reg. The PHS Policy and AWAR require:

- Each member of the IACUC must be provided with, at the minimum, a list of protocols to be reviewed (PHS Policy IV,C,2; AWAR §2.31,d,2).
- Both the PHS Policy (IV,C,1,a–IV,C,1,g; IV,D,1,a–IV,D,1,e) and the AWAR (§2.31,d,1,i–§2.31,d,1,xi; §2.31,e,1–§2.31,e,5) have very specific requirements about information which must be included in the protocol as well as the items the IACUC must consider as described below.
- No member of the IACUC may participate in the review or approval of a research project in which the member has a conflicting interest (e.g., is personally involved in the project) except to provide information requested by the IACUC (PHS Policy IV,C,2; AWAR §2.31,d,2). Some IACUCs require this member to leave the room during discussion or voting on the protocol in question.

- Approval of a protocol considered by the full committee "may be granted only after review at a convened meeting of a quorum of the IACUC and with the approval vote of a majority of the quorum present" (PHS Policy IV,C,2). The AWAR(§2.31,d,2) have the same requirement.
- A member who has a conflicting interest in a protocol under review may not contribute to the constitution of a quorum (PHS Policy IV,C,2; AWAR §2.31,d,2). This person may not vote on the protocol in question.
- Institutions must maintain written documentation of committee deliberations (PHS Policy IV,E,1,2; AWAR §2.35,a,1,2). While most IACUCs include this information in the minutes, such documentation can be maintained instead on file with the protocol. Filing records of deliberations along with the protocol is of importance to institutions subject to state open records laws. It is up to the IO, IACUC Chair, or staff to explain to either APHIS/AC, NIH/OPRR, or AAALAC the rationale for the institution's recordkeeping methods.
- Both the investigator and the institution must be notified in writing of the committee's decision (PHS Policy IV,C,4; AWAR §2.31,d,4).

9:30 Is it important for the IACUC to know whether a protocol is new or a resubmission with a change in title?

Opin. Whether a protocol is new or a resubmission with a change in title may determine the type of review depending upon the policies established by the IACUC. For example, an IACUC may decide that a resubmission with a change in title *only* could receive designated review if it was reviewed by the IACUC within the past year. Alternatively, a change from a funding source that conducts peer-review of proposals to one that does not may trigger full committee review.

9:31 What procedures can the IACUC use to determine if a protocol is new or a resubmission with a change in title?

Opin. If a protocol is designated as a resubmission with a change in the title only, the IACUC may decide to obtain verification by comparing the resubmission with the previously approved protocol on file. Investigators sometimes inadvertently modify protocols during the course of resubmission without recognizing that the IACUC must approve all proposed significant changes (PHS Policy IV,B,7; AWAR §2.31,C,7).

9:32 Should IACUC decisions be influenced by the source or size of a research grant?

Reg. The IACUC must review grants under the requirements of PHS Policy IV,C,1, regardless of the size of the grant. Non-PHS-supported research may or may not be covered by the PHS Policy, depending on the statement of applicability in the institution's Animal Welfare Assurance.

Opin. On reflex, most IACUC members would probably answer "no" to this question. In general, all protocols should be reviewed with the same rigor regardless of how much money is involved or the source of funding. While political pressures for IACUC approval of large or prestigious grants are real, studies should be approved based on appropriateness of the proposed research. Certainly, animal welfare should never be compromised in the interest of increasing grant funds. Nevertheless, in one study,[11] results of a survey indicated that IACUC deliberations potentially can be influenced by the size of a grant as well as by pressure and perceptions that the committee may not even recognize.

9:33 Should the decisions of the IACUC be influenced by the potential scientific importance of a project proposed in a protocol?

Reg. The IACUC has the authority to approve a proposed project. NIH/OPRR and APHIS/AC have stated that under no circumstances is an IACUC required to approve a project against its will.[12]

Opin. IACUC decisions should not be influenced by the investigator, species, funding source, or "hotness" of the research topic. Review should be based on animal welfare issues in consideration of the regulatory requirements. Nevertheless, both the dollar value of the grant, as well as a species-specific view of an animal's societal worth potentially can affect IACUC deliberations.[11]

9:34 If a protocol is completely original and contains untested surgical or experimental procedures, how could such a protocol be reviewed and approved when the procedures cannot be referenced?

Opin. Few protocols actually contain untested surgical procedures. Those that do can be handled in one of two ways. In some cases, the IACUC could recommend that a pilot study be conducted involving only the part of the protocol using the untested procedures. With these pilot data derived from successful use of the previously untested procedures, the investigator could then obtain IACUC approval for the complete protocol. Pilot studies must be reviewed and approved by the IACUC.

Alternatively, the IACUC could approve the protocol as submitted but with a reduced number of animals, the remainder being approved when the investigator has tested the new procedure. Both of these actions allow the researcher to continue the study while protecting the animal subjects. (See 13:6, 13:9.)

9:35 What are some possible actions an IACUC can take with respect to protocols that have been reviewed?

Reg. Both the PHS Policy (IV,B,6) and the AWAR (§2.31,C,6) allow the IACUC to review and approve, require modifications in (to secure approval), or withhold approval of proposed activities related to the care and use of animals.

Opin. In determining the disposition of a protocol submitted for review, an IACUC has several options including:

- Approval
- Approval with conditions (see 9:37)
- Tabling the protocol with a request for major revisions
- Withheld approval

9:36 What constitutes "approval" of a protocol? Does this action necessarily mean that no further changes or information are needed?

Reg. For PHS Policy and the AWAR, IACUCs either approve, require modifications in (to secure approval), or withhold approval of protocols (AWAR §2.31,c,6; PHS Policy IV,B,7). Designated reviewers may approve, require modifications to secure approval, or request full committee review. Anything short of final approval is not adequate for initiation of animal activities or submissions of an IACUC approval date to the NIH as part of a grant application.

Opin. Generally, an approved protocol is one which contains all the required information and has been judged by the IACUC to be acceptable. Approval as submitted also may be granted when additional information required or requested is incidental to the proposed study. For example, the committee may simply request that a "lay description" of the proposed work be rewritten so that the language is more elementary, or that the investigator comply with a simple format requirement.

9:37 What constitutes "approval with conditions?"

Reg. (See 9:36.)

Opin. The IACUC may grant conditional approval of a protocol when it is determined that no major revisions or clarifications are required. Usually, the Chair or the AV is assigned to review the PI's response or revised protocol, and are empowered by the IACUC to approve the protocol without further review by the full committee. Conditional approval does not, however, mean that the study can be immediately initiated. The PI must first comply fully with all conditions arising from the IACUC's review and then a final approval is given. The reader is cautioned that terms such as "conditional approval," "provisional approval," or "approved pending clarification" frequently cause confusion. NIH/OPRR and APHIS/AC prefer that IACUCs either avoid these terms or describe them in sufficient detail to be fully understood. (See 9:42.)

IACUCs might determine that a protocol is approvable, contingent on receipt of a very specific modification (e.g., receipt of assurance that the PI will conduct the procedure in a fume hood). The IACUC can handle this modification (or clarification) as an administrative detail so that an individual, such as the Chair, can verify the same. On the other hand, protocols that are missing substantive information necessary for the IACUC to make a judgment (e.g., justification for withholding analgesics in a painful procedure) are incomplete. If the protocol is incomplete, it is not possible to satisfy the first step in all protocol review processes, i.e., a complete description of the proposal and the opportunity for the entire committee to call for full review. NIH/OPRR and APHIS/AC recommend that IACUCs devise effective ways of differentiating between substantive omissions and administrative issues.[13]

9:38 What constitutes a "tabled" or "postponed" protocol?

Opin. The "tabling" or "postponing review" of a protocol are terms often used interchangeably by IACUCs, but are not used as in the more formal structure of Roberts' Rules of Order. The IACUC may "table" or "postpone" further review of a protocol pending receipt of additional substantive information or a significant revision of the protocol. The action of tabling a protocol is generally used during the process of full committee review when the IACUC decides it is necessary for the whole committee to re-review the protocol before further action can be taken. Tabling a protocol is usually reserved for proposals that do not contain sufficient information or those in which the IACUC has identified a serious animal welfare concern.

9:39 What might cause a protocol to be "disapproved" (approval withheld)?

Reg. Approval may be withheld if any of the PHS Policy (IV,C,1) or AWAR (§2.31,d,1; §2.31,d,2,e) criteria are not met.

Opin. Withheld approval of protocols is usually infrequent, at best. Generally, approval is withheld when the PI and the IACUC cannot agree on fundamental aspects of the proposed study such as the protocol design, animal welfare issues, or the PI will not agree to comply with the IACUC's requirements.

Both the PI and the institution must be notified in writing of the committee's decision. This written notification should include a statement of the reasons for the decision and provide the PI with an opportunity to respond in person or in writing (PHS Policy IV,C,4; AWAR §2.31,d,4).

9:40 How should an investigator respond to questions or conditions from the IACUC?

Opin. Responses from the PI to questions or concerns raised by the IACUC should be in such a format so as to result in a protocol that contains a complete, easy to discern description of the proposed activities. The response from the PI should clearly answer each issue raised. This can be achieved via a point-by-point letter that serves as an amendment to the protocol or by a revised protocol which incorporates all required changes and clarifications. Requiring the PI to respond to the IACUC's review by submitting a revised protocol has the advantage of ensuring that a complete and accurate protocol is contained in one document. Copies of the final, approved IACUC protocol should be maintained in the IACUC administrative office, the animal facility, and in the investigator's laboratory.

9:41 Should there be a time limit to receive responses to queries raised by the IACUC?

Opin. Establishment of time limits and other constraints should be considered when the IACUC develops an SOP. Specific time constraints and deadlines can benefit both the investigator and the committee when both know the expectations. Ironclad deadlines and inflexible staff, however, can present a major point of contention for faculty already juggling multiple projects and who are attempting to comply with myriad paperwork and deadlines.

9:42 Once a response from an investigator is received, can the IACUC Chair or his designee approve protocols if conditions are met?

Opin. Generally, this should be determined when the decision is made to grant "conditional" approval. That is, approval is granted upon the condition that the PI meet certain requirements, and the committee agrees that the assigned reviewers or the Chair can determine if the conditions have been met satisfactorily and then grant approval (PHS Policy IV,C,2; AWAR §2.31,d,2). It should be noted that "conditional approval" is not mentioned in either the AWAR or PHS Policy. Technically, there initially would be "withhold approval" and subsequent approval via the designated member review process. A distinction should be made between conditions based on administrative details and information that has a bearing on the IACUC's decision. (See 9:36, 9:37.)

9:43 Is a majority vote required for any IACUC actions related to protocol approval?

Reg. Approval of a protocol considered under guidelines for full committee review "may be granted only after review at a convened meeting of a quorum of the IACUC and with the approval vote of a majority of the quorum present" (PHS Policy IV,C,2). The AWAR (§2.31,d,2) have the same requirement. (See 9:19 for designated member review considerations.)

9:44 At a full committee meeting of a 20-member IACUC, 15 members are present. Six vote to approve a protocol, six abstain, and three vote against approval. Is the protocol considered approved as per the AWAR and PHS Policy?

Opin. No. In this example, the 15 members present constitute the quorum. In order to approve a protocol, a simple majority of those members present, i.e., eight, would have to vote for approval of the protocol. Since only six members voted to approve, the vote count fell short by two votes and the protocol cannot be approved per the AWAR and PHS Policy.[14] (See 9:43.)

9:45 How should the IACUC handle minority opinions to IACUC actions on the review of protocols?

Reg. PHS Policy (IV,E,1,d) requires institutions to maintain copies of minority views of semiannual reports (not protocol reviews), but does not specify how the IACUC should handle them. There is also a PHS Policy (IV,F,4) requirement that minority views filed by IACUC members be forwarded (via the IO) to NIH/OPRR along with the annual report to NIH/OPRR (but again, this does not refer to protocol reviews). If the minority views relate to an IACUC action that is required to be reported promptly (PHS Policy IV,G,3), they should be provided to NIH/OPRR at that time. The AWAR (§2.31,c,3; §2.35,a,3) also require maintaining records of semiannual reports, including minority views.

Opin. As neither the PHS Policy nor the AWAR address this issue relative to specific IACUC protocols, it should be considered when the IACUC develops an SOP. A common practice is to record dissenting opinions in the minutes. However, another option is to allow the dissenter to write a letter expressing her opinion. This is then filed along with the IACUC protocol.

9:46 A protocol has been approved with changes requested by the IACUC. The protocol is associated with a grant application to the NIH. Does the approval letter sent by the Research Administration Office to the NIH have to detail the changes made in the animal use portions of the grant?

Reg. The PHS Policy (IV,D,2) requires "verification of approval (including the date of the most recent approval) by the IACUC of those components related to the care and use of animals. ... If verification of IACUC approval is submitted subsequent to the submission of the application or proposal, the verification shall state the modifications, if any, required by the IACUC." For competing applications or proposals, verification of IACUC approval may be filed at a time not to exceed 60 days after the proposal application deadline (PHS Policy IV,D,2).

Opin. While the 60-day grace period is helpful to investigators, it creates a situation where an IACUC may require changes after the grant has been submitted to the NIH. It is likely not to be in the best interest of an investigator to have to amend a grant proposal under such circumstances.

9:47 Should the IACUC review just the IACUC protocol or should the IACUC also review the animal care and use sections of an associated grant proposal?

Reg. Verification of IACUC approval submitted to the NIH means that the IACUC has reviewed and approved those components of the grant application related to the care and use of animals. The signature of the institutional representative on the PHS 398 form is a legally binding statement that the IACUC has approved all animal activities covered in the grant application.

Opin. If the IACUC protocol accurately reflects the information contained in the grant application, then the verification is obviously valid. The IACUC, however, cannot be assured of such validity unless it also reviews the animal care and use sections of the associated grant application.

9:48 Should the IACUC accept an approval statement from an IACUC at another institution?

Reg. An IACUC may accept the approval of an IACUC at another institution if that institution has an approved NIH/OPRR Animal Welfare Assurance. This practice is normally limited to collaborations or sub-grant/subcontracting involving performance sites other than the awardee institution. In most instances, the IACUC of the performance site assumes responsibility for animal activities in its facilities. Both institutions should have a clear understanding of what their respective responsibilities are in this situation, particularly if one IACUC agrees to abide by the determinations of another IACUC.[15] (See 8:11.)

Opin. This is an area for which the IACUC should establish an SOP. The IACUC should have some "comfort level" about the other institution, its committee, and the quality of its research. To avoid problems, a collaborative protocol approved by an IACUC at another institution should probably receive at least the equivalent of designated review by the local IACUC. In some instances, full committee review is warranted. Issues to consider in accepting an approval statement from an IACUC at another institution include:

- Does the institution have an approved NIH/OPRR Animal Welfare Assurance?
- Is the institution a USDA-registered research facility?
- Is the institution accredited by AAALAC International?

9:49 Can an investigator receive "conditional approval" for a protocol approved at another institution until the IACUC can review and approve the protocol? What should be the conditions and limitations for this conditional approval?

Reg. (See 9:36.)

Opin. An IACUC may choose to grant conditional approval for a protocol approved by a different institution. However, any conditions regarding acceptance of an approval statement from an IACUC at another institution should be described in the IACUC's written policies. For example, in order to facilitate research, an IACUC may choose to allow animals to be purchased or transferred to its institution, but procedures which involve animals cannot be performed until the local IACUC has unconditionally approved the protocol. (See 9:37.)

9:50 An investigator subcontracts part of a research project to another institution where animals will be used. What oversight and paperwork responsibilities does the primary institution have relative to the PHS Policy and the AWAR?

Reg. NIH/OPRR requires that any subcontracted work involving animals that is supported by PHS funds be conducted only at other NIH/OPRR Assured institutions (PHS Policy V,B). If the performance site (collaborating institution or subcontractor) is not NIH/OPRR Assured, then NIH/OPRR requires that an Assurance for the performance site be negotiated in order for the work to go forward (PHS Policy V,B). Alternatively, an institution can choose to accept responsibility for the work under their own Assurance with a written agreement between the institutions. It should, however, be recognized that the latter arrangement means that the NIH/OPRR-Assured institution assumes full responsibility for ensuring that all the animal work conducted in the collaborating institution is in full compliance with the PHS Policy and the AWAR. This arrangement necessitates that the NIH/OPRR-Assured institution's IACUC conduct semiannual program reviews and facility inspections of the collaborating facility. (See 8:11.)

Opin. When an institution subcontracts part of a research project to another institution, there should be a clearly defined written agreement concerning the responsibilities of each institution to comply with the PHS Policy and AWAR regardless of whether both institutions have approved Assurances on file with NIH/OPRR. If a serious compliance problem arises at a subcontract site, it will likely impact the primary institution. However, if the subcontract site is registered with the USDA and is the facility whose IACUC approved the protocol, that site would be held primarily responsible. Written agreements can help minimize and resolve potential problems.

9:51 Can an investigator appeal the decision of the IACUC relative to a protocol review decision?

Reg. The PHS Policy (IV,C,4) and the AWAR (§2.31,d,4) require that written notification of withheld approval include a statement of the reasons for the decision, "to give the investigator an opportunity to respond in person or in writing." Nevertheless, IACUC decisions to withhold approval may not be overturned by a higher authority (PHS Policy IV,C,8; AWAR §2.31,d,8).

Opin. The above regulatory information suggests that an appeal should be an option. Furthermore, the "Supplementary Information" that accompanies the August 31, 1989 *Federal Register* containing 9 CFR Parts 1, 2, and 3 Final Rules (page 36132), states that "on the basis of the response, the committee may reconsider its decision." This is another area for which the IACUC is advised to establish a written SOP.

9:52 Can a protocol for which approval has been withheld be resubmitted with modifications?

Reg. Yes, neither PHS Policy nor the AWAR preclude resubmission.

Opin. Resubmission of a modified protocol should be encouraged. A resubmission provides the investigator an opportunity to respond to the IACUC's review as allowed per PHS Policy IV,C,4 and AWAR §2.31,d,4. The IACUC also may consider assigning a member to work with the investigator to develop a protocol that can be approved by the committee and allow the investigator's research program to progress. (See 9:51.)

9:53 What is an appropriate and effective mechanism for appeal of IACUC decisions?

Opin. Some suggestions to consider in developing an SOP for appeals include:

- Assignment of IACUC members to work with the investigator to facilitate the process.
- Assignment of an IACUC member to be a "hearing" officer.
- Involvement of the "Dean for Research" or institutional equivalent as a "hearing" officer.
- Participation of the IO as a "hearing" officer.

(See 29:30.)

9:54 Is there any institutional authority that can reverse the decision of the IACUC?

Reg. No. The AWAR (§2.31,d,2) state that "officials of the research facility ... may not approve an activity involving the care and use of animals if it has not been approved by the IACUC." The PHS Policy (IV,C,8) is nearly identical. (See 9:51, 29:32.)

References

1. AVMA Panel on Euthanasia, Report of the AVMA Panel on Euthanasia, *J. Am. Vet. Med. Assoc.*, 202, 229, 1993. Available on the World Wide Web at: *http://www.nal.usda.gov/awic/pubs/noawicpubs/avmaeuth.htm*
2. NIH/OPRR Reports, Dear Colleague Letter. Available on the World Wide Web at: *http://www.nih.gov/grants/oprr/tutorial/index.htm* (Documents also can be ordered by fax: (301) 594-0464.)

3. Prentice, E.D., Crouse, D.A., and Mann, M.D., Scientific merit review: the role of the IACUC, *ILAR News*, 34, 15, 1992.
4. Office for Protection from Research Risks, National Institutes of Health, The Public Health Service responds to commonly asked questions, *ILAR News*, 33, 68, 1991.
5. Federal Register 54, 36114, 1989.
6. Miller, J.G., personal communication, 1991.
7. Black, J., Letter to the editor, *ILAR News*, 35, 1, 1993.
8. Prentice, E.D., Crouse, D.A., and Mann, M.D., Letter to the editor, *ILAR News*, 35, 2, 1993.
9. Garnett, N. and Potkay, S., Use of electronic communications for IACUC functions, *ILAR J.*, 37(4), 190, 1995.
10. OPRR Report, "Dear Colleague Letter," May 21, 1990. Available on the World Wide Web at: *http://www.nih.gov:80/grants/oprr/faxcall.htm*
11. Silverman, J., Do pressure and prejudice influence the IACUC? *Lab. Anim.*, 26(5), 23, 1997.
12. Garnett, N.L. and DeHaven, W.R., Protocol review: a word from the government, *Lab. Anim.*, 27(3), 19, 1998.
13. Garnett, N.L. and DeHaven, W.R., Protocol review. OPRR and USDA commentary, *Lab. Anim.*, 27(8), 1998.
14. Silverman, J., Majority rules? *Lab. Anim.*, 25(4), 22, 1996.
15. Garnett, N.L., personal correspondence, 1999.

10
Amending IACUC Protocols

Diane J. Gaertner and Kathleen D. Moody

Introduction

Biomedical research plans evolve as data are gained and analyzed. The plans for experiments utilizing animals must frequently be revised during the course of an approved protocol. Also, animals used to model human diseases may show unexpected clinical signs of illness. Many IACUCs have developed a mechanism that allows investigators to amend an approved animal use protocol without having to completely rewrite the protocol. This chapter describes the mechanisms being used to amend protocols and suggests when amending an approved protocol may be useful and appropriate.

10:1 What is the purpose of an amendment to an existing protocol?

Reg. PHS Policy (IV,B,7) requires PIs to seek IACUC approval for significant protocol changes. The AWAR (§2.31,c,7) have similar language.

Opin. The purpose of a protocol amendment is to allow a PI to change an approved protocol at times other than when regularly scheduled continuing protocol reviews occur. A PI's research plan may change and evolve while a study is in progress, especially since protocols are often written and approved long before the funding is obtained to commence work using animals. As research needs change, the protocol amendment process provides an opportunity for an investigator to refine his experiment. From the PI's perspective, a relatively simple, straightforward, and timely method to make significant changes to a previously approved protocol facilitates research while keeping the IACUC apprised of changes from the original animal use plan.

10:2 What are the regulatory requirements for protocol amendments?

Reg. Both the AWAR and PHS Policy require that the IACUC review and approve proposed significant modifications to ongoing activities using animals prior to initiation. The AWAR (§2.31,d,1) state that "… the IACUC shall determine that the proposed activities or significant changes meet the following requirements," which are then detailed at §2.31,d,1,i–§2.31,d,1,xi. These requirements include addressing pain and distress, alternatives to painful procedures, animal housing and veterinary care, personnel training and qualifications, surgical standards, and appropriate euthanasia techniques. PHS Policy (IV,C,1,a–IV,C,1,g) states that the IACUC should review the animal-related components and determine that the proposed research projects are in accordance with PHS Policy, the AWA, the *Guide*, and the institution's Assurance with the NIH/OPRR using criteria similar to the AWAR "unless acceptable justification for the departure is presented." (See 10:1.)

10:3 What activities can be considered appropriate for an amendment?

Opin. The IACUC must consider its regulatory responsibilities when considering the best methods for reviewing proposed amendments to approved protocols. Whether the activity proposed can be a simple amendment or requires a rewritten protocol with regular IACUC review depends upon whether the proposed change is considered to be minor or significant. Federal regulations and policies are not specific concerning what is considered to be significant, although NIH/OPRR has cited some examples of changes they consider to be significant, including:[1]

- The objectives of a study.
- Switching from nonsurvival to survival surgery.
- The degree of invasiveness of a procedure or discomfort to an animal.
- Species or the number of animals used.
- Personnel involved in animal procedures.
- Anesthetic agent(s), the use or withholding of analgesics.
- Methods of euthanasia.
- The duration, frequency, or number of procedures performed on an animal.

Table 10.1 lists major and minor protocol changes that might necessitate a revision process and the suggested level of review needed.

TABLE 10.1

Examples of Major and Minor Changes to Protocols and the Suggested Mechanism for These Types of Changes

Type of Change	Major (Significant)	Minor
Examples	Change in purpose or specific aim of study	Substitution of a qualified student or technician
	Change in principal investigator	Addition of a faculty collaborator
	Change of species	
	Addition of USDA-regulated species	Addition of another strain of the same animal species
	Large increase in animal numbers	Change in sex of animal to be used
	Addition of survival surgery	Small increase in animal numbers
	Addition of painful procedure	Need to repeat an experiment
	Unanticipated marked increase in clinical signs or proportion of animal deaths	Addition of minor surgery
		Addition of sample collection times
		Additional noninvasive sampling
Suggested mechanism	Rewrite protocol and have IACUC review	Submit as amendment

As previously indicated, the IACUC must distinguish between minor protocol amendments (changes) and significant proposed changes to the protocol which necessitate a complete revision and resubmission of the entire protocol. Thus, all substantive protocol changes must be reviewed and approved by the IACUC, although less extensive changes may utilize the amendment mechanism rather than necessitating a complete rewrite of the protocol. An institution can permit alterations in administrative information to be changed without IACUC review (e.g., electronic mail address of the PI or a change in the funding agency).

Since each institution's research and animal use program is unique, the NIH/OPRR has suggested that each institution develop uniform institutional guidelines regarding significant protocol modifications and have them available to investigators to clarify the amendment process.[1,2]

Others have suggested assessing the potential or actual reduction in animal welfare and the change in the overall ethical cost–benefit ratio when considering whether an amendment poses a significant change. In this view, the proposed amendment must be considered in conjunction with the original protocol to adequately determine if a significant change has been proposed.[3,4]

Protocol changes which are very complex or submissions of multiple sequential amendments to an existing protocol are not well suited to the amendment mechanism. Ideally, revisions submitted as amendments should be relatively simple and should not affect the existing documentation. Otherwise, the amended protocol can confuse rather than clarify. Complex changes to an approved protocol should be documented by the submission of a completely rewritten protocol which must go through the same approval mechanisms as the original protocol before animal experiments can proceed. The rewritten protocol either may be considered a new protocol (and given a new identification number) or its relationship to an earlier protocol may be maintained by utilizing a numbering system that indicates the original protocol number.

Surv. What activities can be considered appropriate for an amendment? More than one response is possible.

• Request for a few more animals	11/11
• Addition of any new species	8/11
• Request for many more animals	7/11
• Changes in personnel	6/11
• Any change in protocol for an APHIS/AC regulated species	7/11
• More severe clinical signs seen than expected	6/11
• Use of a more painful procedure	5/11

10:4 What format should be used to submit an amendment?

Opin. Amendments may be submitted through formal or informal mechanisms provided that all the necessary information is supplied. Institutions with large numbers of approved protocols often use an amendment form to document proposed changes in protocols. Depending on the specific changes requested, questions to be answered for each protocol amendment should include:

- What is the purpose or rationale for the protocol amendment?
- Will different people use the animals? If personnel have changed, their qualifications (e.g., education, training, and experience) must be documented.
- Will the species, sex, or strain of animal change?
- Will there be specialized housing requirements (e.g., housing that is not standard for the species, or housing which does not meet the animals' physiological or behavioral requirements)?
- Will more animals be needed? If so, justification for the increased number must be provided.
- Will additional minor surgical procedures or sampling of body fluids or tissues occur?

- Are animals expected to experience more clinical illness, pain, or distress as a result of the procedures proposed in this amendment? Are there alternatives to the use of animals in painful procedures? How will any pain or distress from this new procedure be minimized?
- Will there be new procedures involving the animals? Such changes must be described with the same level of detail as is required in a new protocol.
- Will there be a change in the methods of anesthesia, analgesia, or euthanasia?
- Will prolonged restraint of conscious animals be required?
- Will new hazardous agents be used?
- Will the surgical plans change (minor to major, multiple survival surgery, additional procedures)?

Surv. Although standardized protocol amendment forms can be used, there are other methods to initiate the process. Indicate what formats are acceptable to your IACUC for amendments (indicate all that are used).

- Using a letter or memo — 9/11
- Amendment form or other means, but not mandatory — 4/11
- Amendment form mandatory — 3/11
- Accepts amendments by electronic mail — 2/11
- Accepts new protocol submission form as an application for a protocol amendment — 1/11

10:5 Does the designated member review and full committee review processes also apply to amendments?

Reg. PHS Policy (IV,C,2) and AWAR (§2.31,d,2) require the same review procedure for proposed significant changes in ongoing activities (protocols) as they do to new activities.

Opin. The procedures used for new protocol review also can be applied to amendments. Whether a full review mechanism or a delegated reviewer mechanism is used (see Chapter 9), amendments must be reviewed in the context of the complete protocol, which may necessitate discussing the entire approved protocol. This means that the designated reviewer or the full committee may have to review the original protocol, and the full committee may have to discuss the proposed amendment in the context of the complete protocol. IACUC approval of protocol amendments must be documented either in the IACUC minutes or designated reviewer approval must be documented in the protocol file via a dated signature.

10:6 Can the IACUC Chair or an IACUC subcommittee review and approve significant changes or must the full committee review and approve such submissions?

Reg. The AWAR (§2.31,d,2) and the PHS Policy (IV,C,2) identify full committee review and designated member review as the only approved methods to evaluate animal protocols and proposed significant changes to ongoing protocols. In designated member review (see 9:19 to 9:24), all IACUC members must be provided with a list of the proposed significant changes to a protocol so that each member has the opportunity to request full committee review. If no member calls for full committee review, the IACUC Chair may designate one or more IACUC members to review the proposed changes, and have the authority to approve, require modifications in, or request full committee review. In cases where full committee review is appropriate, the PHS has determined that the voting mechanism of a quorum of a convened IACUC cannot be preempted by serial one-on-one meetings, telephone, fax, or electronic mail to poll or obtain members' votes.[5]

Opin. Institutional policy will determine which types of protocols and protocol amendments are eligible for designated member review and which require full review at a convened meeting of the IACUC. Any member of the IACUC, including the Chair, may serve as the designated reviewer. *The reader is advised that the NIH/OPRR has determined that IACUC members, not institutional policy, determine whether new protocols or significant changes to ongoing protocols will be reviewed by the full committee. This is because each member must always have the opportunity to call for full committee review.* The exception to this is if an institution adopts a policy that *all* protocols and significant changes to ongoing protocols must be reviewed by the full IACUC, which would obviate the need for an opportunity to call for full committee review.

Surv. Which IACUC processes does your IACUC use to amend protocols? Indicate all used.

- All amendments must go through the full committee process — 1/11
- Some amendments must go through the full committee process — 9/11
- All amendments go through the designated reviewer process — 1/11
- Some amendments go through the designated reviewer process — 4/11
- An IACUC subcommittee reviews and approves some amendments — 3/11
- All amendments are approved by a less formal process — 6/11

- Some amendments are allowed to be exceptions to these policies   7/11

Various exceptions were granted by the above IACUCs, including allowing personnel changes or administrative changes to be approved by the Chair without the protocol undergoing a formal delegation mechanism. Institutions utilize various mechanisms to distribute information about protocols eligible for delegated review and these mechanisms can be used for protocol revisions and initial submissions.

10:7 How many animals may be added to an ongoing study using an amendment mechanism?

Reg. PHS Policy (IV,D,1,a) and the AWAR (§2.31,e,1–§2.31,e,2) require that proposals to the IACUC specify and include a rationale for the *approximate* number of animals proposed to be used. This is an implicit requirement that institutions establish mechanisms to monitor and document the number of animals acquired and used in approved activities.

Opin. The cutoff for new animal orders may be the exact number of animals approved by the IACUC or a small percentage (e.g., 5%) in excess of the approved number of rodents.[6] Similarly, some institutions allow an amendment mechanism for ordering up to 10% more animals, with any additional animal requests requiring submission of a new protocol form.[7] In the authors' opinion, requests for additional nonrodent mammalian species require review and approval by the full committee mechanism. Whatever mechanism is used to add animals, it must satisfy the PHS Policy that the number of animals approved must be limited to the number needed to obtain valid results.[6]

Surv. Can a request for a few more animals be approved by your IACUC as an amendment without necessitating a complete protocol revision?

- Amendment is acceptable   10/11
- Complete protocol revision is required   1/11

References

1. Potkay, S., Garnett, N.L., Miller, J.G., Pond, C.L., and Doyle, D.J., Frequently asked questions about the Public Health Service Policy on humane care and use of laboratory animals, *Lab. Anim.*, 24, 24, 1995.
2. Division of Animal Welfare, National Institutes of Health, Office for Protection from Research Risks, Frequently asked questions about the Public Health Service Policy on care and use of laboratory animals, *ILAR News*, 35, 47, 1993.

3. McLaughlin, R., What constitutes significant changes? *Protocol Review Panel I*, Presented at IACUCs: Improving the Efficiency, Annual meeting of the Applied Research Ethics National Association (ARENA), Boston, March 26, 1998.
4. Oki, G.S.F., Prentice, E.D., Garnett, N.L., Schwindaman, D.F., and Wigglesworth, C.Y., Model for Performing Institutional Animal Care and Use Committee: continuing review of animal research, *Contemp. Topics Lab. Anim. Sci.*, 35, 53, 1996.
5. Garnett, N. and Potkay, S., Use of electronic communications for IACUC functions, *ILAR J.*, 37, 190, 1995.
6. Potkay, S., Garnett, N., Miller, J.G., Pond, C.L., and Doyle, D.J., Frequently asked questions about the Public Health Service Policy on humane care and use of laboratory animals, *Contemp. Topics Lab. Anim. Sci.*, 36, 47, 1997.
7. Doyle, D.J., Oki, G.S.F., and Prentice, E.D., Conducting continuing review and IACUC review of amended protocols, Presented at *Innovative Biomedical Technologies: The IACUC Response*, Annual Meeting of Public Responsibility in Medicine and Research (PRIM&R), Boston, March 28, 1998.

11

Continuing Review of Protocols

Gwenn S. F. Oki*

Introduction

Initial review and approval of a project by an IACUC represents an informed judgment by the Committee that when the protocol is initiated as written it will be conducted in full compliance with all applicable federal requirements contained in the AWA, AWAR, and PHS Policy. The IACUC review responsibilities do not, however, end with initial approval of the protocol. The IACUC is obligated by both the AWAR and the PHS Policy to conduct ongoing review.

The purpose of this chapter is to present general concepts of continuing review that hopefully will be useful to IACUCs. It should be noted that continuing review is just as important as initial review, since it not only serves to ratify the decisions of the IACUC, but also helps the investigator maintain compliance.

11:1 What is meant by "continuing review"?

Opin. APHIS/AC interprets continuing review, performed no less than annually, as a monitoring process in an effort to determine that the study remains in compliance, that the activities have been "conducted in accordance with the approved protocol," that significant modifications receive prior IACUC approval,[1] and "to ensure that any new requirements of PHS, USDA, or the institution are transmitted to the investigator."[2] The NIH/OPRR requires triennial continuing review to be a *de novo* process, meeting all the new proposal review criteria as set forth in PHS Policy IV,C,1 to IV,C,4.

---

* Ernest D. Prentice contributed to this chapter.

11:2 What is the intent of continuing review?

Opin. In general, the intent of a regulation or policy can be found in the preamble. For example, the preamble to the AWAR makes a single statement indicating that the intent of continuing review is "to provide current information to the research facility regarding all ongoing activities so that it can remain in compliance."[3] Although the intent of continuing review was not articulated in the form of a preamble to the PHS Policy, NIH/OPRR has interpreted continuing review performed no less than annually as a monitoring function.[1,4]

11:3 Is a protocol initially considered approved on the date when the research funding begins or on the date when the IACUC has taken final action on the protocol?

Reg. PHS Policy (IV,D,2) considers a protocol approved on the date of the IACUC approval.

Opin. While it appears to make sense to synchronize the IACUC initial protocol approval date with that of the funding start date, this practice is not permissible. If the IACUC approves a protocol and submits a letter to the funding agency certifying that the protocol was approved, the actual date of that approval must be indicated.[5] This is not a future date with final approval being contingent on actual funding. In general, NIH will allow a 60-day grace period during which the investigator must obtain IACUC approval in order for the grant proposal to go forward through the peer review process. In the majority of cases this precedes possible funding by 9 to 12 months.

11:4 What is the maximum life of an approved protocol?

Opin. The AWAR and PHS Policy do not limit the life of a protocol; however, both require that the IACUC conduct continuing reviews of approved studies at specified intervals. (See 11:6, 11:12.)

11:5 What do the AWAR and PHS Policy indicate with regard to continuing review of protocols?

Reg. The frequency of IACUC consideration of approved, ongoing activities is one of the few areas in which the NIH/OPRR and APHIS/AC have differing requirements. The AWAR (§2.31,d,5) state that "the IACUC shall conduct continuing reviews of activities covered by this subchapter at appropriate intervals as determined by the IACUC, but not less than annually." PHS Policy (IV,C,5) states that "the IACUC shall conduct continuing review of each previously approved, ongoing activity covered by this Policy at appropriate

Opin. intervals as determined by the IACUC, including complete review in accordance with IV,C,1–4 at least once every three years."

Opin. Other than the maximum time interval between continuing reviews, the AWAR makes no further statement regarding the form or substance of continuing review. The PHS Policy (IV,C,5), however, indicates that review criteria set forth at PHS Policy IV,C,1 to IV,C,4 must be satisfied at least triennially. This is the same criteria that IACUCs must use to conduct an initial review of a proposed protocol. Hence, triennial review of a protocol constitutes a *de novo* assessment of a currently approved study. (See 11:6.)

11:6 How often should continuing review of protocols be performed?

Reg. As mentioned in 11:5 the AWAR requires that continuing review be performed at intervals determined appropriate for the study being conducted but no less than once annually. The PHS Policy requirement is the same, except that the interval is no less than triennially. (See 11:5.)

Opin. Each IACUC should establish the frequency of continuing review at the time of initial review of the protocol. Many institutions have divided continuing reviews, based on review intervals, to separately meet the AWAR on an annual basis and to require a new protocol submission at the time of triennial review. Since the PHS Policy necessitates *de novo* review utilizing the same assessment criteria established for new protocol submissions and because there will more than likely be changes in the experimental methods and procedures, new institutional policies and IACUC membership, it may be more efficient to require the investigator to submit a completely new protocol at the time of triennial review. It should, however, be noted that there is no AWAR or PHS Policy time frame requirement for resubmission of a new IACUC application for an ongoing study.

11:7 What is an effective and appropriate way to conduct continuing review of a protocol?

Opin. Most IACUCs require investigators to use an institution-specific continuing review form. This standardizes the information required for all continuing studies and streamlines the review process. The form should be designed in consideration of the intent of continuing review described previously. One model developed in collaboration with NIH/OPRR and APHIS/AC has been described.[6] In addition, NIH/OPRR has published the following advice:[4]

> "A relatively simple mechanism to meet both federal requirements (AWAR and PHS Policy) is to circulate annually to all in-

vestigators with IACUC-approved activities a standard form giving current basic information, such as IACUC approval number, IACUC approval date, title of project, and species used. The investigator then notes that either no changes have taken place or he/she describes any changes that have occurred. Responses are reviewed by an IACUC designee for assessment of the changes reported. Any changes to the approved activity that are deemed of sufficient magnitude to merit further consideration may then be presented to the IACUC. All of these dispositions should be documented as official IACUC actions."

11:8 Is full committee review by the IACUC necessary in order to reapprove a protocol?

Reg. Review procedures under PHS Policy (IV,C,2) apply to the initial and triennial review and IACUC review of proposed significant changes to the use of animals in ongoing activities. Review procedures under the AWAR (§2.31,d,2) apply to the initial and annual review and IACUC review of proposed significant changes to the use of animals in ongoing activities.

Opin. Full committee review is not necessary. Some IACUCs use the "designated reviewer process" as indicated in the AWAR (§2.31,d,2) and the PHS Policy (IV,C,2).

- Each IACUC member is provided with a list of continuing protocols to be reviewed. Any IACUC member shall have available the written descriptions of the research protocols.
- Any IACUC member can request full committee review. If full review is *not* requested, then one IACUC member (the designated reviewer), selected by the IACUC Chair, is provided with the specific continuing review form(s), the corresponding approved protocol files, and is responsible for conducting the continuing review.
- The designated reviewer is authorized to approve require modifications or to request full IACUC review.

In most situations, continuing studies involving no changes or minor proposed changes are most efficiently handled in this manner. Alternatively, some IACUCs conduct full review of all continuing protocols utilizing a primary reviewer. In this scenario, all IACUC members receive a copy of the continuing review form. The primary reviewer, IACUC Chair, AV, and IACUC administrator also are provided a copy of the currently approved protocol. At the time of the convened meeting, the primary reviewer provides a summary of his findings with a recommendation for action. Following discussion, a vote is taken.

Depending on the size of the institution's animal program, either the designated reviewer process or full committee review may be most appropriate. For institutions with smaller animal programs, the latter may be the chosen method; while for larger institutions, the former may be most efficient in consideration of time constraints and the volume of continuing reviews.

11:9 If a designated reviewer is used (see 11:8, 9:19 to 9:24), must the full IACUC still give final approval to the decision made by that person?

Opin. The designated reviewer process utilized by various IACUCs is meant to save time and to promote efficiency, especially for high volume IACUCs. As indicated in the AWAR (§2.31,d,2) and PHS Policy (IV,C,2), the designated reviewer is authorized to approve, to require modifications, or to request full IACUC review. If the latter is not requested by the designated reviewer or by another IACUC member, then the designated reviewer has the authority to approve the continuation report or protocol. Requiring the full IACUC to give final approval of the designated reviewer's approval is an unnecessary duplication of effort which defeats the purpose of the "designated reviewer process."[7] If the IACUC insists on final ratification of the designated reviewer's approval then continuation reviews should be conducted at the time of a convened meeting. (See 9:19 to 9:24.)

11:10 How do investigators change procedures or personnel on their IACUC-approved protocol?

Opin. IACUCs at many institutions require that investigators submit a protocol amendment form when a protocol alteration is needed. This includes modifications in experimental procedures, an increase in the number of animals required, a change in personnel involved in the study (i.e., direct handling and use of animals, etc.). Any protocol modification requires IACUC approval prior to implementation and can be submitted at the time of continuation review. (See Chapter 10.)

Utilizing the designated reviewer process (see 11:8), a copy of the original protocol is given to the reviewer who will review the protocol and the continuation review form. Ideally, any questions that arise during this review are addressed through review of the original protocol. Depending on the number of amendments, it may be advisable for the PI to submit a completely rewritten protocol for future reference. When a protocol has undergone a number of amendments, even if minor in nature, the reviewer may have difficulty discerning what constitutes the most recent version of the current protocol.

11:11 What action should the IACUC take if an investigator does not respond to a request for information concerning the continuing review of his protocol?

Reg. If an investigator permits the IACUC approval to "expire" (i.e., the investigator does not respond to a request for updated protocol information in order for the IACUC to fulfill its responsibility for continuing review) and animal work is ongoing, then the research is no longer in compliance with PHS Policy or the AWAR. IACUCs are required to report serious or continuing noncompliance to NIH/OPRR (PHS Policy IV,F,3,a) or APHIS/AC (AWAR §2.31,d,6).

Opin. The AWAR (§2.31,d,5) and PHS Policy (IV,C,5) require that reviews of animal protocols are conducted at least annually and triennially, respectively. If the IACUC's request for information is necessary in order for the continuation report to be approved, the IACUC has no choice but to temporarily halt further use of new animals in the study. If the continuation reapproval date has been exceeded with no response from the investigator, appropriate individuals in the institution should be notified (e.g., the animal facility, the institution's AV, the PI's department head) with a copy to the PI. Operationally, IACUCs should have established policies and procedures for handling these types of situations.

11:12 When should an entirely new protocol be submitted? Why?

Opin. Many institutions require investigators to submit a new protocol every 3 years. Although this fits in well with PHS Policy (IV,C,5) for complete *de novo* review at least triennially, it should be remembered that the AWAR and PHS Policy do not limit the life of a protocol. Therefore, it is possible to allow a protocol to continue indefinitely without a new submission as long as both the AWAR and PHS Policy requirements for continuing review on an annual and triennial basis are met. For PHS purposes, the triennial review must include a review of all current information relevant to the protocol, although it is not required that an entirely new protocol be submitted. (See 11:4, 11:5.)

11:13 How much information should the IACUC require concerning results of a study at the time of continuing review?

Opin. Reporting the results of an ongoing study is not an AWAR or PHS Policy requirement. However, at the time of triennial review, the IACUC is required to perform a *de novo* review of the continuing study (see 11:5). This necessitates that the protocol meet all review criteria and study conduct requirements as stipulated in PHS Policy

IV,C,5. It, therefore, makes sense for the IACUC to require that the investigator provide the results of the study to date in order to facilitate IACUC review. Indeed, the only way the IACUC can justify the animals already used to date in the research project is to obtain information on the progress of the study. Admittedly, investigators encounter "blind alleys" and other problems that may affect study results. In some cases, animal experiments must be repeated. The IACUC should be advised of such problems arising during the course of continuing review in order to fulfill its responsibilities.

References

1. Division of Animal Welfare, Office for Protection from Research Risks, National Institutes of Health, Issues for Institutional Animal Care and Use Committees (IACUCs): Frequently asked questions about the Public Health Service Policy on Care and Use of Laboratory Animals, *ILAR News*, 35(3-4), 47, 1993. Available on the World Wide Web at: *http://www.nih.gov/grants/oprr/ilar93.htm#6*
2. U.S. Department of Health and Human Resources, Public Health Service, Institutional Animal Care and Use Committee Guidebook, NIH Publication No. 92-3415, 1992, Chapter A-3.
3. Office of the Federal Register, 9 CFR Parts 1, 2, and 3, *Federal Register*, 54 (168): 36133.
4. Division of Animal Welfare, Office for Protection from Research Risks, National Institutes of Health, Issues for Institutional Animal Care and Use Committees (IACUCs), The Public Health Service responds to commonly asked questions, *ILAR News*, 33(4), 68, 1991.
5. Public Health Service Form 398, pp. 8–9, 1995.
6. Oki, G.S.F., Prentice, E.D., Garnett, N.L., Schwindaman, D.F., and Wigglesworth, C.Y., Model for performing Institutional Animal Care and Use Committee continuing review for animal research, *Contemp. Topics Lab. Anim. Sci.*, 35(4), 53, 1996.
7. Garnett, N.L. and DeHaven, W.R., OPRR and USDA commentary, *Lab. Anim.*, 27(8), 18, 1998.

12
Justification for the Use of Animals

Joseph S. Spinelli

Introduction

Within a scientific institution various committees review research protocols. There may be one to determine which projects may be submitted to extramural funding agencies while another, the Institutional Review Board, considers the ethical use of humans in research. Biosafety, radiation safety, and chemical safety committees also deal with human safety. Only one committee, the IACUC, deals with animal welfare.

During the IACUC review process, the committee must consider whether or not the research question is worth asking, if the materials and methods involving animals are appropriate, if the project might harm humans, and if issues of animal welfare are adequately addressed. Investigators must satisfy the IACUC that animals, if their used is justified, will be used properly from legal and ethical points of view.

The survey conducted in this chapter reflects responses primarily from academic institutions.

12:1 What is the definition of an animal?

Reg. PHS Policy (III,A) defines an animal as "any live, vertebrate animal used or intended for use in research, research training, experimentation, or biological testing or for related purposes."

The AWAR (§1.1, Animal), based on the AWA (§2,g), define an animal as "any live or dead dog, cat, nonhuman primate, guinea pig, hamster, rabbit, or any other warm-blooded animal, which is being used or is intended for use for research, teaching, testing, experimentation, or exhibition purposes, or as a pet. This term excludes birds, rats of the genus *Rattus* and mice of the genus *Mus* bred for use in research, and horses not used for research purposes and other farm animals, such as, but not limited to, livestock or poultry used or intended for use as food or fiber, or livestock or poultry used or intended for use for improving animal nutrition, breeding, management, or production efficiency, or for improving the quality of food or fiber. With respect to a dog, the term means all dogs, including those used for hunting, security, or breeding purposes." (See 8:8, 12:5.)

12:2 Is it required that the use of animals be justified on an IACUC protocol?

Reg. Institutions using animals regulated by the AWAR (§1.1, Animal; and see 12:1) must assure that:

- All proposals to the IACUC contain a rationale for involving animals, and for the appropriateness of the species and numbers of animals to be used (AWAR §2.31,e,2).
- The principal investigator has considered alternatives to procedures that may cause more than momentary or slight pain or distress to the animals (AWAR §2.31,d,ii).
- The investigator has provided a written narrative description of the methods and sources (e g., the Animal Welfare Information Center) used to determine that alternatives to animal use are not available (AWAR §2.31,d,1,ii).
- The principal investigator has provided written assurance that the research activities do not unnecessarily duplicate previous experiments (AWAR §2.31,d,1,iii). For a more in-depth discussion of alternatives, see 12:15 to 12:20.

The HREA (§495,c,2) requires applicants for grants, contracts, or cooperative agreements involving research on animals to submit "a statement of the reasons for the use of animals in the research to be conducted with funds provided under such grant or contract." Applications and proposals (competing and noncompeting) for awards submitted to PHS that involve the use of animals are required to specify the species and approximate number of animals to be used, the rationale for involving animals, and the appropriateness of the species and numbers to be used (PHS Policy IV,D,1,a; IV,D,1,b).

The *Guide* (page 10) states, "The following topics should be considered in the preparation and review of animal care and use protocols:

- Rationale and purpose of the proposed use of animals.
- Justification of the species and number of animals requested. Whenever possible, the number of animals requested should be justified statistically.

- Availability or appropriateness of the use of less-invasive procedures, other species, isolated organ preparation, cell or tissue culture, or computer simulation."

12:3 Should the IACUC perform an ethical review of protocols?

Opin. As noted in 12:1, the AWAR and the PHS Policy place conditions on the use of animals. Those conditions are likely to become more restrictive if the average citizen is convinced that animals are badly treated in research, whereas they are likely to remain as they are if the average citizen believes that institutions using animals do so in an ethically responsible way. It is in the interest of both institutions and animals if the IACUC regards itself as a bioethics committee rather than as an organ to merely comply with federal policy. While considerations of a bioethics committee include compliance with federal regulations, the considerations of the committee are likely to be much broader. (See 12:4.)

12:4 How can an IACUC perform an ethical review of protocols?

Opin. IACUC members often have divergent views on what is ethical. This complicates the review of protocols because individuals use different standards of morality to judge whether a project is "ethical." The process can be facilitated by having the IACUC adopt an ethical assessment tool to evaluate projects. The use of such a tool does not necessarily lead to uniformity of opinion, but it can facilitate a systematic review of each project. This allows the Chair to guide the discussion in a way that maximizes the probability that people with divergent views discuss the same issues. The use of an ethical assessment tool also can contribute to the training of new members on how projects should be reviewed.

Humans often completely or partially control the way animals live and die. Interactions between animals and humans are referred to as "human dominance of animals" or "human/animal interactions." The ethical assessment tool is summarized in Table 12.1. While this table can be used as a guide to evaluate the ethics of each type of human dominance of animals, the factors that determine whether or not a particular human/animal interaction is ethical largely depends upon the type of human dominance of animals. Therefore, one would emphasize one or another factor in considering whether it is ethical to maintain a pet dog in one's house vs. raising dairy cattle vs. the use of laboratory animals.

The value of the interaction relates to the benefit derived from the human dominance of animals. Issues to be explored include the value for the human or humans participating in the human domi-

TABLE 12.1

A Reference for Reviewing Human Dominance of Animals

Value of the Interaction	Quality of Life for the Animal
Benefit for the Animals or Humans: • The human in the interaction • The animal in the interaction • Other humans • Other animals	• Nutritional state • Physical surroundings • Physical health • Mental health • Freedom from pain and discomfort • Freedom from suffering
Quality of Death for the Animal	**Contextual Features**
• Freedom from fear • Freedom from pain	• Compliance with laws • Preservation of species • Effect on those working with the animal • Effect on the environment

nance of animals, the value to the animal or animals involved, or the value to other humans or other animals. The human dominance of animals tends to be more ethical when there is a demonstrable benefit than when there is little or no benefit. Human dominance of laboratory animals benefits the humans who are designing and responsible for the procedure. Many who oppose the use of animals in research assert that this benefit drives the interaction to an unethical plane. That is one of the reasons to use a generalized ethical assessment for evaluating *all* human dominance of animals. The benefit that humans receive by having dominance over laboratory animals only becomes unethical if the animal is exploited as a result of not having the opportunity for a quality life.

Only rarely will the laboratory animal benefit in a research situation. Individual animals used in research may benefit from their use as pets or other domestic animals participating in a clinical study designed to improve their health. However, in such clinical studies there is also the risk of harm. The reality is that most laboratory animals will be killed at a much younger age than their natural life span. They also may be subjected to procedures that will cause them harm. Thus, for most laboratory animal studies, the animals' participation is likely to be counter productive to their self interest. The crux of the situation relative to the use of laboratory animals and the value of the interaction is how other animals or humans will benefit. This benefit does not have to be a direct clinical benefit. The benefit can be the increase of human knowledge.

The burden of proof is on the researcher to adequately demonstrate that the question he or she is attempting to answer has either not been answered or must be confirmed and that the methods employed are likely to produce an answer to the question being asked. To the degree that an investigator can demonstrate that the question is worth asking, and that the experimental design has a high probability of answering a question with little or no harm to the animal, the procedure could be given high ethical consideration. To the extent that the experimental design has a low probability of answering the question or there is a high level of harm to the experimental subjects, the project would have a low ethical score. Therefore, the quality of the scientific team, the design of the experiment, and the probability for success all are important factors in considering whether or not the use of animals in experiments is ethical.

There are several other issues that affect the ethical value of animal use. As shown in Table 12.1, these include discussions of the quality of life for the laboratory animals, the quality of their death, and various contextual features. Relative to the quality of life of laboratory animals, two major factors are important. One is the quality of the animal care program, as that influences the animal's nutritional state, physical surroundings, and physical and mental health. The second major factor is the design of the research. A major consideration in research design is the training in animal care and use of those handling animals. Questions for the IACUC to ask include:

- In addition to their training, how well will personnel likely execute their knowledge?
- How invasive will the procedure be?
- If surgery is performed, will it be done in an aseptic and professional manner?
- Are there standard operating protocols for the procedures to be performed?
- Is the investigator asking for any exceptions to the standard procedures?
- If there is an exception, has the investigator adequately justified why the exception is necessary and does the committee concur with the methods to be employed?
- What is being done to prevent pain and distress to the animal?
- Is the animal being properly evaluated for pain? If so, how?

An especially important aspect of pain and distress is whether any physical or psychological pain or discomfort can be expected to exceed the animal's tolerance. In this regard, people performing euthanasia must be adequately trained. The method of euthanasia chosen should neither induce fear nor pain in the animal being killed.

Contextual features are broad considerations relating to the animal–human interaction and include issues such as compliance with applicable laws, preservation of a given species, the effect of the animal–human interaction on those humans involved with the animal, and the effect on the environment.

12:5 Since the AWAR definition of an animal includes dead animals (see 12:1), should the IACUC request a justification for the use of dead animals?

Reg. If an animal is not excluded from the AWAR definition of animal (see 12:1), institutions covered by the AWAR are required to have their IACUC request a justification for the use of dead animals. The PHS Policy does not require IACUC review and approval of the use of dead animals unless they are killed for the purpose of being used in PHS-supported activities. (See 8:8.)

Opin. From a legal point of view, justification is not required, but from an ethical point of view, such use should be justified. Justification of the use of dead animals demonstrates the commitment of the institution to provide the same ethical consideration to all sentient species, not just to those designated as animals under a legal definition. While dead animals are obviously not sentient, IACUC considerations should include:

- What is the value of the project?
- Why is the use of dead animals necessary?
- What affect will the use of the dead animals have on the animals' quality of life (before they are killed)?
- What will be the quality of their death?

12:6 Should the IACUC request justification for the use of animals in field studies?

Reg. The AWAR (§1.1) state that a field study is "any study conducted on free-living wild animals in their natural habitat which does not involve an invasive procedure and which does not harm or materially alter the behavior of the animals under study." Under the AWAR (§2.31,d,1) animals involved in field studies are exempt from IACUC review.

PHS Policy does not distinguish between field studies and laboratory studies and, therefore, requires IACUC review and approval of field studies if covered under an Animal Welfare Assurance.

Opin. If invasive procedures occurs in field studies, the AWAR exemption does not apply. Also, assuming the project is performed on vertebrate species, animals in field studies are not excluded from IACUC consideration in institutions covered by the PHS Policy. For consistency, it is a good policy to grant the same ethical review for field studies as for animals in nonfield studies. (See 12:4.)

12:7 What justifications for animal use are considered appropriate?

Reg. The AWAR (§2.31,e) requires that "a proposal to conduct an activity involving animals, or to make a significant change in an ongoing activity involving animals, must contain the following:

- Identification of the species and the approximate number of animals to be used.
- A rationale for involving animals and for the appropriateness of the species and numbers of animals to be used.

PHS Policy (IV,D,1,a; IV,D,1,b) requires the same information. U.S. Government Principle III states, "The animals selected for a procedure should be of an appropriate species and quality and the minimum number required to obtain valid results. Methods such as mathematical models, computer simulation, and *in vitro* biological systems should be considered."

The *Guide* (page 10) expands on this by recommending that "the following topics should be considered in the preparation and review of animal care and use protocols:

- Rationale and purpose of the proposed use of animals.
- Justification of the species and number of animals requested. Whenever possible, the number of animals requested should be justified statistically.
- Availability or appropriateness of the use of less-invasive procedures, other species, isolated organ preparation, cell or tissue culture, or computer simulation."

Opin. If animals are to be used in research, the design of the project should have reasonable potential to answer the question posed. Currently in the U.S., there is much discussion about whether IACUCs should consider the scientific merit of a given protocol. Using the ethics assessment tool described in 12:4, such consideration is mandatory because the quality of the science determines the likelihood of answering the question being asked. The quality of the science becomes one of the major factors in justifying or failing to justify the use of animals. However, there are a variety of options regarding how the scientific review may actually be performed. (See 9:14 to 9:17.)

If a granting agency performs an adequate review of scientific proposals, the IACUC can defer to those agencies for scientific review.

However, this requires some feedback. For example, if the IACUC relies on the NIH to evaluate the scientific merit of a project, then that review should be considered as part of the IACUC's overall review process. (Most IACUC reviews occur prior to or immediately after NIH submissions. The IACUC approval must occur within 60 days of the NIH submission, otherwise NIH will not even initiate scientific review. Therefore, if the IACUC uses the NIH review as scientific justification, an approach may be to provide contingent approval, pending a favorable NIH review.) If the protocol receives a poor review from the granting agency and if the researcher derives funds from another source, the IACUC should consider withholding approval until such time that it is satisfied that the project has scientific merit. *The reader is cautioned that the PHS Policy does not recognize "contingent approval." IACUC approval (not contingent approval) is required by PHS Policy IV,D and by the NIH Grants Policy Statement, Part II, Terms and Conditions of NIH Grant Awards.* IACUCs also have the option of having protocols reviewed for scientific review by consultants having greater expertise in the area of science than members of the IACUC. (See 9:16 to 9:18.)

Surv. What justifications for the use of animals on a protocol are considered appropriate? A small sampling from various institutions shows the diverse methods used to assure that investigators have given adequate justifications for their proposed animal use.

- At University A, the investigator is required to provide a rationale that excludes use of humans, cell culture, etc. They generally regard the determination of the scientific value of the research as being out of the purview of the IACUC, except in rare instances.
- At University B, the IACUC uses government or industry testing standards and requirements, or a literature search showing a lack of acceptable alternatives.
- University C requires a discussion of why nonanimal models are inappropriate or nonexistent and why the proposed animal species was selected over others as the animal model.
- University D requires the justification for using animals to include responses such as: the data to be collected requires complex physiological interactions to be valid; it would be inappropriate to pursue data collection (or conduct training procedures) using humans; or the project requires a unique animal model specifically made to address a research question. Unique models include transgenic, SCID, or athymic mice. An inappropriate justification would be cost.
- University E requires investigators to justify both the use of animals and the species requested. The most appropriate

answer is to provide literature establishing the proposed species as the most appropriate one.
- Commercial Organization A attempts to determine that there is a scientific need for the information, including: development of new procedures or techniques; validation of results from collaborators prior to expansion of studies; further investigations, safety assessment of potential products, or efficacy evaluations.

12:8 Is a requirement by a funding agency to use animals considered adequate justification?

Opin. Adjudication as to what is proper use of animals is not dictated by granting agencies. It is the institution, through its IACUC, that must decide if the justification to use animals is appropriate. If a granting agency requires the use of animals, or a particular species or strain, their reasons for doing so could be submitted to the IACUC for its consideration. However, the IACUC must make that decision based on the parameters previously noted (see 12:7) not just on the policy of the granting agency.

The instructions accompanying the IACUC form of one university succinctly states, "Certain contracts or Requests for Proposals may specify that animals must be used. The IACUC will not accept this in itself as the sole reason for using animals in a study. You still must present a cogent argument why the available alternatives to laboratory animal use, including human studies, are not satisfactory or appropriate for your needs."

Surv. Is a requirement by a funding agency to use animals considered to be adequate justification for your IACUC?

- No 4/7
- Would consider, but would not solely justify animal use 3/7

12:9 How much detail should the IACUC require relative to animal identification in the IACUC protocol and in records? For example, should the IACUC request that rats be identified as "rats" or "*Rattus norvegicus*" or "F344/Crl rat"?

Opin. While there is nothing specific in the AWAR, the PHS Policy, or the *Guide* that requires identification of animal by strain on IACUC protocols, this author believes the spirit of those documents requires it. Justification of animal use is generally required. Often the reason a given species is used is because of the special characteristics of a particular strain or construct. When such strains are to be used, if one is to justify animal use, those special characteristics should be described and the strain or construct should be named using appropriate nomenclature (see *Guide,* page 48). The more detailed the justification, the better. This author's experience is that many investigators understand this need and will list the strains or constructs they wish to use, and state their significance.

IACUCs have broad authority and if they want to require that investigators use official nomenclature of strains on IACUC applications, they are free to do so.

12:10 Are there any federal requirements dictating that animals or certain species of animals must be used in certain forms of research or product safety testing?

Opin. This author is not aware of any federal regulation or law requiring that a certain species be used for a given project. In some instances Requests for Proposals or contract stipulations from the FDA suggest the use of particular species. For example, the FDA *Guidelines for Preclinical and Clinical Evaluation of Agents Used in the Prevention or Treatment of Postmenopausal Osteoporosis*[1,2] state, "Because no animal species duplicates all of the characteristics of human osteoporosis, it is felt that an examination of bone quality in two species is necessary to adequately investigate the effectiveness and safety of drugs for this indication. One study should be conducted in the ovariectomized rat model and the second in a nonrodent model (i.e., larger, remodeling species) which will be left to the discretion of the sponsor. ..."

12:11 What is the position of the Food and Drug Administration (FDA) relative to using animals in the LD_{50} test?

Opin. The FDA position is as follows:[3]

"The term 'LD_{50}' stands for the Lethal Dose 50%. The test requires that about 100 animals be administered varying doses of a substance to find the one that kills 50% of them. The classical LD_{50} test was widely accepted in the 1930s, but it has become apparent that it is unnecessarily precise and that other simpler tests using fewer animals yield acceptable information.

"Safety test results, including toxicity data, are required by several government agencies' regulations to show possible hazards that might be associated with consumer and industrial products. The LD_{50} acute toxicity test is a measure of a substance's ability to cause death within approximately 2 weeks and results in the loss of relatively large numbers of animals, generally rats or mice. Because it is unreasonable for people to be exposed to substances for which safety has not been established, some form of acute toxicity data is required. In many cases, FDA requirements specify that any data in possession of a company relative to the safety of a product must be submitted. In some cases, too, a classical LD_{50} test may have been done several years ago and is only now being submitted, or has been done for reasons not related to FDA requirements. Thus, while regulations do not require or seek classical LD_{50} studies, FDA cannot ignore LD_{50} data if it is submitted. However, FDA has sought to clarify its position that the classical LD_{50} test is outmoded and unwarranted by conducting workshops; publishing reports, articles, and guidelines; and through discussions and meetings with industry officials."

12:12 Should an IACUC approve development of an animal model if another well-established model exists, simply because the investigator has no experience with the established model?

Opin. The IACUC should evaluate each protocol on the basis of how well the investigator has justified the use of whichever species, strain, and/or construct that is proposed. One issue is the appropriateness of the animal model. Based on the justification presented by the investigator, the IACUC should approve the model, require additional information, or disapprove a given model. No formula for that action should, or could, be given here. Much like a jury trial, it will depend on the veracity of the evidence — the justification provided by the investigator.

12:13 Since many people believe that cephalopods are sentient and because they have a well-developed nervous system, does an IACUC have the authority to include such animals under its purview? What about other invertebrates?

Opin. Neither the AWAR nor PHS Policy cover the use of invertebrate animals. Nevertheless, there are no restrictions that either of these agencies have placed on IACUCs relative to oversight of animal care and use activities that exceed the scope of their regulations or policies. Therefore, an institution can confer upon its IACUC the authority to oversee the care and use of invertebrates.

12:14 Must the IACUC approve the use of animals destined to be used as a food source for other animals being used in IACUC-approved studies?

Opin. The AWAR and PHS Policy do not require IACUC approval for this use of animals. It is the proposed use of animals that determines whether IACUC approval is needed, not the destiny of the animals.

If the *only* use is as a food source for other animals, approval is not required. Nevertheless, the use of any sentient animal associated with teaching or research should involve review by the IACUC. The IACUC and the institution are not precluded from doing this. It is this author's belief that doing so strengthens the sincerity of an institution's position that they care about proper animal use when they adopt ethical standards that exceed what is required by current regulations or policies.

12:15 What is meant by "alternatives" to using animals in research, teaching, and testing?

Reg. APHIS/AC Policy #12[4] clarifies that APHIS/AC considers "alternatives" to include refinement, replacement, and reduction in looking for alternatives to painful or distressful procedures. Every protocol using animals must justify their use, the species used, and the number used (AWAR §2.31,e,4). (See 12:16.)

The federal mandate of the HREA and PHS Policy (IV,C,1,A) to avoid or minimize discomfort, pain, and distress in experimental animals, consistent with sound scientific practices, is, for all practical purposes, synonymous with a requirement to consider alternative methods that reduce, refine, or replace the use of animals. Consideration of these issues should be incorporated into IACUC review, investigator training, research proposals, and ongoing monitoring of the institutional animal care and use program.[5]

Opin. Alternatives to the use of animals in research, teaching, and testing is generally thought of as replacement of intact sentient animals by insentient material. Alternatives can consist of nonliving systems such as computer simulations, injection dummies and mannequins, or living systems such as cell cultures.

Any experimental system that does not entail the use of a whole, living animal is considered an alternative. Some of these are relative replacements, as they still entail the humane killing of an animal for obtaining cells, tissues, or organs for subsequent *in vitro* studies. Others are absolute alternatives that do not require any biological material derived from a fully developed vertebrate, nonhuman animal.[6]

A much broader definition of alternatives has been developed.[7] It states that "alternatives are defined as new methods that refine existing tests by minimizing animal distress, reduce animal usage, or replace whole animal tests." It is the opinion of this author that under this definition, an alternative can include a whole insentient animal that is insentient because the areas of the brain that are responsible for perception have been irreversibly damaged or removed. Use of such animals minimizes animal distress. (See 12:16.)

12:16 What do the AWAR and PHS Policy say with respect to use of alternatives to animals in research, teaching, and testing?

Reg. Under the AWAR (2.31,d,1,ii) and the AWA (§13,a,3,B), the IACUC is required to assure that "the principal investigator has considered alternatives to procedures that may cause more than momentary or slight pain or distress" and "has provided a written narrative description of the methods and sources, e.g., the Animal Welfare Information Center used to determine that alternatives were not available."

APHIS/AC Policy #12[4] expands on this requirement, stating that "consideration of alternatives to each procedure which may cause pain or distress must state sources consulted, such as *Biological Abstracts*, *Index Medicus*, Medline, the Current Research Information Service (CRIS), and the Animal Welfare Information Center (AWIC)." The policy also states that "the minimal written narrative should include: the databases searched or other sources consulted, the date of the search and the years covered by the search, and the key words and/or search strategy used by the principal investigator when considering alternatives or descriptions of other methods and sources used to determine that no alternatives were available to the painful or distressful procedure. The narrative should be such that the IACUC can readily assess whether the search topics were appropriate and whether the search was sufficiently thorough." Finally, "reduction, replacement, and refinement (the three R's) must be addressed, not just animal replacement."

Principle III of the U.S. Government Principles states, "The animals selected for a procedure should be of an appropriate species and quality and the minimum number required to obtain valid results. Methods such as mathematical models, computer simulation, and *in vitro* biological systems should be considered." Principle IV states that proper use of animals, including the avoidance or minimization of discomfort, distress, and pain when consistent with sound scientific practices, is imperative. NIH/OPRR has written, "The federal mandate to avoid or minimize discomfort, pain, and distress in experimental animals, consistent with sound scientific practices, is, for all practical purposes, synonymous with a requirement to consider alternative methods that reduce, refine, or replace the use of animals."[5] The attachment to that statement cites all of the PHS statutory and policy bases for consideration of alternatives.

The *Guide* (page 10) recommends that animal care and use protocols should consider the "availability or appropriateness of the use of less-invasive procedures, other species, isolated organ preparation, cell or tissue culture, or computer simulation."

12:17 How can an IACUC be assured that an investigator has considered and rejected the existence and use of alternatives?

Opin. There are strong views on this issue. For example, a colleague recently told the author,

> "It would be much easier if a list of alternatives were made available and then all scientists can relate to this list. For example, all rabbit antibody producers now are aware of the alternatives to Freund's. All the veterinary surgery classes are aware of the alternatives of videos and plastic models. Until then, the IACUC needs to rely on the combined expertise of the members and the cooperation of the researcher to collectively assure that alternatives have been considered and rejected."

Surv. How does your IACUC assure that an investigator has considered and rejected the existence and use of alternatives?

- University A requires that the PI indicate the search strategy used for all protocols in which there are procedures involving any alleviated or unalleviated pain or distress. PIs may be required to repeat the search if it appears mechanical or does not include appropriate search terms likely to reveal alternatives.
- Using its prerogative, the IACUC of Commercial Organization B requires two literature searches with key words and dates, as well as a discussion of replacement, reduction, and refinement of the study. (Authors comment: Although APHIS/AC Policy #12[4] requires but one search be performed, that organization informally suggests that two or more databases be reviewed as part of that search.)
- University B requires a statement that an investigator has considered and rejected the existence and use of alternatives, backed up by the databases searched, date of search, years searched and keywords, one of which must be "alternatives."
- At University C the IACUC ensures that all protocols (including those not covered by the AWAR) list the sources, date the search was performed, dates the search covers, and key words used. Protocols are reviewed to determine if they are inclusive for both alternatives to animal use and alternatives to painful procedures.
- The IACUC at University D requires investigators to list databases, search dates, and key words.

12:18 What sources and methods are useful when searching for alternatives to animal use?

Opin. APHIS/AC Policy #12[4] lists suggested references. They are *Biological Abstracts*, *Index Medicus*, Medline, the Current Research Information

Service (CRIS), and the Animal Welfare Information Center (AWIC). The *Guide* (page 82) has a small list of references.

Altweb[8] is a Website for news, information, discussion, and resources from the field of alternatives to animal testing. The Altweb site is "intended to foster the development of scientifically acceptable *in vitro* and other alternatives to animal testing. Alternatives are defined as methods which reduce animal use, replace whole animal tests, or refine existing tests by minimizing animal distress."

In addition, the not-for-profit PREX[9] online information service at Utrecht University has core databases on laboratory animal sciences and alternatives. These databases give computer access to references and information that may not be available on any other database system currently accessible by the scientific community. And the databases are updated on a continuous basis. Subscription is necessary to access the databases.

The National Library of Medicine publishes *Alternatives to the Use of Live Vertebrates in Biomedical Research and Testing*,[10] where abstracts assist in locating sources to reduce, refine, and replace animals as test systems.

12:19 If the source of information about alternatives to animal use is other than a published article (e.g., a quote or statement from an experienced investigator), how should the IACUC evaluate this?

Opin. From this author's perspective, the report of a verbal communication from an experienced PI may be of great value in determining the presence or lack of alternatives to animal use. The discussion should be reported in detail in the IACUC application and the IACUC members could then evaluate how much weight they want to give to the statement. There must be enough information presented to allow the IACUC to determine if the PI made a good faith effort to search for alternatives.

Surv. If the source of information about alternatives to animal use is other than a published article (such as a quote or statement from an experienced investigator), how does your IACUC evaluate such validation?

- At Commercial Organization A, this is ancillary and does not replace literature searches.
- The IACUC at University A accepts such statements at face value.
- The IACUC at University B does not accept verbal statements.
- At University C, the IACUC evaluates all the responses to the "alternative search" question in relation to the pain classification of the project. For example, if a weak justification for a search for alternatives was given for a study involving the use

of carcasses, parts of carcasses, or simply observing animals in the wild, the IACUC would probably accept it. However, if a weak justification was made where death was being used as an end-point, the committee would spend more time evaluating this "validation."

12:20 Must the IACUC assure that the animal use activity does not duplicate previously conducted work? If so, how is this done?

Reg. The AWAR (§2.31,d,1,iii) require that "the principal investigator has provided written assurance that the activities do not unnecessarily duplicate previous experiments."

Opin. The key word here is *unnecessary*. Duplication is permitted if it serves a valid scientific purpose, such as confirming another's work. To determine if the duplication is or is not necessary, the IACUC needs to evaluate the justification of the investigator who proposes the duplication.

Surv. Does your IACUC assure that the animal activity in the protocol does not duplicate previously conducted work? If so, how?

- University A requires the PI to provide a justification as to why she is proposing to duplicate research.
- At University B, as part of the IACUC form, PIs must attest that their work is not duplicative and demonstrate that each of their projects is distinct and asks different scientific questions. This is checked at the administrative level to make sure that the investigators are not funded by two agencies for studies asking the same scientific question.
- At University C, a written statement by the PI is the basis for this determination. This also is backed up by the specific literature search parameters used.
- At University D, many approved projects have parts that are intended to duplicate previous work. That is usually not questioned because it is justified. Projects that are questioned typically include those that cause pain and duplicate what is already known and will not add any new knowledge to the field. In such cases the committee discusses the necessity and looks at the justification more closely. The justification provided by the PI is used by the IACUC to approve any duplicative studies.
- At University E, the IACUC evaluates the merits of a duplicative project based on information provided by the literature search. Also, if the protocol has undergone peer review for scientific merit by a funding agency and received an award, the IACUC is reassured that the protocol is not likely to unnecessarily duplicate previous work.

- At University F, the IACUC asks the investigator to state on the protocol application whether or not this work duplicates other work. If so, the PI must justify the duplication. If not, the IACUC does not require anything more than a statement that the work is not duplicative.
- Commercial Organization A's IACUC accepts the PI's assurance as valid and truthful. Occasionally, if a committee member knows the study is or may be duplicative, the committee will ask for further justification. This is true especially for safety assessment and validation studies.

12:21 Is the use of animals in teaching protocols considered "necessary duplication?"

Opin. There are no federal laws, regulations, or policies pertinent to this question. This author believes the issue should be left to the discretion of the IACUC. As shown in the survey below, while some IACUCs discourage classroom use of animals, all respondents allow it under special circumstances. The implication is that the IACUCs regard some use of animals as "necessary duplication."

This author further believes that a case can be made for using animals in teaching protocols ("necessary duplication") and that in other instances such use is "unnecessary" duplication. For example, there probably is no impediment to learning if a demonstration involving a live animal is video taped so that it can be shown several years in a row. However, to teach certain surgical procedures to students or practitioners, there may be no reasonable alternative to a living animal.

Surv. Is the use of animals in teaching protocols considered "necessary duplication?"

- At Private Company A, this concern is generally not applicable. However, there may be training protocols used for workshops. The IACUC has approved this duplicative use practice to help assure the quality of research results and humane methods in the future.
- At University A, the use of animals for teaching is something that the IACUC usually rejects unless the requester provides compelling evidence that there are no other alternatives. Even when IACUC approves a protocol, the Chair sometimes talks to the instructor to consider the option of developing a "real-time" video of the procedure that can be shown to the class in lieu of using animals.
- At University B, such use is generally regarded as necessary duplication.

- At University C, such use is occasionally approved. They have disapproved requests and asked that videos or other nonanimal methods be explored.
- At University D, the PI addresses this in the discussion of nonanimal models.
- The IACUC at University E does not consider animals used in teaching as "unnecessary duplication." The responder felt that this is because training involves teaching new students. Thus, although the procedures are performed repetitively, the procedures are new for each student.

References

1. U.S. Food and Drug Administration, Guidelines for Preclinical and Clinical Evaluation of Agents Used in the Prevention or Treatment of Postmenopausal Osteoporosis. Available on the World Wide Web at:
http://www.fda.gov/cder/guidance/osteo.pdf
2. Thompson, D.D., Simmons, H.A., Pirie, D.M., and Ke, H.A., FDA guidelines and animals models for osteoporosis, *Bone*, Oct. 17(4 Suppl), 125S, 1995.
3. Spinelli, J. and Markowitz, H., Clinical recognition and anticipation of situations likely to produce suffering in animals, *J. Am. Vet. Med. Assoc.*, 191, 1216, 1987.
4. U.S. Department of Agriculture, Animal and Plant Health Inspection Service, Policy #12, Written narrative for alternatives to painful procedures, April 14, 1997. Available on the World Wide Web at:
http://www.aphis.usda.gov/ac/policy12.html
5. National Institutes of Health, Office for Protection from Research Risks, OPRR Reports, Dear Colleague Letter, Number 98-01, Nov. 17, 1997. Available on the World Wide Web at:
http://www.nih.gov/grants/oprr/dc98-01.htm
6. Fund for the Replacement of Animals in Medical Experiments. Available on the World Wide Web at:
http://www.frame-uk.demon.co.uk/alternat.htm#Replacement
7. Johns Hopkins Center for Alternatives to Animal Testing. Available on the World Wide Web at: http://www/jhsph.edu/~altweb/caat/caat.html
8. Altweb, available on the World Wide Web at:
http://www.sph.jhu.edu/~altweb/index.html
9. PREX, available on the World Wide Web at: http://131.211.172.21/
10. National Library of Medicine, *Alternatives to the Use of Live Vertebrates in Biomedical Research and Testing*. Available on the World Wide Web at: http://sis.nlm.nih.gov

13

Justification of the Number of Animals to Be Used

Edward J. Gracely

Introduction

Choosing an appropriate number of subjects is an important part of any research project. A proper sample size is essential to obtaining valid results and to minimizing the number of individuals exposed to the potential risks and harms of research. For this reason, the IACUC is interested in ensuring that a sufficient but not excessive number of animals are used. A study with too few animals is ethically problematic because the animals are being subjected to potentially painful procedures or loss of life with relatively little likely benefit to the advancement of knowledge. On the other hand, a study using more animals than are truly needed is also problematic, because it unnecessarily exposes some of them to the same harms.

This chapter wrestles with key questions that may arise in determining animal numbers for IACUC purposes. Although the existing regulations and policy are not very explicit and IACUCs differ on many of these issues, experience, common sense, and the results of an informal survey all help provide direction. The survey results emanate from a questionnaire sent to primarily academic institutions.

13:1 What do the AWAR and PHS Policy state with respect to justifying the number of animals requested in the protocol?

Reg.   The AWAR (§2.31,e,1; §2.31,e,2) state that the proposal to use animals must include "identification of the species and the approximate number of animals to be used" and "a rationale for involving animals and for the appropriateness of the species and numbers of animals to be used." PHS Policy (U.S. Government Principle III) states that "the animals selected for a procedure should be of an appropriate species and quality and the minimum number required to obtain valid results." The PHS Policy (IV,D,1,a; IV,D,1,b) also states that all applications and proposals submitted to the PHS and which involve animals must include "identification of the species and approximate number of animals to be used," and "rationale for ... appropriateness of the species and numbers to be used."

The *Guide* (page 10) states that "whenever possible, the number of animals requested should be justified statistically." This is the most specific statement of how animal numbers should be justified in any of these documents.

13:2 What constitutes sufficient justification for a requested number of animals?

Opin. The AWAR and PHS Policy do not address this issue. Thus, it is left to the institution to develop appropriate policies. Clearly, the IACUC must be able to determine how all of the requested animals will be utilized, what research questions will potentiallly be answered by the number to be used, and why those questions could not be answered with fewer animals.

Describing animal usage is generally straightforward, except for studies that are open-ended with evolving goals. In the latter instance, some attempt at providing a sequence of events and a desired endpoint should be provided. This author is not comfortable with a justification of numbers based solely on an ongoing need for a certain number of animals per year, without an explanation of what is likely to be accomplished in each specific time period.

Justifying the specific animal numbers for individual research questions is often more difficult and there is no perfect way to do it. Previous research, while sometimes helpful, is only a crude guide to the number of animals needed in the present study. Even statistical power analysis requires assumptions (such as the magnitude of effects worth looking for) that are often subjective, or which must themselves be estimated from other data.

Sample size justification should be seen as a means to ensure that researchers and IACUCs are both aware of the issues and are attempting to avoid serious miscalculations. A moderate amount of detail is sufficient to accomplish this. Consider three specific approaches:

- Citation of previous research, with sufficient information provided to indicate that the previous research is similar enough in concept and methodology to make it reasonable to use similar sample sizes in the proposed research project.

- If a power analysis is to be employed, enough information is needed to show that the researcher knows how to analyze the data and use a power analysis. Question 13:4 lists key considerations of a fairly thorough power analysis. A briefer approach may be acceptable, but this author's experience suggests that attempts at an abbreviated power analysis often produces descriptions that are garbled or lack key information.
- Third, if the study has no statistics (e.g., histological characterization of tissues), it must be clear why the specific amount of material is needed, and why the number of animals requested is appropriate to provide that amount of material.

Surv. A Some protocols do not require any statistics. An example is one that focuses on detailed histologic analyses of certain tissue. When your IACUC is presented with such a protocol, which of the following is the *minimum* it will accept for sample size justification?

• A detailed breakdown of assays to be performed, amount of material needed for each, amount available from each animal, and a calculation of total animal numbers from this	3/20
• A statement of the *total* amount of material needed for all assays combined, followed by similar derivation of animal numbers	3/20
• A much simpler statement that N animals are needed to provide enough material for all assays to be performed	11/20
• An answer between the second and third above	2/20
• Answer unclear	1/20

Surv. B Some protocols do require inferential statistical analysis. An example is a study testing whether rats of three different ages have different mean levels of certain plasma proteins or one testing the effect of several treatments on the percentage of mice that develop a disorder. If a statistical power analysis is the primary basis a PI provides for justifying the sample size in a particular protocol, what is the minimum statistical information your IACUC would ordinarily accept in that protocol?

• A full statement of the statistical analyses performed, a detailed description of how the power analysis was tailored to those analyses, a description of basic parameters such as effect sizes and standard deviations, and how they were estimated, etc., plus the results of the power analysis	1/20

- A power analysis statement with effect sizes, standard deviations, etc., along with the results of the power analysis, but *without* the full description of the analyses and some of the explanatory details in the first bullet above   2/20
- Almost any halfway reasonable attempt to describe a power analysis is accepted   11/20
- Response between the second and third bullets above   1/20
- The statement that a power analysis has been done is sufficient, even if no details are provided   5/20

Surv. C Does your IACUC require a statistical power analysis for those protocols involving inferential statistics?

- Yes   2/19
- No   16/19
- Yes (with comment that experience can justify smaller numbers)   1/19

Surv. D Does your IACUC accept a statement based on previous experience or on published research as a basis for sample size justification in a statistical-type study?

- Yes   16/19
- No   3/19

Surv. E If you responded "yes" to Survey D, what is the minimum your IACUC will ordinarily accept for sample size justification?

- A fairly detailed statement describing the previous research, showing points of similarity and dissimilarity with the proposed studies, and indicating the kind of significant or nonsignificant findings that emerged with different sample sizes; plus one or more published references   0/15
- Same as previous response, but references not required   1/15
- A brief description of previous work and what sample sizes were used; much less detail than listed in the first response above, plus one or more published references   1/15
- Same as previous response but references not required   6/15
- One or more references cited as providing the basis on which sample size estimates were made; no details   2/15

- The statement that the sample sizes were chosen based on the previous experience of the researcher, without a description or reference   5/15

Surv. F Will your IACUC accept a protocol with none of the above (Survey E) justifications for sample size. In other words, does your IACUC regard sample size choice as a research-design question out of the purview of the IACUC?

- Yes   2/20
- No   18/20

The reader is cautioned that although two IACUCs responded positively to the question posed in Survey F, the regulatory considerations noted in 13:1 are still required. These two respondents stated that their responses applied to special cases. One provided an example of a teaching protocol in which numbers are given as "one dog per three students." Nevertheless, it is this author's opinion that this *is* a sample size justification and, therefore, must be justified. For example, it is appropriate for the IACUC to ask why one dog per three students is the optimal number.

13:3 Relative to the number of animals used, are there different regulatory requirements for different species?

Reg. Definitions of "Animal" relative to the PHS Policy and AWAR are provided in 12:1.

Opin. There are no species-specific distinctions relative to justifying the number of animals proposed for use. From ethical, humanitarian, and regulatory standpoints, it is as important to properly justify the use of 100 mice as it is to justify the use of 100 dogs. Invertebrates are not included in the definition of an animal noted above, but it seems reasonable that certain apparently sentient and intelligent invertebrate groups (such as octopuses) be given the same consideration as vertebrates.

13:4 What are the key considerations in using a statistical power analysis to justify animal numbers?

Opin. There are several key considerations if a researcher wants to perform and present a fully usable and appropriate power analysis.

- The researcher must have a thorough understanding of the appropriate statistical analysis for the design, or must be willing to seek out and collaborate with someone who does. Much basic science research that uses statistics involves several groups or conditions, and often a number of time points. Such designs require advanced analytic techniques, such as the analysis of variance (ANOVA) and specific follow-up tests (such as the Tukey Test). In some cases the major question involves the interaction of two independent variables, as tested in a two-way ANOVA. These analytic decisions have important implications for the power analysis. For example, a researcher with a complex design who estimates power merely for a two-group comparison may reach a substantially incorrect conclusion.

- The design must be spelled out in sufficient detail for IACUC (or other) reviewers to understand it and to see how the power analysis is configured to match the analysis that will be employed.
- The researcher must present and justify the specific parameters and assumptions required by the power analysis. Typically this will involve estimates of effect sizes (such as the mean difference between two groups) and estimates of variability, such as the standard deviation. Each analytic method has different requirements which must be determined before a power analysis can be run. Prior experience, the theory underlying the research, and pilot data may be useful.

13:5 What role can a biostatistician play in evaluating the justification of the number of animals requested?

Opin. Biostatisticians are trained to perform the kind of calculations used to determine sample size. They can advise the researcher how to properly analyze the data, and can help to write up the analysis and animal number sections of the protocol. To be most helpful, biostatisticians also should be familiar with the IACUC and its task. They should understand the audience that their comments will reach, and understand enough of the science as to be able to recognize what is central and what is not.

13:6 What exactly is a pilot study? How do pilot studies differ from other studies in animal number justification?

Opin. A pilot study is one involving a small number of animals (rarely more than 10) used to demonstrate that a technique can work, or to estimate the variability in the data before performing a statistical analysis. A pilot study does not require the extent of sample size justification required by "full" studies. Nevertheless, a pilot study is part of an IACUC protocol and, therefore, does require a justification, even if the justification is little more than "this is the minimum number of animals needed to perform a pilot study." If pilot studies are to be exempted from all but the most lenient requirements for sample size justification, then common sense suggests that they be narrowly defined and limited to a few special cases.

It is important to note that pilot studies are much more useful for estimating variability in the data than for determining the likely magnitude of effects. The latter requires a study much larger than most pilots. Furthermore, if a pilot study is used to estimate magnitudes of differences or to determine whether or not a particular manipulation is worth pursuing further, then the pilot animals should *not* be used in the succeeding full study (see 13:8). If pilot study animals serve only to estimate variability or to verify that a technique works in your laboratory (e.g., that you can actually get reliable results on a particular piece of equipment), it may be acceptable to include the pilot study animals in the full study.

13:7 Can a study using 4 groups of 10 animals each be considered a pilot study?

Opin. In the author's opinion it is difficult to see how so many animals could be considered a pilot study or why a pilot study would require four groups. Variability can normally be adequately estimated with one group (or, at most, two groups) of animals. Viability of a technique also can typically be tested on one or two groups.

Every possible situation cannot be anticipated. For example, consider a researcher who has four technically difficult manipulations, each of which must be shown to work before a larger study can be undertaken utilizing all four. Conceivably, a pilot study with four groups could be justified here.

13:8 If animals are requested both for a pilot study and the full study to follow, should the IACUC approve both at the same time?

Opin. Assuming the number of animals used in the pilot study are justified (see 13:6) and until the results of the pilot study are known, the IACUC should not approve the entire number of animals requested. The pilot may show that the basic technique is flawed, and the whole study should not progress. In other instances, a pilot has as its purpose the estimation of variability in the data upon which a power analysis will be based. Until these estimates are in hand, any number of animals requested for the full study must be considered unjustified.

Surv. If a person requests animals for a pilot study to be done before the full study and if the practicality of the full study or the sample size required by it may be affected by the pilot study, how does your IACUC proceed?

• Approve all the animals	1/19
• Approve the pilot study, conditionally approve the rest, with some data to be provided before the remaining animals are released	6/19
• Approve only the pilot data, requiring the PI to submit a new request to get the remaining animals	9/19
• Decision varies, based mainly on factors involving the invasiveness risk and severity of the protocol	3/19

13:9 Can the IACUC demand that a pilot study be done?

Opin. The IACUC has the right and responsibility to require an adequate justification of the number of animals requested (see 13:1). If a researcher is unable to provide the same, the IACUC may withhold its approval. If a pilot study can break the impasse, it may be an appropriate choice for both the IACUC and the investigator. The IACUC, by exercising its responsibility to assure that the number of animals to be used is appropriate, may withhold approval of a project until a pilot study is completed. If a pilot study is agreed to, it requires IACUC approval (see 13:6 to 13:8). Actually, the IACUC may be able to demand a pilot study, but there is nothing in the AWAR or PHS Policy that states it can. (See 9:34.)

Surv. In your personal opinion, can the IACUC *demand* that a pilot study be done?

• Yes	18/19
• No	1/19

13:10 Can an IACUC withhold approval of a protocol that requests fewer animals than the IACUC believes are needed for a valid study?

Opin. Yes, if the researcher does not adequately justify the use of that number. The primary question is whether or not the sample size justification itself is adequate (see 13:1). If the justification is invalid or inappropriate, the researcher should be asked to improve it. If the researcher fails to do this acceptably, then approval of the protocol is withheld until the justification is satisfactory to the IACUC. The use of too few or too many animals can be equally deleterious. A researcher who requests fewer animals than one might have expected is not excused from showing how that number is sufficient for research purposes.

Surv. In your opinion, can the IACUC withhold approval of a protocol that appears to be requesting too few animals?

- Yes   19/19
- No   0/19

13:11 When justifying the number of animals requested, should fetuses be counted as vertebrate animals?

Opin. Yes, if the fetus is sufficiently advanced to feel pain. The purpose of the IACUC is to help assure high standards of animal welfare in the institution. If a fetus can feel pain, then its welfare matters and it becomes an IACUC concern. Nevertheless, it is difficult to develop comprehensive and systematic guidelines for avoiding fetal pain, because so little is known about many species and their precise neural development. (See 8:12, 14:20, 16:24.)

Surv. A Does your IACUC count fetuses as vertebrate animals?

- Yes, always   0/20
- Yes, after a certain fetal age (depending on the species)   4/20
- No, only animals actually born are counted by our IACUC   11/20
- No, only animals having been weaned are counted by our IACUC   1/20
- Complex, hard to categorize   3/20
- Never discussed   1/20

Surv. B In order to clarify the concept of "counting" fetuses, a second survey composed of two questions was sent to some people who said they did *not* count fetuses.

In adult animals, a researcher planning to do a potentially painful procedure usually must first anesthetize or euthanize the animal. Does your IACUC require similar protections for fetal animals?

- Yes, always   0/6
- Yes, after a certain fetal age (depending on the species)   4/6
- No, only for animals actually born   0/4
- No, only for animals that were weaned   0/4
- Rarely arise because fetuses share anesthesia with the dam (it is this author's opinion that it is not always true

that an anesthetized dam will have fully anesthetized
fetuses) 2/6

When the researcher lists the total number of animals to be utilized in a study, are fetal animals counted in that total?

- Yes, always 0/6
- Yes, after a certain fetal age (depending on the species) 2/6
- No, we only count animals actually born 2/6
- No, we only count animals that were weaned 2/6

13:12 To reduce the number of animals used, how might an animal be used in more than one biomedical research, teaching, or testing protocol?

Opin. Any sacrificed animal should be utilized in as many studies as can benefit from the use of its tissues. It is appropriate to do so any time several protocols can use an animal without subjecting it to any additional pain or distress.

The real question concerns reuse of an animal in a painful or distressing way after it has already been used in this way once. Although animals are generally not subjected to a second survival surgery procedure, a second use in a nonpainful or nondistressful study can reduce the total number of animals involved in research in a particular institution. For example, after a survival surgery study in which animals need not be euthanized, a simple behavioral study might be appropriate.

13:13 How else can animal numbers be reduced?

Opin. It is worthwhile to carefully consider ways that animal numbers or suffering can be further reduced in experimental settings. Morton[1] describes some of them. For example, he notes that if a certain drug will be considered as too toxic for use if even one of six subjects has severe adverse reactions, it would make sense to test the six in sequence, then stop the study if one of them, in fact, has such a reaction. He notes that in certain kinds of research (such as testing antibiotics for efficacy in an animal model prior to use in a Phase I human study), there may be little purpose in a strict significance level (e.g., setting the cutoff for significance at $p \leq 0.01$) since any findings must still be replicated in humans.

Morton[1] notes that pilot studies can help to avoid problems and the data may be able to be used in the main study. Nevertheless, as noted in 13:6, one needs to be careful with this concept. If pilot studies are used merely to test procedures and estimate variance, it is true that one can use the data in the full study. If, however, pilot studies are used to see, for example, which drugs from a large set of drugs have an effect, one needs *independent* replication of the results with entirely new data. Anyone proposing to mix pilot data with full study data should consult a statistician.

13:14 What role can veterinarians and animal facility personnel play in reducing the number of animals used?

Opin. The IACUC has a general responsibility to oversee the appropriate use of animals in biomedical research. It closely interacts with the AV and the animal care staff. Veterinarians and other animal facility personnel have responsibilities that generally include assuring that animals are obtained from quality vendors, reviewing the health status of incoming animals based on a vendor's reports, assuring quarantine and testing of animals if that is appropriate, and facilitating the rederivation of animal strains when appropriate. They also are responsible for developing the entire program of preventive medicine and often oversee all animal care operations. The use of healthy, well-treated, and well-maintained animals helps decrease the numbers of animals needed by decreasing the need to repeat studies.

Veterinarians and animal facility personnel also can assist with preoperative, operative, and postoperative planning to help assure animal survival, further decreasing the numbers needed.

Reference

1. Morton, D.B., The importance of non-statistical design in refining animal experiments, *ANZCCART News*, VII(1), March 1998, Insert.

14
Animal Acquisition and Disposition

Michael J. Huerkamp and David R. Archer

Introduction

Millions of animals are acquired and used for scientific research annually in the U.S. Much is at stake because animal acquisitions for research and the eventual disposition of these animals come under the stringent regulation of federal law and the oversight of funding agencies. If animals are not acquired and used in accordance with appropriate laws and standards, institutions risk punitive measures or public relations dilemmas. Complicating these issues are myriad details related to the potential sources of animals (e.g., farms, commercial and other breeders, foreign import, pet shops, donations from citizens, transfer from other research projects, interstate movement), the type of research for which animals are acquired, the animal species to be used, stage of development or maturation, and eventual disposition (e.g., euthanasia, slaughter for food, donation to raptor rehabilitation program, adoption). After animals are acquired, the tracking and accounting of their use is arguably one of the biggest administrative challenges at a research institution and is a source of significant interaction and debate between scientists and institutional officials. Consequently, in order to act in a legal, fair, consistent, and rational manner, it is important for IACUCs to be cognizant of the many issues related to acquisition, records, tracking, and disposition of research animals.

The survey information provided in this chapter was supported by a questionnaire sent to 15 academic institutions (12 public, 2 private, 1 foreign) and 1 commercial animal producer.

14:1 What are the usual and ordinary sources of animals for research?

Reg. (See 12:1 for the definition of "animal".)

Opin. The most conventional way to obtain animals for research is to acquire them from outside the institution by purchase from a commercial vendor, licensed dealer, or farm. Other sources, usually requiring a certain level of justification, include pet shops, private pets (such as those used in veterinary clinical studies, see 14:28), and those transferred from other research institutions. In certain geographic areas, dogs and cats may be available from animal control programs, but this is highly variable depending on locality. Animals also may be studied in, or collected from, the wild. Within an institution, sources include breeding or stock colonies and those transferred from, or exchanged with, another research project.

In some cases, scientists may require tissues, rather than live animals, for research. As an alternative to purchasing animals to be immediately euthanized, scientists may obtain tissue from slaughtered livestock, harvested marine animals, or, perhaps, via transfer from a colleague who has euthanized animals for another purpose.

It is important to understand that the breadth of species defined as "animal" vary with regulating bodies and do not necessarily adhere to the classic biological definition. (See 12:1.)

14:2 What are the legal requirements governing procurement of animals and tissues for research?

Reg. The AWAR require that animals used for biomedical research:

- Be acquired by lawful means (§2.30,a,1).
- Be of an appropriate species (§2.31,e,2).
- Be tabulated in the *approximate* number of animals used (§2.36,b,5–§2.35,b,8) or to be used (§2.31,e,1).
- Be used in experiments designed to use the minimum number of animals necessary (§2.32,a,2).

For animals used for food and fiber or in agricultural research, the AWAR does not apply, but recommendations are given in the *Guide for the Care and Use of Agricultural Animals in Agricultural Research and Teaching*.[1] For research funded by entities of the PHS, the salient features of the AWAR are reinforced by the PHS Policy and the *Guide* (pages 8 to 14).

The provisions in the *Guide*, including those that relate to justifying and tracking animal use, also are used by AAALAC in accrediting research institutions. While the AWAR exclude poikilothermic animals and typical laboratory rats and mice from these considerations, all vertebrates are covered under PHS Policy and, by extension, AAALAC. Consequently, in most research institutions in the U.S., federal regulations, AAALAC standards, and granting agencies require that virtually all vertebrate animals be acquired lawfully, used judiciously, and disposed of appropriately. Even non-PHS-funded research may come under PHS requirements if an institution, via its Assurance statement to NIH/OPRR, indicates that it will include activities funded by sources other than PHS. Institutions may exclude non-PHS-funded or agricultural animal activities from its Assurance statement, yet still empower the IACUC to oversee the use of animals in those areas.

Also impacting on research animal acquisition and disposition are regulations and standards covering importation of animals and biologicals,[2] endangered species,[3,4] the safety of the American food supply[5] and the use of infectious agents (biohazards),[6] radioisotopes (usually state regulated), and chemicals and certain toxins.[7] Finally, beyond these bounds, an institution also must act and set policy in ways that serve to protect its good reputation.

14:3 Our institution, a land grant college, uses animals only in agricultural research and training. Are we required to follow either the AWAR or the PHS Policy?

Reg. No. While it is possible to do so and this may be useful from an institutional cost accounting perspective, at this time it is only *recommended* to account for the number of animals used in agricultural research and training.[1] The AWAR (§1.1 Animals; §2.1,a,3,vi) specifically exclude livestock bought, sold, or used in agriculture production, agricultural research, or agricultural training. However, the authors know of at least one institution where local APHIS/AC inspectors have elected to regulate teaching activities with horses under the AWAR. (It should be noted that horses or other farm animals used in teaching activities that are considered "biomedical" would routinely be regulated under the AWA.) PHS Policy also does not normally cover animals acquired and studied for agricultural purposes, unless, of course, those activities are PHS-supported. However, PHS Policy does apply if the agricultural research and teaching component are part of a larger institution with an institutional Animal Welfare Assurance that states that all institutional entities will adhere to PHS Policy. This becomes a germane issue when PHS-funded research is done on institutional farms or other agriculture-related facilities.

Opin. In order to preserve relations with APHIS/AC and a good public reputation, and to obtain and maintain AAALAC accreditation, any institution doing agricultural research with food and fiber species should endorse the principles in the *Guide for the Care and Use of Agricultural Animals in Agricultural Research and Teaching.*[1] Adherence to recommendations in the *Guide* also promotes good science, enhances animal well-being, maintains credibility with the public, provides standardization of research conditions between institutions (and, hence, reduces variability), and ultimately protects the institutional privilege of conducting research with animals.

14:4 Are there any special legal requirements regarding accounting and tracking of specific species used in research?

Reg. The federal government specifically requires that detailed records be kept on individual dogs and cats acquired and used in research (AWAR §2.35,b,1–§2.35,b,7) and that annual animal use be reported by pain/distress category (§2.36,b,5–§2.35,b,8). The species required to be reported are described under "Animal" in §1.1 "Definitions" of the AWAR. Federal[3] and state laws generally require special permits for the use of endangered species in captivity or in field research. Specific recordkeeping can aid the researcher in proving animals were born in captivity and in compliance with applicable laws.

Opin. Annual tracking of animal use is a potentially valuable management resource. Such data can be used to show trends in program size and changes in focus that may be useful in program justification and projections. Such information also may be requested by institutional administrators or public officials and may be helpful in clarifying facts about animal research for the public. Many institutions (such as ours) require accounting of animal use in an annual report.

14:5 Since neither PHS Policy nor the AWAR addresses the research use of amphibians, reptiles, fishes, and other ectotherms in any detail, should the use of these species in research be monitored by the IACUC?

Reg. PHS Policy (III,A) does cover the use of all vertebrates in research including reptiles, amphibians, and fishes, and the PHS Policy (IV,A,1) requires adherence to the *Guide*.

Opin. Although the PHS Policy is intentionally broad in scope and does not prescribe specifics about the care and use of any species, it assigns that task to the IACUC and allows for professional judgment. Many of the principles it advocates generally can be adapted to animal care and use programs for various kinds of amphibians, reptiles, and fishes. Recommendations for specific species are often available from organizations having interest in the appropriate care and use of these species in laboratory and field studies.[8-12] It is clear, however, that individual requirements for these three classes of vertebrates, which contain more than 28,000 species, cannot be addressed in a single set of guidelines. Consequently, NIH/OPRR recommends that the advice of experts be obtained to design and develop studies and suitable housing and care procedures for species not commonly used in research.[13]

14:6 An investigator at our institution uses avian embryos in his research. What is the responsibility of our IACUC, if any, in overseeing this activity?

Opin. Although PHS Policy does not discuss considerations attendant to embryonated bird eggs (or mammalian fetuses, for that matter), non-vertebrate animals (chick embryos within eggs) have the potential to develop into hatched vertebrate animals. While embryonal states of avian species develop vertebrae at a stage in their development prior to hatching, OPRR has interpreted "live vertebrate animal" to apply to avians only after hatching.[14] However, the risk of eggs hatching and producing chicks (requiring food, water, proper housing, and veterinary care and placing them under the purview of PHS Policy) dictates that IACUCs consider developing policies for different aged avian embryos, newly hatched birds, and the point at which bird embryos are considered vertebrate animals.[15] For chickens, the last 3 days of incubation (incubation days 18 to 21) represent the last stage of embryo development and coincide with the chick drawing the yolk sac into the body and having sufficient pulmonary maturation to handle oxygen and carbon dioxide exchange.[16] During this period of time, some chicks may hatch normally and some prematurely hatched chicks could survive outside of the egg with little additional care. (See 8:12.)

14:7 What institutional entity should procure animals for research?

Opin. One of the critical responsibilities of an IACUC is to assure that the institution lawfully procures appropriate animals for research and, in virtually all cases, IACUCs transfer this responsibility to the institutional animal resources program. Veterinarians involved in the animal resources program must have specialized training in laboratory animal medicine (AWAR §2.31,b,3,1; PHS Policy IV,A,3,b,1; *Guide*, page 56) and usually have the expertise to evaluate animal sources and identify those that are appropriate for research given the institution, its resources, and its needs.

It is well within the jurisdiction of the IACUC, especially in consideration of its responsibility to ensure an adequate program of veterinary care and a safe work environment and to provide guidance to the animal resources program concerning the species, sources, genotypes, and microbiological background of animals acquired for research use.

In smaller academic or industrial programs with part-time or consultant veterinary staff, the IACUC may well want or need to have considerable input into this process. Coordination of veterinary care, vendor approval, and a centralized purchasing system will aid the tracking of animal use as discussed in Questions 14:4, 14:8, 14:9, 14:11, and 14:12.

14:8 Should the IACUC become involved in approving an "acceptable" animal vendor?

Opin. The loss of animals or research to preventable infectious diseases has ramifications on the institutional requirement to ensure that research is not *unnecessarily* duplicative (AWAR §2.31,d,1,iii; §2.32,c,3,iii; Guide, page 10). Consequently, it is within the purview of the IACUC to work with the professional veterinary staff to identify acceptable animal health standards. In most instances, IACUCs defer to the professional judgment of the veterinary staff to identify vendors that meet appropriate health and legal standards. Often, this is the most effective means of addressing this obligation. Nevertheless, when rodent use is high, when certain animals are scarce, or when funds are inadequate, veterinarians may be under pressure to permit the entry of animals with sub-optimal health status into the institution In this situation, it may be worthwhile for the IACUC, as the de facto representative of the community of scientists, to work with the veterinary staff to develop and enforce standards that are in the best interests of the community of scientists within the institution.

14:9 What constitutes an acceptable animal vendor? What procedures can the IACUC or animal resources program use to assess vendor quality?

Opin. In general, animal vendors should have a tradition of producing consistently healthy animals of a specific genotype and health status in compliance with applicable laws, statutes and regulations, and institutional needs. Under the AWAR (§2.40,a), licensed vendors are required to have a program of veterinary care. Specifically, the AWAR requires licensing of any of the following supplying warm-blooded animals, other than rats and mice, to research institutions: pet shops (§2.1,a,3,i), breeders selling 25 or more dogs or cats annually (§2.1,a,3,iv), persons or commercial enterprises deriving more than $500 annually from such sales (§1.1 "Dealer," §2.1,a,3,ii), and any person or entity (Class B dealer) selling any dog or cat not born and raised on its premises (§2.1,a,3,iv). Additionally, institutions should make a dedicated effort to ensure that all transactions involving animal procurement are done in a lawful manner (Guide, page 57). Other considerations are whether the vendor has an adequate program of veterinary care and a history of consistently meeting consumer needs and expectations. Ideally, vendors should be AAALAC-accredited or provide assurance that they meet standards for their industry (i.e., health quality, animal care, customer service) and applicable federal laws. For rodent vendors, the health status of the animals should be documented through a regular program of health surveillance and verified by a rational means that includes sampling technique, sampling strategy, appropriate diagnostic tests, sufficiently frequent testing, a reliable laboratory, and prompt reporting of changes in health status. In most instances, the evaluation and consideration of animal vendors is handled within the program of veterinary medical care with IACUC oversight.

14:10 What precautions should the IACUC and animal resources program require if animals of unknown health status must be acquired?

Opin. Acquisition of animals of unknown health status should be limited to those instances when the animals are only available from one source. Nevertheless, this process still has several risks. The primary threat is the transmission of infectious diseases to pathogen-free animals. In addition to causing otherwise preventable pain and distress, research may be compromised or invalidated and, therefore, deemed as being unnecessarily duplicative (see 14:8). The IACUC should ensure that the institution has the components of a program of veterinary medical care that provide for the stabilization, isolation, and health characterization of animals of an undefined health status (*Guide*, pages 58 and 59).

Quarantine programs should be suitably long to permit the incubation stage of a disease to become clinically manifested or detected by culture or the presence of serum antibodies, pathogen DNA, or pathognomonic lesions. Where bacteriology or polymerase chain reaction (PCR) technology can be employed, results often can be obtained quickly. Nevertheless, responsive antibody production may not be detectable for 3 weeks or more depending on the amount and timing of the pathogen inoculum. Although investigators may feel a stringent quarantine and testing program is a hindrance to their research, data are easily compromised when using infected animals.

An effective quarantine program also should provide for the surgical rederivation of diseased animals and other relevant services, such as pathogen testing of cells and cell lines that may carry infectious agents that could potentially affect the health of animals within the program.

14:11 Should the IACUC attempt to track the number of all animals acquired or only those used under approved protocols?

Opin. Although neither the PHS Policy nor the AWAR explicitly require an institutional mechanism to track animal usage by investigators under IACUC-approved activities, both require that proposals to the IACUC specify and include a rationale for the approximate number of animals proposed to be used (AWAR §2.31,e,1; §2.31,2,2; PHS Policy IV,D,1,b; *Guide,* page 10). These provisions implicitly require that institutions establish mechanisms to monitor and document the number of animals acquired *and* used in approved activities.[13] Additionally, §2.31,d,1 of the AWAR states that the IACUC must approve significant changes to ongoing activities, and changing the number of approved animals is typically a significant change which should be tracked.

Tracking is particularly important for institutions that maintain large breeding colonies of nonhuman primates where animals produced from this population may be held for lengthy periods of time before assignment to a research protocol. Other institutions, using random source dogs and cats must maintain acquisition and disposition records in compliance with federal law (AWAR §2.35,b). These resources also may operate, depending on the institutional need, by maintaining pools of unassigned animals and allotting them to research protocols only after a suitable stabilization period.

Finally, tracking of animals held and used is needed for the institution to comply with the annual reporting requirement under the AWAR (§2.36,b) and for other possible requirements, such as the AAALAC annual report for AAALAC-accredited institutions.

14:12 How should the number of animals acquired and used be tracked by the IACUC?

Reg. The AWAR requires that the common names and numbers of warm-blooded animals (other than rats and mice) used for teaching, research, experiments, or tests or held for such purposes be given in the annual report to APHIS/AC (AWAR §2.36,b,5–§2.36,b,8).

Opin. Although it is implicitly required that institutions establish mechanisms to monitor and document the number of animals acquired and used in approved activities (see 14:11), IACUCs approach this consideration with some lack of unity. Some committees attempt to track all animal use with precise accuracy while others attempt to track use with approximate accuracy. The major reason for this diversity in accounting essentially relates to discrepancies in tracking and reporting requirements between the AWAR and PHS Policy. Species not currently regulated by the AWAR (i.e., rats, mice, birds, fish) are often accounted using approximate rather than precise figures (PHS Policy IV,D,b; *Guide,* page 10). Thus, institutions that track and report the *approximate* number of certain species used in research may be in full compliance with the AWAR and PHS Policy. However, many internal and external factors (including financial management, public opinion, and the spirit of the law) support an accurate accounting of the animals used in research. The advent of commercial animal ordering software allows for most integrated animal resources programs to track the number of animals purchased from commercial sources or imported to the institution through a quarantine program with relative ease. This provides accurate figures with relatively little investment in time and personnel. When acquisitions are linked to IACUC approval numbers, investigators can be automatically informed when their ordering has reached a preset percentage (e.g., 80 to 90%) of the animals approved on that protocol. More flexibility is often found in practices concerning rodents and ectotherms than nonrodent mammals. Small institutions that use limited numbers of animals may choose to maintain a hard-copy log of each IACUC-approved activity, merely subtracting the number of animals acquired for each order from the number approved, with verbal or written notification to the investigator as the number of animals approved is approached.[7]

The real challenge in animal tracking is found in large breeding colonies of prolific animals such as ectotherms (e.g., fish), chickens, or rodents. Dealing with the sheer volume of animals produced can be a full-time job for animal caretakers and users, and attempts to accurately count the animals are often overwhelming. Activities involving the production of genetically manipulated animals are of low yield, but require the production of unpredictably large numbers of animals, the vast majority of which are unsuitable for research.[17] Additional challenges are presented where animals are used or collected in the field or bred in fisheries or when animals are transferred from one investigator to another. Studies involving fish and their offspring may involve tens to hundreds of thousands of animals. In such studies, it is not ordinarily possible to accurately count the precise number of animals used. While it is not outside the realm of possibility to do this, the cost-benefit ratio is not conducive to the practice and doing so arguably would have little impact on the humane care or use of animals. Although of less significance, extra animals included in a shipment may not be counted against ceilings approved by the IACUC.

There also is no agreement on how to count animals used on more than one study or animals kept over the years on the same study.[18] Animals used in more than one study usually fall into one of two categories: studies where the animal is killed for a specific purpose but tissues taken for another study (discussed in 14:22) or where the longevity or rarity of the animal, the duration (e.g., aging) or innocuous nature of the primary study (e.g., behavioral), or the need to study disease pathogenesis/treatment over a lengthy period allows for the animals to be used in long-term or justified sequential or serial experiments. The latter scenarios usually occur with nonhuman primates or higher vertebrates and as such it is especially important that each use of an animal is justified in an application, and that multiple experiments do not jeopardize the quality of the data. For reporting purposes, animals should only be included once except where specific procedures are required to be reported.

Surv. Does your IACUC track or intend to track the number of animals acquired for research with precise or approximate accuracy?

- Track all animal use with precise accuracy   7/16
- Track animal use with approximate accuracy   8/16
- No attempt to track   1/16

14:13 What factors prevent the precise prediction and accounting of the number of animals acquired and used for research?

Opin. Although precise accounting of animals is not always required (see 14:12), investigators must still satisfy the requirement that the number of animals used be limited to the appropriate number necessary to obtain valid results (AWAR §2.32,c,2; PHS Policy IV,1,g; U.S. Government Principles III). This often produces conflicting pressures, on the one hand, to justify animal numbers (including the use of statistical analysis and power calculations (see Chapter 13), but on the other to account for the inherent unpredictability of science. This conflict can often lead to discrepancies between the number of animals approved and the number necessary for the completion of the work. Thus, failed experiments, unforeseen technical challenges, unexpected disease, or other complications may cause more animals to be used than originally expected (or requested). Therefore, it is not reasonable to expect scientists to be able to project the exact number of animals needed to do the work. Consequently, scientists usually do not see precise accounting as valuable and, thus, do not commit precious human resources to the activity. IACUCs, in turn, are loathe to stringently enforce what has been conveyed as a loose regulatory standard. In addition, the peer review groups of most funding agencies do not micromanage projects down to details such as animal numbers. Instead, most rely on the powerful influence of funding, an influence that is often underestimated by the public and largely ignored by the vocal opponents of research. Still, the expectation of most organizations is that there will be a good faith effort made in tracking the number of animals used.[13]

Avian and rodent breeding colony accounting may be facilitated if the colony is managed by a centralized animal resources program, but there are trade-offs in this practice, such as insulation of researchers from their animals and inequity between production and demand. Historically there has often been an "us and them" relationship between investigators and many IACUCs, which can lead to less than accurate accounting of animal usage. Generally, IACUCs need to communicate their role to investigators in an effective way and be responsive to the sometimes rapidly changing needs of the investigators. One effective mechanism is workshops explaining the relevant laws and the obligation of the investigators.

14:14 Should the number of animals generated by breeding colonies be tracked?

Reg. The AWAR (§2.36,b,8) require annual reporting of the species and number of regulated animals that are bred, conditioned, or held for use.

Opin. The expectation of legal, regulatory, and accrediting bodies is that research facilities make a good faith effort at counting the number of progeny not only used in research, but the total number of animals produced.[13] When a breeding colony is a centralized core, supports multiple projects, or even individual investigators, animal production should be tracked by a dedicated protocol. Generally, progeny are counted at weaning and debited against a numeric ceiling approved by the IACUC. The number of animals approved by the IACUC should include both those that are used in research as well as those that are not used. Unfortunately, the unavoidable production of unusable offspring is an integral part of the research enterprise. In the case of dogs and cats, federal law requires precise accounting at the time of weaning (AWAR §2.35,b; §2.38,g,3).

Surv. Does your IACUC require that animals produced from an in-house rodent breeding colony be accounted?

- Yes   16/16
- No   0/16

14:15 Is there a maximum percentage of unused breeding colony animals that are euthanized (without having been used as breeding stock or experimental subjects) that should be tolerated by the IACUC?

Opin. The production of unused animals that are subsequently euthanized is almost unavoidable in large breeding colonies of ectotherms and rodents. This is a significant issue for commercial and core production colonies which must operate at a production excess (approximately 30% or more) in order to meet unexpected increases in market demand or that produce animals at an excess when demand unexpectedly takes a down turn. A number of user factors such as newly funded projects, changing experimental initiatives, failure to plan, cessation of standing orders, and periods of relative research dormancy, alone or collectively may have a dynamic impact on production. For animals from a production colony, matching supply with need is a moving target and, as such, production cannot be expected to match demand.

The case of creating genetically unique animals using homologous or nonhomologous recombination presents many examples of animals that are produced and euthanized without further use. In the creation of a transgenic (nonhomologous insertion) mouse using the pronuclear microinjection technique, only 10 to 40% of the mice born will carry the transgene, thus, 60 to 90% cannot be used.[19] Although most transgenic founders will subsequently transmit the foreign gene in 50% of their offspring, approximately 20 to 30% are mosaic and transmit the gene at a lower frequency, i.e., 5 to 10%.[20] Consequently, in the breeding of transgenic founder animals anywhere from 50 to 95% of the offspring will not carry the gene of interest, will have no research value in many instances, and must be euthanized.

Whether producing genetically unique mice or breeding small colonies used by a single investigator, mating must continue in order to perpetuate the colony even when progeny are not needed for experiments (it is possible to freeze embryos for later rederivation, but this is not a realistic approach for most investigators). When this happens, the offspring that are not selected as breeders must be euthanized. In managing inbred animals by a single line system, genetic divergence is prevented by euthanizing breeders, their progeny, and grand-progeny that cannot be traced within a minimum number of generations to a common ancestor. As one might imagine, in a large production operation involving multiple genotypes of animals, this results in the euthanasia of large numbers of animals.

For these reasons, it is recommended that IACUCs permit animals to be produced from breeding colonies in reasonable excess of projected or historical need.

14:16 Should investigators be allowed to acquire, but not use, animals prior to IACUC approval of the protocol?

Opin. It is not advisable to allow advanced purchase for several reasons. First, if animals are acquired prior to consideration of a research proposal and the proposed use of the animals is subsequently rejected by the IACUC or the number of animals approved by the IACUC is reduced from the original request, this may present problems in managing the population[21] and would be in violation of legal requirements to use the minimum number of animals necessary (see 14:13). Second, there may be unintentional use or temptation on the part of the scientist to use or to prepare to use the animals without appropriate authorization.[21]

14:17 Can animals be transferred between studies having IACUC-approved protocols?

Opin. Providing certain requirements are met, it is permissible to transfer animals between studies. Such requirements should include informed consent of the transaction by the IACUC, approval by the IACUC of the use of the species in the research, and a mechanism to debit animals from one protocol and transfer them to another. Unrestricted trading of animals between research protocols could lead to unauthorized use, defeat attempts at tracking animal utilization, put legal records pertaining to dogs and cats in jeopardy (AWAR §2.35,c,1; §2.35,c,2), and may be confusing for veterinarians trying to determine the ownership of animals requiring medical care.

Surv. Does your IACUC permit animals to be transferred between studies having IACUC-approved protocols?

- Permit transfer   16/16
- Prohibit transfer   0/16

14:18 Under what circumstances or conditions may animals be transferred between studies and how can the IACUC track this activity?

Opin. At institutions where there are breeding colonies of animals supporting more than one research project, there must be a mechanism to transfer animals from the breeding colony to IACUC-approved research protocols. An example of the latter is an institution having a core transgenic animal facility. Animals of a desired genotype are produced under one IACUC-approved activity (transgenic core) and then are transferred to the IACUC-approved activity of a specific scientist for research use. Other cases involve simple transfers where one investigator receives a few breeders of one-of-a-kind genetically manipulated mice from the breeding colony of a colleague. Nonhuman primates used in operant conditioning or behavioral paradigms present another example. These animals are usually purchased from a commercial vendor and are quarantined in the animal research facility. During the quarantine phase (or shortly thereafter), the investigator may become aware of behavioral considerations that render the animal less appropriate for the intended research and more appropriate for use by a colleague. In this instance, the investigators may wish to exchange animals. The same may be true for dogs or cats obtained for use in research involving surgical procedures.

Demeanor and size in this circumstance may promote scientists to trade animals between acute and chronic studies or where the animals have anatomical or other attributes that are better suited for certain studies than others.

The most important consideration is that the IACUC employ a mechanism to ensure that the total number of animals used and the species of animal are not at a variance from the approved protocol[22] and that there are not multiple survival surgeries in unrelated projects (AWAR §2.31,d,1,x,A; *Guide,* page 12). The IACUC should develop some guidelines for the transfer of animals between approved protocols if this is a circumstance that might be faced repeatedly.[21] Important considerations in setting guidelines include obtaining the surgical history of the animals from either the investigator or veterinary staff so that animals are not unknowingly subjected to multiple major survival surgical procedures, unless there is an approved exception (AWAR §2.31,d,1,x,A; *Guide,* page 12). The movement of animals from one protocol to another ideally should involve debiting or crediting the number of animals against those approved for the study. IACUCs frequently refer these activities to the animal resources program that often has the staffing and computer resources to handle the task.

14:19 If a protocol is being renewed, how can the IACUC ensure that it is not unintentionally renewing the original number of animals requested?

Opin. This is an important issue from the perspective of avoiding unnecessarily duplicative experiments and providing a justification of the total number of animals used in research. An example of this situation is useful for illustrative purposes.

> An investigator is funded for a 5-year study involving 30 macaques. He purchases them over the first 3 years of the study and then submits a renewal after 3 years which requests 30 macaques. How can the IACUC be sure that the 30 animals requested on the renewal are new monkeys and not just a relisting of those already on study and still in the investigator's colony? As an extension of this problem, what if the investigator originally was approved for 30 animals, but only acquired 20 by the time of the 3 year renewal? If 10 animals are now requested in the renewal, how can the IACUC determine whether these are the remaining 10 from the original approval, or 10 more above the original request of 30 (for a total of 40)?

These problems can be addressed with careful formulation of the renewal and modification requests. For an annual renewal (end of first or second year), the application should not include animal numbers as the request has already been approved for the first 3 years. The annual accounting of animals for APHIS/AC should be monitored at the level of purchasing or when animals are assigned from a breeding colony to a particular protocol. Requests for additional animals before the end of the original protocol should be handled in a different manner to new or 3-year renewals to specifically indicate that additional animals are required. One mechanism to help eliminate the issues raised in the examples is to close out the original protocol regardless of how many animals were used or not used and to consider the continuation as an essentially new proposal with a new set of experiments and requirement for justification of animal numbers. Obviously there are circumstances, as in the example, where further questions need to be asked. Examples are

- Is this a new protocol or a 3rd-year renewal?
- If the protocol is a renewal, what is the previous protocol number?
- If the protocol is a renewal, are there *existing* animals that are already assigned to the study?
- If so, how many? (Do not include this number in your answer to the following question.)
- Justify the number and use of animals specific to this application.

Specifically, special attention should be paid to the type of application (new or renewal), the numbers of animals used, the number requested, and the number to be carried over from the previous protocol. Although it is beyond the jurisdiction of individual IACUCs, many of these problems could be alleviated, in the opinion of the authors, if periods covered by IACUC protocols and extramural grants could be matched in duration. (See 11:3.)

Many IACUCs do not consider this issue with meticulous detail. Instead, they rely on faith that good science is not going to be repetitious or wasteful, that the vast majority of scientists can be trusted, that economic factors (as much as anything) have a limiting effect on animal use and dictate judiciousness, and that the IACUC application will show that the research requires a certain number of animals and is novel and important.

14:20 Should unweaned animals (particularly rodents) or animals used or harvested *in utero* be accounted for?

Opin. At our institution, the use of fetuses or animals up to the age of weaning is not tracked, but the adult breeding animals needed to produce them are accounted for. However, if painful procedures are to be done and the animal is at a stage of development so as to perceive pain, it is clearly within the jurisdiction and responsibility of the IACUC to address this issue (*Guide*, page 10). (See 8:12, 13:11, 16:24.)

14:21 Is IACUC approval needed for the use of animal tissues acquired from a slaughterhouse?

Opin. IACUC approval is not required if the collection takes place postmortem and as a by-product of the commercial enterprise.[21-23] However, it is in the institution's best interest to ensure that issues such as occupational health and safety, potential disease transmission to colony animals, and related matters are addressed.[22] Thus, it is recommended that the IACUC and investigators maintain appropriate documentation, and institutions tailor their policy to meet their needs. If the samples are obtained antemortem and are not incidental to slaughter (e.g., blood) or if they dictate the slaughter procedures in any way, then a protocol may be necessary depending on the nature of the study.[23]

Some IACUCs have short, one-page forms for the purpose of documenting animal tissue acquisition and use. A useful mechanism is to have a form for tissues from any source including slaughterhouses, intramural projects where animals are euthanized and tissues are made available to others, or extramural transfers. Completed forms can be given an expedited review by the committee chair or designee acting under the auspices of the IACUC (e.g., the AV). (See 8:8.)

Surv. Does your IACUC approve the use of animal tissues acquired from a slaughterhouse?

- No requirement/not addressed   10/16
- No requirement except for tissues from nonhuman primates   1/16
- Require formal IACUC approval   0/16
- Require IACUC approval via expedited review, notice of intent, or memorandum of understanding   5/16

14:22 Is IACUC approval needed for use of animal tissues acquired from animals euthanized as part of an unrelated experiment?

Opin. As with 14:21, this is an issue that divides IACUCs. The harvest of tissues from dead animals itself is not considered research under the AWA and, therefore, is not a regulated activity.[23] Committees that elect not to oversee this activity generally do so as long as the tissues collected do not alter the approved procedures in any way, that the animals were used exclusively for the research of others (and not for the individual receiving the tissue), and that the tissues are collected after the animal is dead. In institutions where IACUC approval is required (see 14:21), the process often involves the use of a brief form and an expedited review by the IACUC Chair, AV, or other designee.

NIH/OPRR considers the use of shared tissues and slaughterhouse material to be an effective application of Reduction, Refinement, and Replacement, and it encourages this practice.[24] However, proposals involving the killing of animals to use some or all of their tissues or one that involves specific antemortem manipulation, requires IACUC review. In all cases, a "paper trail" is recommended to show that an institution has applied the appropriate standards to the acquisition, use, and disposition of animals.[24]

PHS grant applicants using shared animal tissues or slaughterhouse specimens are advised to specify the origin of tissues when describing their proposed use in an application. If the "No" box is checked on the vertebrate animal block on the face page of the PHS application form, any reference to the use of animals in the application is likely to trigger questions about IACUC approval.[24] To avoid delays in peer review, applicants should be advised to explain in the PHS grant application that tissues come from a slaughterhouse or as a by-product of other IACUC-approved activities. (See 8:8.)

Surv. Does your IACUC require its approval for the use of tissues acquired from animals euthanized as part of an unrelated experiment?

- Require approval   7/16
- Do not require approval   9/16

14:23 Are there reasons why IACUCs should review the acquisition of animal tissues acquired from animals euthanized as part of an unrelated experiment?

Opin. Assuming the animals and the method of euthanasia were covered under an approved protocol, IACUC approval is not generally needed.[23] However, approval is obviously necessary if the investigator receiving the tissue, in the course of meeting research needs, causes more animals to be used or different procedures to be done on live animals. Review by the IACUC also should be considered when there is a risk of pathogen transmission from one facility to another within a research institution. For example, a number of rodent viruses may be transferred in tissues. Direct or indirect contact with such tissues may serve to infect previously disease-free rodents in another animal research facility, invalidate the research in which the animals were used, and cause additional animals to be used unnecessarily as a replacement for those infected. Occupational health and safety issues also dictate IACUC consideration of tissue transfer and use. Some rodent pathogens, such as lymphocytic choriomeningitis virus, are transmissible to humans. Tissues from nonhuman primates, wild-caught animals, livestock, and unconditioned dogs or cats also may serve as a source of pathogens for humans. Additionally, there may be experiments done that could lead to public relations challenges for the institution. Examples that come to mind are research using fetal tissue that may not be consistent with the institutional mission or philosophy or research on tissues acquired from animals euthanized at an animal shelter. (See 8:8.)

14:24 Is IACUC approval needed for acquisition of animal tissues from abroad?

Opin. Investigators that collaborate with scientists in other countries may receive animal tissues from these foreign sources and may sometimes go abroad to collect samples from captive wild animals maintained in zoological collections and in research colonies. Just as animal-related activities supported by the PHS require review by the IACUC of the domestic awardee institution, foreign institutions that serve as performance sites also must have Assurances on file with NIH/OPRR.[13] NIH/OPRR considers institutions whose scientists are engaged in such collaborative work accountable for the animal-related activities from which they receive animals or animal parts. When a foreign institution holds a PHS Assurance, it also is expected that the institution will conduct the study in accordance with the applicable host nation's policies and regulations. In the specific case of sample collection, the review should take into account the species involved, the nature of the specimen, and the degree of invasiveness of the procedure, giving appropriate consideration to the use of anesthetics and analgesics. When samples are obtained directly by citizens of a foreign country for subsequent shipment, the recipient PHS-supported investigator should determine the proposed methods of collection and present that information to their IACUC for review. Prior to sample collection and regardless of whether specimens are obtained by an awardee institution's investigator directly or by persons in a foreign country, NIH/OPRR strongly recommends that each awardee institution consult with other agencies of the U.S. government concerning importation requirements.[13] Depending on the species involved and the nature of the specimen, the following may be of assistance: the U.S. Fish and Wildlife Service, Department of the Interior (for compliance with the International Convention on Trade in Endangered Species of Fauna and Flora [CITES]),[4] APHIS, U.S. Department of Agriculture (regarding potential animal pathogens),[2] and the Centers for Disease Control and Prevention (concerning importation of nonhuman primates and potential pathogens of human beings).[6]

14:25 Under what conditions can the IACUC permit the use of animals purchased from a retail pet store?

Reg. The AWAR (§2.1,a,3,i) require licensing of pet shops that supply warm-blooded animals, other than rats and mice, for research purposes.

Opin. As a general rule, this is a practice that should be avoided. It may be permissible for ectotherms, such as fish, for rare strains of rodents or rabbits, or for certain birds or reptiles when such animals cannot be acquired from a more traditional or conventional source. Where regulated species (see 14:1) are to be obtained from a pet shop, the institution has an obligation to obtain the informed consent of pet shop management and to procure the animals in a lawful manner (*Guide*, page 57). Purchase of animals from pet shops also should fall under the auspices of the IACUC because of the potential for abuse and the risk to the institutional reputation should this practice become public knowledge. Animals that are obtained from pet stores may not be acclimated to a laboratory environment and are generally of unknown genetic and health backgrounds. With respect to the latter, the veterinary staff should be called upon to evaluate the health status of the animals and quarantine, isolate, and otherwise manage these animals to protect other research colonies.

14:26 Are there legal requirements that must be met for purchase of animals from a commercial farm?

Reg. (See 12:1 for the APHIS/AC and PHS Policy definition of "Animal.") Traditional farm animals, such as sheep, swine, or cattle, are regulated if used in research for purposes other than improving animal nutrition, breeding, husbandry, production efficiency, or improving the quality of food and fiber (AWAR, §1.1 "Animal").

Opin. While APHIS/AC can regulate the sale of farm animals for research purposes, it has chosen not to do so at this time. The only persons required to be licensed for selling farm animals to research are those who sell *exclusively* to research.[23] Thus, the livestock producer or livestock marketer who sells an occasional animal is not required to be licensed, nor do they have to meet the transportation requirements, even if transporting animals to a research facility. However, the research facility becomes responsible at the point they assume custody of the animal. Therefore, if an employee of the research facility goes to a livestock market or farm and purchases livestock, then as an agent of the research facility he must satisfy the transportation requirements (AWAR §3.136–§3.142) and the institution must satisfy health and husbandry standards (AWAR §3.125–§3.133).

14:27 What guidelines can the IACUC establish for the use of endangered species in research?

Opin. The use of endangered species in research must conform with all federal and state laws and international conventions to which the U.S. subscribes. This includes the Lacey Act, the Endangered Species Act, the Marine Mammal Protection Act, and the Convention on International Trade in Endangered Species (CITES) treaty. It is desirable for IACUC investigator guidelines to repeat this requirement.[22]

14:28 Does the IACUC need to approve studies using privately owned animals, such as those used in a clinical trial in a school of veterinary medicine?

Opin. There is no clear path to follow. Neither the AWA, AWAR, nor PHS Policy distinguish between privately owned animals and those owned by the institution. The law and regulations give the IACUC room to deliberate justifiable exemptions. Ultimately this involves doing what is appropriate by all parties and documenting the exemption and its rationale. While one could arguably conclude that the IACUC specifically is not required to review the use of pets in research (unless PHS supported), where privately owned animals may be involved in the research enterprise, it is recommended that institutions have a written policy.[22] A number of factors impinge on this situation, which include but are not limited to:

- The specific type of work.
- The need to board or retain the animals for a period of time.
- The nature and resources of the facility.
- Involvement of federal funding.
- Requirement for IACUC review by a nonfederal granting agency.
- The content of the institutional Animal Welfare Assurance.

Pets used in clinical trials are privately owned animals recruited from volunteers and, as such, usually remain the property of the owners who maintain responsibility for their upkeep, housing, and transportation. However, once these animals appear on institutional property, rules should be established as to how the animals are managed and temporarily housed. Such rules should take into consideration facility operations, informed consent, institutional liability, public safety, and animal welfare. Where IACUCs oversee research with private pets, review may be done by the IACUC itself or by a designated subcommittee of members representing (for example) the veterinary school faculty and staff. Should client-owned animals be donated to the institution, the full range of compliance issues take effect. At veterinary schools where the IACUC does not review the research use of client-owned animals, teaching hospitals typically have policies governing the conduct of research. Typically, the proposed research is reviewed by the department and then a hospital board or committee akin to an Institutional Review Board for human subject experimentation.

Surv. Does your IACUC approve studies using privately owned animals, such as those used in a clinical trial in a school of veterinary medicine?

- Require approval   11/16
- No oversight   1/16
- Not applicable   4/16

14:29 What are the legal requirements regarding the disposition of research animals?

Opin. Animals used in research may be disposed of by euthanasia, adoption, retirement (e.g., endangered species), return to production agriculture (food and fiber species), sale for slaughter, release to the wild, or transfer to another institution or research project. In most instances, animals used in research are euthanized. Considerations when disposing of these animals include the experimental history (e.g., method of euthanasia and exposure to chemical, toxins, radiation, and drugs) and may involve burial in a landfill, incineration, alkaline hydrolysis with discharge into the sewage system, or rendering into animal feeds. In some instances, particularly in the case where radioisotopes have been given to an animal, the carcass must be held for a period of time until the radioactivity decays to safe levels.

14:30 Can animals used in research, teaching, or testing be adopted as pets once they have been used in an IACUC-approved project? What is an appropriate mechanism for facilitating this process?

Opin. Providing the animals are healthy and have not been disfigured or disabled by the research, adoption of animals that make suitable pets is appropriate. This practice, like many others, may have legal ramifications and guidelines for adoptions should be formally developed in consultation with legal counsel. In institutions were it is permitted, adoptions to a reliably suitable and permanent home occur on a limited basis after review and approval by the researcher, approval of the request by the IACUC, and confirmation of the good health of the animal by a veterinarian. There may be other restrictions, such as limiting adoptions to employees and students of the institution and not to the general public. Some public institutions do not permit adoption per se as adoption entails a change in ownership that involves divestiture of public property. However, at least one pragmatic institution skirts this issue by permitting animals to leave under the status of a long-term loan. At the authors' institution, healthy animals are released after the approval process and the completion of relevant records (i.e., USDA Veterinary Services Form 18-6: Record of Disposition of Dogs and Cats). Copies of all correspondence related to adoptions and the permanent medical records of adopted animals are maintained in the animal resources program office.

Surv. Does your IACUC permit adoption of animals as pets once they have been used in an IACUC-approved project?

- Yes, under limited circumstances (e.g., approval by the PI and IACUC, proof of normal health as determined by veterinary examination, and the absence of disfiguring or potentially debilitating conditions) — 13/16
- No — 2/16
- Not applicable — 1/16

14:31 Should the IACUC permit farm animal species that have been used in teaching, testing, or research to be sold for slaughter? What guidelines would be useful to the IACUC in this regard?

Opin. Domestic livestock, usually raised for food or fiber, also are commonly used in research and training. In some cases, it may be appropriate to return live animals that have been used for research or training purposes back to production agriculture or to render carcasses into livestock feed. However, if this is done, it should be done with caution and with full awareness of the potential for institutional and individual liability.

When animals used in research are subsequently slaughtered for human or animal food, the major concern is adulteration of the meat from residues (e.g., antibiotics, hormones, toxins, radioisotopes, carcinogens) or infectious agents that would render the meat or its food products unwholesome. The agencies that regulate slaughter are either USDA's Food Safety and Inspection Service (for federally inspected facilities) or state agencies for state-inspected slaughter facilities. Endangerment of the nation's food supply by failing to adhere to food safety standards constitutes a felony in the U.S. which could be punishable by fines or sanctions to the institution or institutional officials, imprisonment of convicted parties, and revocation of the license of the AV. Owing to these potential legal ramifications, an institutional policy on the sale, slaughter, consumption, or rendering of research animals is strongly encouraged.[21,23] The policy should be developed with appropriate legal counsel and should be consistent with the institution's Animal Welfare Assurance. Veterinary care of animals sent to market or slaughter and drug withdrawal periods must be consistent with the Animal Drug Use Clarification Act of 1994 (AMDUCA).[5] In the event that the animals are given a new animal drug as defined by the Food and Drug Administration (FDA),[25] no meat, eggs, or milk from those animals may be processed for human food without FDA approval. If animals are being slaughtered at the research facility for subsequent consumption, the method of euthanasia must be stated in the approved protocol and should be in compliance with the AWAR (§1.1 "Euthanasia") and the American Veterinary Medical Association recommendations on euthanasia.[26] For animals that have had invasive surgical procedures, the general practice should be for the level of surgical asepsis and postoperative care to be consistent with current veterinary standards for the species. (See 18:7.)

Surv. Does your IACUC permit farm animals which have been used in teaching, testing, or research to be sold for slaughter or released back to production agriculture?

- Permit the subsequent sale of research animals for slaughter or return to production agriculture   7/16
- Do not permit subsequent sale or return to agriculture   4/16
- Not applicable   5/16

14:32 Should institutional programs related to veterinary care (such as a rodent sentinel program, quarantine programs involving diagnostic sample collection from animals, necropsy service, and therapeutic surgical interventions) be approved via the protocol review system?

Reg. The AWAR (§2.31,a) require that all animal activities for regulated species be reviewed by the IACUC. The AWAR (§2.31,c,1) and the PHS Policy (IV,B,1) state that the entire "program" must be reviewed by the IACUC at least once every 6 months. The *Guide* (page 9) states that the IACUC should do the same every 6 months.

Opin. The authors interpret this to mean that the IACUC should oversee sentinel and breeding programs, pathogen screening, surgical rederivation program, stock animals, etc. in some manner, probably through SOPs or IACUC-approved protocols. All clinical activities (e.g., diagnostic and therapeutic interventions, sentinel animal usage) involving existing research or instructional animals come under the purvey of the AV, should be covered under the semiannual institutional program assessment, and be approved and monitored by mechanisms other than protocol review (e.g., through SOPs).

In a "gray zone" there are activities such as blood donor animals used in veterinary schools or rodent colonies used as a source of food for raptors in rehabilitation programs. In these instances, one can argue that if there are no teaching or research components to these activities, they do not fall under IACUC jurisdiction. Even in an instructional setting, one must differentiate clinical teaching using clinical cases, from using laboratory animals as teaching aids. It is suggested that the IACUC or the institution have a written policy about "gray zone" cases, and the IACUC or some other institutional entity have oversight responsibility. This permits the development of standards for the humane care and use of these animals and avoids pitfalls that may be unnecessarily harmful to the animals or embarrassing to the institution. For rodent colonies, this arrangement complements the program of adequate veterinary care by enabling the colony to be enrolled in a health surveillance and management program or programmatically isolated from other colonies. (See 8:13, 8:15.)

Surv. Which of the following institutional programs related to veterinary care are reviewed by the IACUC and approved via the protocol review system (more than one response is possible)?

• Rodent sentinel program	11/16
• Rodent quarantine program (including surgical rederivation)	3/16
• Necropsy service	3/16
• Veterinary therapeutic surgical interventions	2/16
• Genetic quality control	1/16

14:33 Our institution's veterinarians or physicians occasionally are asked to provide veterinary services to a pet either for free or for reduced cost. How should our IACUC address these situations?

Opin. Although research institutions have great resources and clinicians with unique skills, engaging in these activities is a legal "slippery slope" fraught with risk. There is great potential for both institutional and veterinary liability should something go awry.

All states have veterinary practice acts that regulate the practice of veterinary medicine in that state and protect veterinary private practice as a business. Many states permit veterinarians specializing in laboratory animal medicine to practice without a license providing that certain conditions are met (e.g., licensed in at least one state, specialty board certification, limit activity to specialty area). When unlicensed specialists, including physicians, undercut fees or provide veterinary services for individuals who otherwise would seek care from private practitioners, the institution and veterinarian assume a legal risk.

Some institutions and individuals believe that they can protect themselves by rendering free service or having clients sign consent forms. Nevertheless, the judicial system most likely would not enforce provisions of consent forms in which clients agree not to hold an institution or individual liable for the negligent treatment of an animal.[27] Nor will any consent by a client to have a veterinary procedure done by a nonveterinarian protect the nonveterinarian from prosecution for practicing veterinary medicine without a license.[27]

In some instances, veterinarians and institutions must be pragmatic. Acutely ill or dying animals on or near institutional property, such as stray cats in end-stage disease, dogs that have been hit by cars, or injured birds may require humane intervention and cooperation with local animal control authorities. Whatever the reason, veterinarians are bound by oath to relieve animal suffering, but as noted previously, must also act in ways that protect research animals from those of an undefined health status. Consequently, institutions may want to develop policies that address situations where emergencies involving animals that are not university property require a prompt and humane intervention. The unwritten policy of our institution is to stabilize animals requiring critical care with the understanding that they will be transported by a county animal control officer immediately to a local veterinary hospital for subsequent care.

14:34 My laboratory animal facility is associated with a human hospital. What policy, if any, should our IACUC have relative to housing seeing eye or other animals that might accompany patients admitted into the hospital?

Opin. At our institution with a large medical center and several affiliated hospitals, seeing eye dogs and animals involved in pet-facilitated therapy programs are permitted in the wards and public areas of the hospital. These issues fall under the jurisdiction of hospital policy and practice and not under the oversight of the IACUC as the animals are privately owned and not used for research, testing, teaching, or training. Under the Americans with Disabilities Act, hospitals, as public accommodations, are expected to provide reasonable access to services and facilities (healthcare) for those using service animals,[28] but are not required to provide care or supervision while they are on site.[29] Likewise, at our medical center, pet-facilitated therapy is made available to patients through a relationship with a bonded, outside provider. The provider ensures that the privately owned animals participating in the program are of the appropriate age, size, and disposition, are acclimated to patients and the hospital environment, and are in apparent good health (including parasite control and full immunizations) as certified annually by a contract veterinarian.

References

1. Committee to Revise the Guide for the Care and Use of Agricultural Animals in Agricultural Research and Teaching, *Guide for the Care and Use of Agricultural Animals in Agricultural Research and Teaching*, 1st revised ed., Federation of Animal Science Societies, Savoy, IL, 1999, p. 4.
2. Office of the Federal Register, Code of Federal Regulations, Title 7; Part 371.4(b)(2), Animal and Plant Health Inspection Service–Veterinary Services, Washington, D.C., 1989 (revised 1998).
3. Office of the Federal Register, Code of Federal Regulations, Title 50; Part 17; Section 17.22, Permits for Scientific Purposes, Enhancement of Propagation or Survival, or for Incidental Taking of Endangered Wildlife, Washington, D.C., 1975 (revised 1997).
4. Office of the Federal Register, Code of Federal Regulations, Title 50; Part 23, Endangered Species Convention, Washington, D.C., 1977 (revised 1997).
5. Office of the Federal Register, Code of Federal Regulations, Title 9, Part 530, Extralabel Drug Use in Animals, Washington, D.C., 1997 (revised 1998).
6. Office of the Federal Register, Code of Federal Regulations, Title 42, Part 71, Foreign Quarantine; Subpart F, Importations; Section 51, Dogs and cats; Section 52, Turtles, tortoises, and terrapins; Section 53, Nonhuman primates; Section 54, Etiologic agents, hosts and vectors, Washington, D.C., 1985 (revised 1997).
7. Office of the Federal Register, Code of Federal Regulations, Title 29; Part 1910, Occupational Safety and Health Standards, Subpart Z, Toxic and Hazardous Substances; Section 1450(e), Chemical Hygiene Plan, Washington, D.C., 1990 (revised 1998).
8. Institute for Laboratory Animal Resources, National Academy of Sciences, Amphibians, in *Guidelines for the Breeding, Care and Management of Laboratory Animals*, Washington, D.C., 1974.
9. Institute for Laboratory Animal Resources, National Academy of Sciences, *Recommendations for the Care of Amphibians and Reptiles in Academic Institutions*, Washington, D.C., 1991.
10. Schaeffer, D.O., Kleinow, K.M., and Krulish, L., Eds., *The Care and Use of Amphibians, Reptiles, and Fish in Research*, Scientists Center for Animal Welfare, Greenbelt, MD, 1992.
11. American Society of Ichthyologists and Herpetologists, American Fisheries Society, American Institute of Fisheries Research Biologists, Guidelines for the use of fishes in field research, *Fish J.*, 13 (2), 1, 1987.

12. American Society of Ichthyologists and Herpetologists, The Herpetologists' League, and Society for the Study of Amphibians and Reptiles, Guidelines for the use of live amphibians and reptiles in field research, *J. Herpetol.* (Suppl) 4, 1, 1987.
13. Potkay, S., Garnett, N., Miller, J.G., Pond, C.L., and Doyle, D.J., Frequently asked questions about the Public Health Service Policy on Humane Care and Use of Laboratory Animals, *Contemp. Topics Lab. Anim. Sci.*, 36 (2), 47, 1997.
14. Division of Animal Welfare, Office for Protection from Research Risks, National Institutes of Health, *ILAR News*, 33 (4), 68, 1991.
15. Saif, Y.M. and Bacon, W.L., Protocol review. Simply stated, *Lab. Anim.*, 25(5), 22, 1996.
16. Swayne, D.E., Protocol review. Preventative measures, *Lab. Anim.*, 25 (5), 21, 1996.
17. Hogan, B., Beddington, R., Constantini, F., and Lacy, E., *Manipulating the Mouse Embryo: A Laboratory Manual*, 2nd ed., Cold Spring Harbor Laboratory Press, Plainview, NY, 1994, 117.
18. Tomson, F.N., personal communication, 1998.
19. Hogan, B., Beddington, R., Constantini, F., and Lacy, E., *Manipulating the Mouse Embryo: A Laboratory Manual*, 2nd ed., Cold Spring Harbor Laboratory Press, Plainview, NY, 1994, 120.
20. Wilkie, T.M., Brinster, R.L., and Palmiter, R.D., Germline and somatic mosaicism in transgenic mice, *Devel. Biol.*, 118, 9, 1986.
21. Garnett, N.L., personal communication, 1998.
22. Bayne, K.A., personal communication, 1998.
23. DeHaven, W.R., personal communication, 1998.
24. Garnett, N.L., OPRR and USDA/Animal care response on applicability of the animal welfare regulations and the PHS policy to dead animals and shared tissues, *Lab. Anim.*, 26(3), 21, 1997.
25. Office of the Federal Register, Code of Federal Regulations, Title 21, Part 511, New animal drugs for investigational use, Washington, D.C., 1975 (revised 1998).
26. American Veterinary Medical Association, Report of the AVMA Panel on Euthanasia, *J. Am. Vet. Med. Assoc.*, 202, 229, 1993.
27. Tannenbaum, J., personal communication, 1998.
28. Office of the Federal Register, Code of Federal Regulations, Title 28, Part 36, Nondiscrimination on the Basis of Disability by Public Accommodations and in Commercial Facilities; Section 301(a), Eligibility Criteria, Washington, D.C., 1991 (revised 1998).
29. Office of the Federal Register, Code of Federal Regulations, Title 28, Part 36, *Nondiscrimination on the Basis of Disability by Public Accommodations and in Commercial Facilities; Section 301(c)(1-2), Service Animals,* Washington, D.C., 1991 (revised 1998).

15
Animal Housing and Use Sites

Cynthia Gillett

Introduction

This chapter focuses on what is and is not acceptable housing or holding locations for animals used in research, teaching, and testing. The AWAR, *Guide*, and PHS Policy are the primary references utilized. There are few absolutes when considering animal housing and use sites. Professional judgment, institutional policy, and protocol goals and needs must all be considered when determining what is and is not an acceptable location for the care and use of animals.

Twenty-three individuals from throughout the U.S. and Canada, including the author, responded to a questionnaire about animal housing and use sites. Their responses are tabulated and summarized where appropriate to provide further insight into the breadth of policies currently in practice.

15:1 What is the definition of an animal facility?

Reg. The AWAR (§1.1) has several definitions. *Housing facility* is any land, premise, shed, barn, building, trailer, or other structure housing or intended to house animals. Separate definitions also are provided for *indoor housing facility, outdoor housing facility, sheltered housing facility,* and *research facility.*

PHS Policy (III,B) defines an animal facility as "any and all buildings, rooms, areas, enclosures, or vehicles, including satellite facilities, used for animal confinement, transport, maintenance, breeding, or experiments inclusive of surgical manipulation."

The *Guide* (pages 72 and 73) states that an animal facility consists of functional areas for animal housing, care, and sanitation; receipt, quarantine, and separation of animals, separation of species, or isolation of individual projects; and storage. Some facilities may encompass space for surgery, intensive care, necropsy, radiography, diet preparation, procedural space, clinical treatment, supply receiving, cage washing, carcass storage, facility administration, training, locker rooms, and break rooms.

Opin. The author's working definition of an animal facility is an integrated concept that is inclusive of the many programmatic aspects which relate to animal housing, care, and use in an institutional setting. However, for the purposes of this chapter, animal facility shall refer to the primary housing location and support space for animals and animal studies overseen by the IACUC. It is important to note that the AWAR definition of a Research Facility broadly encompasses institutions or persons who use or intend to use live animals in research, tests, or experiments. Thus, a Research Facility, as defined in the AWAR, should not simply be thought of as those areas where animals are housed or used, but the institution as a whole.

15:2 What are the circumstances and length of time that animals may be removed from their primary housing site?

Reg. The AWAR (§2.31,c,2) require that any location where animals are housed for more than 12 hours be inspected by the IACUC at least once every 6 months. PHS Policy (III,B) states that "a satellite facility is any containment outside of a core facility or centrally designated or managed area in which animals are housed for more than 24 hours." Nevertheless, PHS Policy II requires compliance with the AWA. Therefore, the PHS Policy definition of a satellite facility and the duration of stay requirements may be applied to species that are not covered by the AWAR, such as typical laboratory rats and mice.[1]

The *Guide* (page 9) states that the IACUC is responsible for oversight of animal-activity areas; however, it does not define that term. It also states (page 71) that "animals should be housed in facilities dedicated to or assigned for that purpose and should not be housed in laboratories merely for convenience. If animals must be maintained in a laboratory area to satisfy a protocol, the area should be appropriate to house and care for the animals."

Opin. The author's interpretation of the term "animal-activity area," as stated in the *Guide* (page 9) is that it implies research procedural areas such as a laboratory or testing site, etc. Neither the *Guide,* the AWAR, nor the PHS Policy prohibit animals from being removed from the animal holding facility for procedures or for housing elsewhere, such as an investigator's laboratory or a satellite facility. However, doing so may trigger additional regulatory oversight responsibility on the part of the IACUC or bring into play individual institutional requirements. Although one could apply different time standards based on whether or not an AWAR-covered species is involved, applying the most restrictive criterion (12 hours) will result in compliance in all situations.

Surv. Under what circumstances and for how long may animals be removed from their primary housing site?

- Less than 12 hours outside of animal holding facility permitted with no prior IACUC approval 9/23
- More than 12 hours out of the primary site becomes a satellite facility requiring justification to and prior approval by the IACUC and approval of husbandry plans and physical plant 10/23
- Up to 24 to 48 hours out of primary site permitted with no prior approval 1/23
- No overnight out of primary area housing allowed without IACUC approval 2/23
- Never allowed 1/23

15:3 What are examples of acceptable circumstances for keeping animals outside of the primary animal facility?

Opin. The AWAR, PHS Policy, and *Guide*, while providing definitions and oversight recommendations for animal facilities (including satellite facilities, see 15:1), do not provide examples of acceptable circumstances or rationale for keeping animals outside of the primary animal facility.

There are other considerations when housing animals outside of a centrally managed facility. For reasons of disease control, some facilities have a policy stating that if animals leave the facility they may not be returned. This is particularly true for microbiologically defined (specific pathogen free) rodents and transgenic mice. The method and route of transportation to the study area also must be monitored to minimize potential for disease transmission and for conformance with transportation guidelines.

Surv. What are some examples of acceptable circumstances and justification for keeping animals outside of the primary animal facility? More than one response per respondent is possible.

- Biohazard studies requiring specialized containment 3/17
- Necessary equipment cannot be easily relocated to the animal facility 3/17
- Behavioral or stress testing that would be affected by daily transport from animal holding room 3/17
- Postoperative monitoring 1/17

- Insufficient procedure space in animal facility 2/17
- Continuous administration of test substances via highly technical, automated procedures (e.g., automated administration of intravenous drugs) 2/17
- Frequent, repeated observations are part of protocol 3/17

15:4 Are there regulatory considerations regarding the actual act of transportation that the IACUC must consider when animals are moved from their primary housing site to a secondary site?

Reg. The AWAR (Part 3–Standards) has extensive transit regulations for each covered species. They include primary enclosures, primary conveyances, animal identification, food and water requirements, care in transit, handling, sanitation, ventilation, shelter, and temperature. The AWAR (§2.38,f) also covers the transit of animals as part of an adequate "handling" consideration. PHS Policy II requires compliance with the AWAR, where applicable.

The *Guide* (page 57) states that "All transportation of animals ... should be planned to minimize transit time and the risk of zoonoses, protect against environmental extremes, avoid overcrowding, provide food and water when indicated, and protect against physical trauma."

Opin. The AWAR on transport of animals apply to commercial transportation, but many of those requirements are common sense and designed to ensure the well-being of the transported animals. These regulations should be consulted when planning intrafacility and interfacility institutional animal transfers. While animal transport is not required to be directly reviewed and approved by the IACUC, transport activities should be reviewed during the semiannual program evaluation to ensure appropriate procedures are being followed. Animal transport issues also should be raised during protocol review when justification for off-site housing is being requested.

15:5 Are there practical, nonregulatory considerations regarding the actual act of transportation that the IACUC must consider when animals are transported from their primary housing site?

Reg. The AWAR (§2.38,f,1) states that handling of all animals shall be done as expeditiously and carefully as possible in a manner than does not cause trauma, overheating, excessive cooling, behavioral stress, physical harm, or unnecessary discomfort.

Opin. The practical, nonregulatory considerations which can arise during animal transport tend to be situational and are best handled by discussions with the laboratory animal facility professional staff, rather than with the IACUC. Examples include situations such as whether or not a box of mice can be taken outside to transport them to a laboratory in another building; whether or not a cardboard box can be used to transport a rabbit to the laboratory; or which elevator can be used to transport animals to a laboratory.

The answers to these and similar issues depend on factors such as climate, time of year, facility design, institutional policies, etc. The IACUC, during the course of its semiannual facility inspection, may come across evidence of unacceptable transportation practices (e.g., a soiled transport enclosure, inappropriate elevator or corridor use) which can then be addressed with the PI's and laboratory animal care staff.

15:6 If animals are being housed in a research laboratory setting, are there species differences to consider?

Reg. The housing requirements for various species are addressed in Part 3 of the AWAR. These standards apply if the animals are approved by the IACUC to be "housed" in a laboratory. The IACUC may make exceptions to these standards if there is scientific justification to do so (AWAR §2.31,c,3).

Opin. The need for laboratory housing should be protocol driven and not motivated by convenience or economics. A comprehensive husbandry plan for the animals should be submitted to the IACUC. The IACUC should keep records of approved laboratory housing sites and perform, or delegate, periodic inspections of these areas when animals are present, in addition to the scheduled semiannual facility inspections.

Surv. If animals are permitted to be housed in a research laboratory setting, are there species differences to consider? List all species that can be housed outside of your facility.

Survey respondents indicated that decisions on laboratory housing were made on a case-by-case basis, with nonmammals (e.g., fish, birds) being more likely to be permitted to be housed in a laboratory setting.

- Rodents can be housed in a laboratory   6/20
- Nonhuman primates can be housed in a laboratory   2/20
- Nonmammals can be housed in a laboratory   11/20
- Rabbits can be housed in a laboratory   2/20
- Species differences not a factor, case-by-case assessment   3/20

15:7 May animals be housed at sites not belonging to your institution? Under what circumstances?

Reg. The AWAR (§2.38,i) state, "If any research facility obtains prior approval of the APHIS, AC Regional Director, it may arrange to have another person hold animals: provided that:

1. The other person agrees, in writing, to comply with the regulations in this part and the standards in Part 3 of this subchapter, and to allow inspection of the premises by an APHIS official during business hours
2. The animals remain under the total control and responsibility of the research facility
3. The Institutional Official agrees, in writing, that the other person or premises is a recognized animal site under its research facility registration."

APHIS/AC suggests that APHIS Form 7009 should be used for approved off-premises housing sites.

PHS Policy (IV,B) requires that awardee institutions and other participating institutions that receive PHS support have an approved Assurance on file with NIH/OPRR. NIH/OPRR has stated that vendors supplying custom antibodies to investigators must either have an Animal Welfare Assurance on file with NIH/OPRR or must be included in the receiving research institution's Assurance statement.[2] Additionally, if AWAR-covered species, such as rabbits, are used to produce antibodies at the vendor's site or are in other research studies, then the vendor must be registered with APHIS/AC as a research facility. APHIS/AC Policy #10 clarifies the need for licensing or registration of a facility using covered species for antibody production.[3] (See 8:11.)

Opin. The *Guide* does not address this issue. Additional concerns include the need or desirability for the noninstitutional site to be either APHIS/AC registered, have an Assurance on file with NIH/OPRR, or be AAALAC accredited. These considerations are dependent on the primary institution's status and on the individual housing situation. Affiliate Veterans Administration Medical Center facilities are an example where housing animals at a noninstitutionally owned site might be routinely permitted. In the latter example, this should be listed as a "site" of the research institution.

Surv. A May animals be housed at sites not belonging to your institution?

- Never allow it 2/18
- Allow it if justified to and approved by the IACUC 15/18
- Allow it with minimal prerequisites other than inclusion in the IACUC semiannual inspections 1/18

Surv. B Under what circumstances can animals be housed at sites not belonging to your institution? More than one response may be given.

- Inability to house the animals at the institution due to space constraints or lack of specialized housing (e.g., farm animals, nonhuman primates) 6/9

- Collaborations with neighboring biomedical institutions, zoological parks, marine biological centers 5/9

15:8 Are there any circumstances under which animals may be housed in a person's home or on a private farm?

Reg. Regulatory requirements are the same as for 15:7. The home or farm must be listed as a site of the institution.

Opin. Prior approval by the IACUC for private housing is considered by surveyed respondents to be essential. When animals remain privately owned, as opposed to institutionally owned, they most often remain housed in that person's home or farm, e.g., blood donor animals or veterinary clinical studies on companion and farm animals.

Surv. What are the circumstances under which animals may be housed in a person's home or on a private farm?

- Private party housing was not applicable to their situation or not allowed 6/16
- Large animals housed at a private farm under contract 7/16
- Fish and songbirds housed privately 2/16
- Lease-back agreements with private farms or homes; land is leased back to the university during the research project, maintaining site authority for the university 2/16

15:9 If animals are wild-trapped, are there conditions the IACUC should place on that activity in terms of type of trap, frequency of checking the trap, and euthanasia of injured or to-be-collected animals?

Reg. The AWAR definition of animal (§1.1; see 12:1) includes most warm-blooded mammals; therefore, wild mammals are subject to IACUC protocol review. PHS Policy II requires compliance with the AWA, as applicable. AWAR §2.31,f,1 also applies for handling wild animals. The IACUC may make exceptions to these standards if there is scientific justification to do so (AWAR §2.31c,3).

The *Guide* (page 5) states that "investigators conducting field studies with animals should assure their IACUC that collection of specimens or invasive procedures will comply with state and federal regulations and this *Guide*."

The PHS Policy (III,A) defines an animal as any live, vertebrate animal, and therefore, all vertebrate animals used in field studies fall under this Policy.

Opin. Since both the AWAR and PHS Policy include wild animals in their definition of an animal, the IACUC is required to review field studies and animal trapping for appropriate animal care and use. Thus, protocols must provide sufficient detail on methods to be used (e.g., tagging, collaring, blood collections, euthanasia), frequency of observations, and a contingency plan for animals hurt in the collection process. There are several references providing advice on review of field studies.[4,5] One such book[4] reviews protocol aspects which IACUCs should assess for field research. It discusses species selection, site selection, and methodologies to be employed. (See 16:21.)

Surv. What conditions should the IACUC place on field activities in terms of type of trap, frequency of checking the trap, and euthanasia of injured or to-be-collected animals? More than one response is possible.

- Not applicable due to no field studies   5/19
- Prior IACUC approval required   14/14
- For small mammals — setting traps in evening and checking in the morning   3/14
- For mist traps — constant observation and immediate checking upon capture   1/14
- Frequent observation, daily at minimum   6/14
- Food and water should be available, depending on timing   2/14
- Shade or cover should be available   1/14
- Trap is appropriate size for the animal   2/14
- Traps disabled when not in active use   1/14
- Identification systems atraumatic (i.e., minimize pain and discomfort and not likely to result in subsequent injury or increased predation)   1/14
- Researcher safety (i.e., personnel made aware of potential zoonoses and preventive actions, properly trained in handling to minimize stress to animal and maximize personnel safety)   3/14
- Euthanasia provisions (i.e., animals will be euthanized if required by study or due to accidental injury, and method meets field study guidelines)   5/14
- Full description of trap function and use of least-stressful trap available   4/14
- Permits if necessary   2/14
- Case-by-case analysis   2/14

15:10 Should field sites be included in the semiannual IACUC inspections?

Reg. The AWAR (§2.31c,2) state that "animal areas containing free-living wild animals in their natural habitat need not be included in such inspections." Neither the *Guide* nor PHS Policy address the issue of inspecting field sites.

PHS Policy does not specifically discuss wild animals or field studies. Since PHS Policy III,A applies to all live vertebrate animals, field sites could be considered to be animal study areas by the NIH/OPRR and, therefore, require semiannual inspection by the IACUC.

Opin. It is the author's opinion that the question of whether field study sites are to be included in the IACUC semiannual inspection must be decided on a case-by-case basis. Strict interpretation of PHS Policy requires inclusion, but in practical terms, very few field studies are funded by the PHS. Certainly, field study sites which include housing enclosures should technically be inspected, as the animals involved are no longer truly free-living and are dependent on adequate monitoring and oversight. The IACUC should try to inspect such sites when they are actually in use. However, practical issues must also be considered. If the field site is geographically distant, it may not be feasible to send an inspection team. Alternative review methods for such sites can include photographs, videos, and submission of comprehensive husbandry procedures.

Surv. Does your IACUC include field sites in its semiannual IACUC inspections?

- Not applicable 5/19
- Yes 1/14
- No 5/14
- Yes, if close by 1/14
- Yes, if institutionally owned or affiliated 1/14
- Yes, if institutionally owned or affiliated and there is an invasive procedure 1/14
- Yes, if invasive procedures are used 3/14
- Yes, if there is a temporary holding enclosure 2/14

15:11 Are animals that are used in field studies included on the APHIS/AC Annual Report?

Reg. AWAR-regulated species should be reported as noted in the AWA (§13,7,A, requirement for an annual report from research facilities) and the AWAR (§2.36, also a requirement for an annual report). The AWAR (§2.36,b,4) do not exclude animals used in field studies and the AWAR definition of an animal (§1.1) does not exclude wild animals (and, therefore, does include wild rats and mice such as a deer mouse).

Opin. The answer to the above question seems straightforward, but because field studies tend to have less direct oversight by way of

IACUC inspections, animal ordering, etc., including the animals on the APHIS/AC annual report, was occasionally not done by some survey respondents. However, the institution and the IACUC should provide a mechanism for reporting (on the APHIS/AC annual report) the AWAR-regulated species used in field studies. For example, the IACUC can send each investigator having an approved field study protocol an annual request for summary information on the number and species used during the reporting period (October 1 to September 30). These numbers can then be compiled for the institutional Annual Report which is due before December 1 of each year.

Surv. Are animals used in field studies included on your APHIS/AC Annual Report?

- Not applicable 5/21
- Include AWAR-covered species used in field studies 9/16
- Include field study animals only if euthanized during the study 3/16
- Do not include field study animals in annual report 4/16

15:12 With reference to housing requirements, should the IACUC make an attempt to proactively distinguish between biomedical research and agricultural research when reviewing proposals using large farm animals?

Reg. The definition of an animal in the AWAR (§1.1) excludes "... farm animals, such as, but not limited to, livestock or poultry used or intended for use as food or fiber, or livestock or poultry used or intended for use for improving animal nutrition, breeding, management, or production efficiency, or for improving the quality of food or fiber." (See 12:1.)

The *Guide* (page 4) states that "housing systems for farm animals used in biomedical research might or might not differ from those in agricultural research. Animals used in either biomedical or agricultural research can be housed in cages or stalls or in paddocks or pastures. ... The protocol, rather than the category of research, should determine the setting (farm or laboratory)."

PHS Policy II does not make reference to a particular type of research. It states that it is applicable to all PHS-conducted or -supported activities involving animals. Section 495a of the HREA, upon which the PHS Policy is based, does refer to "animals to be used in biomedical and behavioral research" which, by inference, might be construed as not applicable to agricultural research. Nevertheless, it is prudent to follow PHS Policy II.

Opin. The AWAR language has become the working definition of agricultural research. The unstated converse is that farm animals used in studies whose goal is the advancement of biomedical science *are* regulated by the AWAR. Therefore, the practical reason why an IACUC might wish to distinguish between biomedical and agricultural research is that the latter need not be included in AWAR requirements such as annual reporting, search for alternatives, etc. and need not conform to recommendations in the *Guide* with particular reference to the space and physical plant recommendations that are focused on biomedical research animals. (See 15:13.)

15:13 What criteria are used to distinguish biomedical from agricultural research? How are the standards different for biomedical research housing sites vs. agricultural research housing sites?

Reg. (See 15:12.)

Opin. It is clear that farm animals such as sheep, cows, goats, swine, and horses used in biomedical research are covered by both the AWAR and the *Guide*. It is also clear that the same animals used for research on livestock production are not ordinarily covered by the AWAR. PHS-funded agricultural projects (which would be covered by the *Guide*) are very unlikely, given that the focus of PHS funding is biomedical research.

What is not clear are the circumstances under which farm animals used in biomedical research must be housed according to the *Guide*'s recommendations as compared to when they may be housed in more farm-like settings (such as paddocks, pastures, and barns). The *Guide* states that the housing system chosen should be IACUC protocol driven. This author interprets that to mean that farm animals used in studies in which the minimization of variables is vital to the study's outcome (e.g., transplant research) should be housed in the more environmentally controlled setting of a traditional laboratory animal facility. Conversely, farm animals intended for use in studies for which each animal's environment need not be finely controlled (e.g., antibody production) may be housed in farm-type settings.

Surv. What criteria are used to distinguish biomedical from agricultural research? How are the standards different for biomedical research housing sites vs. agricultural research housing sites?

• Use AWAR definition for agricultural research	2/9
• If the end goal is to benefit human health, it is biomedical research	2/9
• If end goal is to affect production practices, it is agricultural research	2/9
• The funding agency determines the category of research, e.g., NIH funding means biomedical research, Pork Producers Council means agricultural research	2/9

- Housing standards for agricultural research are those of a well-managed farm 3/9
- Housing standards for agricultural research are those in the *Guide for the Care and Use of Agricultural Animals in Agricultural Research*[6] 3/9
- Housing standards are protocol driven, biomedical research animals could be housed in farm setting and agricultural research animals may need controlled environment of biomedical research 2/9

Note: One IACUC had an agricultural animal subcommittee, while another institution had a separate IACUC for wildlife and agricultural research.

15:14 Is it acceptable to house different animal species in the same holding room?

Reg. AWAR adequate veterinary care standards (§2.33,b,1; §2.33b,2) require appropriate facilities for housing animals and appropriate methods to be used to prevent diseases.

The AWAR (§3.33,b) prohibit the co-housing of hamsters or guinea pigs in the same primary enclosure with any other species, and §3.58,a does the same for rabbits. The AWAR (§3.33,c) require the separation of guinea pigs or hamsters, that are under quarantine or treatment for a communicable disease from other guinea pigs, hamsters, or other susceptible species. The AWAR (§3.58,b) provides the same protection for rabbits. The AWAR (§3.7,d) also prohibit the cohabitation of dogs or cats with other species in the same primary enclosure unless they are compatible. Similarly, AWAR §3.133 (which covers animals other than dogs, cats, guinea pigs, hamster, rabbits, nonhuman primates, and marine mammals) states that "animals housed in the same primary enclosure must be compatible. Animals shall not be housed near animals that interfere with their health or cause them discomfort."

PHS Policy is silent on the subject of mixing species, although PHS Policy II requires compliance with the AWAR, where applicable.

The *Guide* (page 22) states that "the environment in which animals are maintained should be appropriate to the species, its life history, and its intended use." The *Guide* (page 58) also states that "physical separation of animals by species is recommended to prevent interspecies disease transmission and to eliminate anxiety and possible physiologic and behavioral changes due to interspecies conflict. … In some instances, it might be acceptable to house different species in the same room, for example, if two species have a similar pathogen status and are behaviorally compatible."

Opin. Professional judgment usually dictates that mixing of certain species in the same room is inappropriate to their life histories (e.g., cats and dogs). The potential for disease and parasite transmission is an additional consideration regarding housing different species in the same room or enclosure. As examples of recommended separate housing by species, the *Guide* (page 59) notes that nonhuman primates should be separated by geographical origin and rabbits should be housed separately from guinea pigs.

The NIH has an Intramural Position Paper on Housing Multiple Species of Large Laboratory Animals.[7] The content is directly applicable only to the NIH campuses; they are not recommendations promulgated by the *Guide* or any regulatory agency.

Surv. Which species do you house together in the same holding room?
Note: Several respondents commented that mixing species is a veterinary, not an IACUC, decision. Institutions not responding to housing certain species together may have not done so because they do not house that species at all, or they have not encountered a need to house species together.

- Species mixing in an animal holding room prohibited — 2/16
- Rodent species, particularly rats and mice housed in Microisolator™-type cages — 8/16
- Amphibians and reptiles, e.g., frogs and turtles, multiple amphibian species — 3/16
- Ungulates, such as goats and sheep — 3/16
- Pigs and sheep, pigs and dogs — 3/16
- Multiple macaque species, e.g., cynomolgus, rhesus, stump tail — 4/16
- Baboons and macaques — 2/16
- Guinea pigs and chinchillas — 1/16
- Ferrets and cats — 1/16

References

1. Division of Animal Welfare, Office for Protection from Research Risks, National Institutes of Health, The Pubic Health Service responds to commonly asked questions, *ILAR News*, 33:4, 69, 1991. Available on the World Wide Web at: http://www.nih.gov/grants/oprr/ilar91.htm

2. Potkay, S., Garnett, N., Miller, J., Pond, C., and Dole, D., Frequently asked questions about the Public Health Service policy on humane care and use of laboratory animals, *Lab Anim.*, 24(9), 24, 1995. Also see *OPRR Reports*, #95-02, March 8, 1995. Available on the World Wide Web at: *http://www.nih.gov/grants/oprr/dc95-3.htm*
3. U.S. Department of Agriculture, Animal and Plant Health Inspection Service, Policy #10. Licensing and registration of producers of antibodies, sera and/or other animal parts and pregnant mare urine (PMU), April 14, 1997. Available on the World Wide Web at: *http://www.aphis.usda.gov/ac/policy10.html*
4. U.S. Department of Health and Human Services, Public Health Service, National Institutes of Health, Institutional Animal Care and Use Committee Guidebook, NIH Publication No. 92-3415, 1992, Chap. B.
5. American Society of Mammalogists, Acceptable field methods in mammalogy: preliminary guidelines approved by the American Society of Mammalogists, *J. Mammal.* 68(4), Suppl., 1, 1987. Available on the World Wide Web at: *http://asm.wku.edu/announcements/*
6. Committee to Revise the Guide for the Care and Use of Agricultural Animals in Agricultural Research and Teaching, *Guide for the Care and Use of Agricultural Animals in Agricultural Research and Teaching*, 1st revised ed., Federation of Animal Science Societies, Savoy, IL, 1999, 4. (Available from the Federation of Animal Science Societies, 1111 N. Dunlap Ave., Savoy, IL 61874.)
7. National Institutes of Health, Intramural Position Paper on Housing Multiple Species of Large Laboratory Animals. Available on the World Wide Web at: *http://oacu.od.nih.gov/arac/index.htm*

16
Pain and Distress

Paul Flecknell and Jerald Silverman

Introduction

A great diversity of views are held concerning the acceptability of using animals in biomedical research, but many agree that it is desirable to reduce to a minimum any pain or distress associated with that research. This approach has been described most fully in Russell and Burch's classic text, *The Principles of Humane Experimental Technique*,[1] and then popularized as the "3 Rs" of animal experimentation: **R**eduction of the number of animals used, **R**eplacement of animals with nonsentient alternatives (or with human subjects), and **R**efinement of experimental design to minimize pain and distress. Aside from its attraction on ethical grounds, the alleviation of unnecessary pain or distress also may improve the quality of scientific data obtained. Pain and distress may represent uncontrolled experimental variables that can have significant effects on research. Reducing the magnitude of this variation, therefore, can benefit both the welfare of the animals used and the scientific output from the project. It follows that a consideration of methods of minimizing or eliminating pain and distress should be of central concern to IACUCs.

In many instances we lack the scientific data necessary to make objective judgments concerning the significance of animal pain and distress. Often it is necessary to weigh the available evidence and try to strike a balance between a concern for animal welfare and the requirements of a particular research project. When attempting to achieve this balance in circumstances where there is limited data, it is helpful to establish whether adoption of measures thought likely to improve animal well-being will interfere with a study, or will simply require modest additional work on the part of an investigator. If measures that may improve animal welfare can be implemented without compromising the scientific integrity of a project, then it is preferable for this to be done.

16:1 What is the difference between stress and distress?

Opin. Many attempts have been made to differentiate stress and distress[2,3] and these vary in their approach. In general, stress is a normal biological event, whereas distress has some aversive or unpleasant component. Stress arises as a result of stressors in the animals' environment; for example, a change in environmental temperature. Normal homeostatic mechanisms are triggered and the animal adapts to the changes and restores its system to a normal state. These adaptive responses occur continuously and are a feature of all living organisms. As the effort required to adapt to stressors increases, a point may be reached where the animal fails to adapt fully, and this is considered by some authors to represent distress. Other opinions require that the animal becomes aware of the effort of its adaptive responses, and perceives this effort as something it wishes to avoid, leading to a subjective state of "distress."

16:2 What is the difference between distress and suffering?

Opin. Whether there is a difference between distress and suffering depends primarily upon how distress and suffering are defined. These definitions are still debated and, at present, IACUCs should determine which working definitions are acceptable to them. Useful discussions can be found in References 1, 4, and 5. (See 16:1.)

One possible definition is that distress is frequently used as a phrase to encompass the state produced by exposure to stressors such as excess heat, cold, or lack of food (see 16:22). Those who advocate use of the term "suffering" usually require that the animal perceives the threat these stressors pose to its integrity, and often include a temporal dimension to this emotion requiring the state to last for some time before it is defined as suffering.

16:3 What is the difference between pain and nociception?

Reg. Pain is not directly defined in the AWAR, but a painful procedure (§1:1, Painful procedure) is defined as one "that would reasonably be expected to cause more than slight or momentary pain or distress in a human to which that procedure is applied. ..." The example provided is pain in excess of that caused by injections or other minor procedures. Nociception is not defined. Neither the PHS Policy nor the *Guide* provides definitions of pain or nociception, although the definition of a painful procedure in U.S. Government Principle IV, which is incorporated into the PHS Policy, is similar to the AWAR definition.

Opin. Nociception represents only the detection of potentially damaging stimuli, and does not include the subjective experience of these stimuli. Pain in humans is described as having both a sensory and emotional component. For example, the IASP[6] definition is as follows:

> Pain is an unpleasant sensory and emotional experience associated with actual or potential tissue damage or described in terms of such damage.

Defining animal pain is difficult because of uncertainties relating to the emotional component of pain in nonhumans. This has been avoided in some definitions by interpreting pain in relation to its effects on animal behavior. For example:

> Pain in animals is an aversive sensory experience that elicits protective motor actions, results in learned avoidance, and may modify species-specific traits of behavior, including social behavior.[7]

A useful summary of nociception and pain assessment and recognition is given in the guidelines of the Association of Veterinary Teachers and Research Workers.[8]

16:4 What should be the minimum expectations of the IACUC related to the relief of pain and distress?

Reg. The IACUC, under the AWAR (§2.31,d,1,iv,A–§2.31,d,1,iv,C; §2.31,d,1,v) must assure that procedures causing more than slight pain or distress will be performed with appropriate sedatives, analgesics, or anesthetics unless certain specific criteria are met (e.g., IACUC-approved scientific justification to withhold analgesia). They must involve the AV or the AV's designee in their planning, and not use paralytics without anesthesia. Animals experiencing severe or chronic pain or distress that is not alleviated must be euthanized during or at the end of the study.

The PHS Policy (IV,C,1,a–IV,C,1,c) is worded similarly to the AWAR. PHS Policy (IV,D,1,d) notes that applications and proposals to the PHS must contain a description of the procedures that are designed to assure that discomfort and injury to animals will be limited to that which is unavoidable in the conduct of scientifically valuable research, and the proper drugs will be used where indicated to minimize animal pain and discomfort. U.S. Government Principles IV to VI largely reiterate the AWAR and PHS Policy.

The *Guide* (page 10) states that the IACUC should consider the use of appropriate sedation, analgesia, and anesthesia for animals. It also notes (page 64) that the recognition of pain in different species is a key to its prevention or alleviation, and that professional judgment should be used to determine the appropriate analgesics or anesthetics. The *Guide* (page 65) warns that some drugs, such as sedatives,

Opin. anxiolytics, and neuromuscular-blocking agents are not analgesics or anesthetics.

Opin. Causing pain and distress are considered undesirable and form the basis for much public disquiet about the use of animals in research. It is appropriate for an IACUC to require all protocols to be designed to avoid or minimize pain or distress. It also may be considered appropriate for an IACUC to require investigators to demonstrate that they have considered all appropriate alternatives in their use of animals.[9] Each of the so-called "3 Rs" can result in a reduction in pain and distress.[1] Alternatives that **R**eplace the use of sentient animals largely eliminate pain and distress since they use nonsentient alternatives, e.g., tissue culture. **R**eduction alternatives (e.g., appropriate statistical methodology leading to a reduction in total animal use) result in fewer animals potentially experiencing pain and distress. Finally, **R**efinement alternatives seek to reduce to a minimum the pain and distress experienced by those animals that are still necessary to use. Perhaps it is this last area that the IACUC will see as its greatest concern, and it should review not only the use of analgesic and anesthetic drugs, but also the training and competency of those involved in the procedures. For example, an inexperienced investigator may handle an animal with excessive force and cause significant pain or distress which could have been avoided by proper training. It also is likely that an investigator skilled in surgical procedures will carry out a project using fewer animals (because of fewer technical failures), causing less postoperative pain (because of less tissue trauma), with more rapid recovery (because of shorter anesthesia time), than an inexperienced or less competent investigator. All of these factors should be integrated into the IACUCs' expectations when assessing the efforts an investigator intends to make to reduce pain and distress.

16:5 What alternatives must be considered for protocols which will involve pain or distress? How can this be done?

Reg. The AWAR (§2.31,d,1,ii) require the PI to consider alternatives to procedures that may cause more than momentary or slight pain or distress to the animal. The PI must provide a written description of the methods and sources used to determine that alternatives are not available. This requirement is clarified in APHIS/AC Policy #12[10] which gives examples of sources that can be used (e.g., Medline, Animal Welfare Information Center). Policy #12 states that the minimal written narrative should include the databases or other sources used, the date of the search, the years covered by the search, and the key words or search strategy used.[10] More important, Policy #12 notes that Reduction, Replacement, and Refinement (see 16:4) must be addressed, not just animal replacement.

The *Guide* (page 10) states that the availability or appropriateness of the use of less invasive procedures, other species, isolated organ preparations, cell or tissue culture, or computer simulations should be considered.

U.S. Government Principle III, which is part of the PHS Policy, states mathematical models, computer simulation, and *in vitro* biological systems should be considered. (See 12:15 to 12:19.)

Opin. From 16:4 it can be seen that all protocols which have the potential to cause pain and distress should be reviewed with reference to reduction, replacement, and refinement. This requires significant effort on the part of both the IACUC and the investigator. Investigators should at least explain and justify their use of animals, indicating what literature sources they have reviewed to show that there is no practical alternative to animal use. As the number of databases on alternatives grows, reference to searches on these can be included in a submission.

Investigators also should demonstrate to the IACUC that an appropriate experimental design has been used which maximizes the information gained and minimizes the number of animals used. It is helpful to indicate whether expert advice from a statistician has been sought, particularly when the experimental design is complex. A useful practical approach to estimating numbers of animals required has been published.[9] Finally, a range of options are available for reducing pain and distress, and those appropriate to the investigation should be considered. For example, it might be possible to conduct a study entirely under general anesthesia or reduce pain by using appropriate analgesics. It also may be possible to limit the distress experienced by animals, or the duration of that distress, by requiring defined endpoints for a study (e.g., placing upper limits on total tumor burden in studies of carcinogenesis, or defining a set of criteria that will be used to determine when an animal should be removed from study and humanely killed).

16:6 Since the actual assessment of pain cannot be made until after a procedure is performed, is it appropriate for the IACUC to attempt to assess the potential for pain when first reviewing a protocol?

Reg. (See 16:3 for the definition of a painful procedure.)

Opin. Although the degree of pain cannot be determined for any particular individual animal until after a procedure has been performed, it is frequently possible to make an informed estimate of the likely degree of pain that may be caused. If the procedure, or a similar procedure, has been carried out previously at the institution, then this can be used to help predict the likely consequences. If the technique is new to the institution, then colleagues at other institutions where the procedure has been performed might be consulted. Extrapolations of pain experienced by humans undergoing similar procedures can act as a rough guideline. There should be few occasions when some indication cannot be obtained as to the likely consequences in terms of pain and distress. This initial assessment should certainly be sufficient to determine, for example, what level of analgesic use would be appropriate, and what type of aftercare the animal might require. In all instances, measures for the control of pain require monitoring of animals to ensure they are effective.

16:7 Can the IACUC approve protocols in which animals will experience pain or distress not relieved by the use of anesthetics or analgesics? Under what circumstances?

Reg. If an investigator can justify, in writing, that withholding anesthesia or analgesia is required for scientific reasons, and will only continue for the necessary period of time, then the IACUC can potentially approve such an activity (AWAR §2.31,d,1,iv,A). The PHS Policy (IV,C,1,b) has similar wording.

Opin. IACUCs may frequently be asked to approve protocols in which animals experience pain or distress which cannot be alleviated by the use of anesthetics or analgesics. For example, animals used in studies of arthritis or inflammatory conditions may experience pain, but analgesic administration may interfere with the protocol to such an extent that its use would invalidate the data obtained. Pain and distress also may occur in animal models involving neoplasia, chronic organ failure, infectious diseases, toxicity testing, and a wide range of other circumstances. In many of these studies, alleviation of pain and distress by pharmacological means may not be possible because of interactions between the drugs used and the research protocol. In these circumstances, efforts must be made, whenever possible, to reduce the number of animals used in such a study, replace the study with alternative techniques that do not cause pain, and to refine the study design so that pain and distress are minimized. Although these basic tenets should be applied to all protocols, they clearly are of particular importance in circumstances where significant pain and distress can be anticipated.

16:8 What are some typical criteria an IACUC or an investigator can use to determine if an animal is in pain or distress?

Opin. Most attempts to recognize pain or distress in animals rely on observing a combination of behavioral and physiological variables, and looking for deviations from normality. For example, an animal in pain may change its spontaneous behavior so that it becomes less active, decreases grooming activity, and reduces its food and water consumption. It also may change its responses to handling, for example, by increased aggression. Conversely, some animals may become apathetic and unresponsive to handling. The various criteria that might be used have been discussed at length by various authors.[3,8,11] See Reference 12 for an extended literature review.

The key points to note are that signs of pain and distress may be very subtle and their recognition almost always requires a detailed knowledge of the normal behavior of the animal species. Recognition also requires that sufficient time be allocated for observation of the animal and, in some circumstances, it may be necessary to observe the animal's behavior in such a way that it is unaware of the presence of the observer, e.g., by using a video camera.

16:9 An IACUC protocol indicates that a procedure is moderately painful to its canine subjects for approximately 1 hour, then never repeated. During that time, pain-relieving drugs cannot be used, but an alternative pain relief mechanism will be used. That is, calming music, a darkened room, and human petting will constantly occur. The technique has not been previously attempted. Should the IACUC consider this to be alleviated or unalleviated pain?

Reg. See 16:3 for the definition of a painful procedure and 16:4 for related regulatory information. The AWAR (§2.36,b,7) specifically require painful procedures, for which appropriate anesthetic, analgesic, or tranquilizing drugs are not used, to be placed in Category E on the APHIS/AC Annual Report. The *Guide* (page 65) notes that in addition to anesthetics, analgesics, and tranquilizers, nonpharmacological control of pain is often effective.

Opin. Practical experience in clinical veterinary practice has shown that many companion animals, especially dogs and cats, respond positively to human contact. For example, it is well accepted that good nursing, which includes stroking and verbal reassurances, plays a role in reducing pain and distress in the postoperative period. This accords with human clinical experience, in which it has been shown that the emotional state of the patient influences the degree of pain experienced and the analgesic dose required to alleviate that pain. Practical evidence that similar mechanisms can be involved in animal pain is provided by the changes in behavior of animals that are believed to be in pain postoperatively when they are given nursing attention. Such attention may silence an animal that is vocalizing, may trigger eating or drinking, or may result in a previously agitated animal becoming calm and resting, or sleeping. Obtaining a positive response to such contact depends critically upon the previous experience of the animal and the nursing skills of the personnel involved. If an animal has not been adequately socialized to accept (and welcome) human contact, then attempts to provide reassurance may be counterproductive and increase the animal's distress. If such a technique is to be considered as a replacement for conventional pain-alleviating techniques, then it is important that the IACUC require the personnel involved and their previous relationship with the animals used to be specified. The personnel should have experience with the techniques, and the technique should (ideally) be familiar to the animals. It is important to carry out a pilot study to ensure that the methods proposed to reduce pain or distress are adequate. Finally, it is appropriate to consider other, nonpharmacological methods of pain relief, such as acupuncture or transcutaneous nerve stimulation.

16:10 Can ketamine alone or with xylazine be considered adequate anesthesia for major surgery, such as abdominal surgery in laboratory rodents, rabbits, or cats?

Opin. Ketamine, when used alone, immobilizes most animals and produces some analgesia. The degree of analgesia, and degree of immobilization, varies considerably in different species.[13] Ketamine is relatively ineffective in providing analgesia for the viscera,[14] and when used alone does not provide adequate anesthesia for abdominal surgery in rodents, rabbits, cats, or indeed any species except (possibly) nonhuman primates.[15,16] The degree of muscle relaxation is so poor that this often renders ketamine unsuitable even for superficial procedures. The addition of drugs with sedative effects such as xylazine, medetomidine, acepromazine, or diazepam greatly improves the quality of anesthesia. In general, combining ketamine with drugs which have analgesic and sedative properties (e.g., xylazine and medetomidine) produces more effective surgical anesthesia than combinations such as ketamine/acepromazine or ketamine/diazepam.[13] When combined with xylazine (or medetomidine), ketamine provides surgical anesthesia in most strains of rodents, and in rabbits and cats.[17,18] The degree of analgesia provided by ketamine/medetomidine or ketamine/xylazine is usually sufficient for abdominal surgery.

16:11 Is chloralose considered to be an anesthetic or a hypnotic?

Opin. Chloralose is a hypnotic like many other agents that are used to anesthetize animals (e.g., pentobarbital). It is often assumed, incorrectly, that because a drug is hypnotic it cannot be used to provide surgical anesthesia. The definition of hypnosis is "a condition of artificially induced sleep, or a state resembling sleep, resulting from moderate depression of the central nervous system from which the patient is readily aroused."[19] Although drugs classed as hypnotics produce this effect as the dosage is increased, progressively greater depression of the CNS occurs, so that animals progress from sedation to hypnosis to general anesthesia. Some hypnotics cause such severe cardiovascular depression at the doses necessary to achieve general anesthesia that they cannot be used in this way without risking the death of the animal. Others, such as chloralose and the barbiturates, can be given at doses sufficient to produce general anesthesia in many species. Different species, and different strains of the same species, vary in their response and, in some circumstances, the depth of anesthesia will be inadequate for surgery. Providing the anesthetic depth is assessed (e.g., by evoking a response to a painful stimulus such as a toe pinch), then dose rates can be adjusted as necessary or additional analgesia can be provided by drugs such as morphine or fentanyl.[20]

16:12 Should the IACUC request special safety precautions when urethane is used as an anesthetic?

Opin. Urethane is carcinogenic[21] and, if it is to be used as an anesthetic, appropriate precautions should be adopted to avoid any safety hazard to personnel. The exact nature of the precautions may vary, but they generally mirror those required to ensure safe handling and use of carcinogens. The IACUC should consider utilizing the services on the institution's Biosafety Committee. (See Chapter 20.)

16:13 Should the use of urethane be allowed as an anesthetic for recovery surgery?

Opin. Although urethane is a useful long-term anesthetic, it has many side effects which usually preclude its use for recovery anesthesia. The dose used for anesthesia is sufficient to cause neoplasia in some species (see 16:12); the drug is an irritant when given intraperitoneally and recovery can be very prolonged. Urethane also causes hemolysis.

16:14 Is a measured decrease in food or water consumption a reasonable indicator of pain in laboratory rodents?

Opin. Numerous factors can influence food and water consumption in rodents, and pain is only one of these. If a rodent reduces its food and water intake, pain should be included as a possible cause along with infectious disease processes, changes in environment, alteration in the husbandry regimen, and other factors. Nevertheless, in the postoperative period, immediate changes in food and water consumption seem especially useful as indicators of postoperative pain.[22,23] Following surgery, rats and mice almost invariably show a small fall in body weight (between 5 and 15%, depending upon the nature of the surgical procedure) as a consequence of reduced food and water consumption. If analgesics are provided, this fall in consumption is reduced. Since both opioids (e.g., buprenorphine) and nonsteroidal antiinflammatory drugs (e.g., carprofen) have this effect, and since administration of these drugs to normal animals does not increase food and water consumption, it seems reasonable to conclude that some of the reduction is due to postsurgical pain.[22,23] The degree of reduction ranges from 10 to 15% after minor procedures (e.g., skin incision) to 100% after major invasive surgery.

As rodents are often group-housed, measuring individual food and water consumption can be difficult. It is often more convenient to record body weight. It is important to note that many studies are carried out on growing animals, so establishing average weight gain for a few days before an operative procedure is essential. Although this measure has been shown to be useful, it is essentially a retrospective index of how much pain or discomfort was produced over the preceding 12 to 24 hours. Behavioral assessment, if carried out effectively, allows dose rates of analgesics to be adjusted depending upon the animals' responses.

16:15 What percentage of weight loss over what period of time might indicate that an animal is experiencing pain or distress?

Opin. As discussed in 16:14, loss of weight does not necessarily imply an animal is experiencing pain and distress. However, chronic weight loss, particularly if unexpected, should be investigated carefully. In many institutions, limits are placed on weight loss, based on the assumption that a fall in weight must reflect some adverse effects and some degree of distress. Brief periods of inappetence will cause weight loss of 5 to 10% in rodents and, once weight loss exceeds 20%, general loss of body condition and fat or muscle mass becomes clinically apparent. Most guidelines are based on consensus, not on published data, and it is advisable to examine the issue on a case-by-case basis. It is important to establish why the weight loss is occurring. Is this due, for example, to lowered feed intake, increased metabolism, or decreased absorption of nutrients? Is the weight loss an unavoidable consequence of the research protocol, or could supplemental feeding, use of analgesics, antianxiety drugs, or other methods prevent or ameliorate it? It also is important to compare an animal's weight with that of an untreated, normally growing control. Finally, it is worth noting that adults of several strains of laboratory animals when fed *ad libitum* become obese. This emphasizes the importance of adopting a reasoned, logical approach to interpreting the significance of weight loss in an animal.

16:16 What are useful guidelines for the IACUC to consider when the use of neuromuscular blocking (NMB) agents is requested?

Reg. The AWAR (§2.31,d,1,iv,C) state that neuromuscular blocking agents (paralytics) should not be used without anesthesia. The PHS Policy (U.S. Government Principle V) states that surgical or other painful procedures should not be performed on unanesthetized animals paralyzed by chemical agents.. The *Guide* (page 65) states that when these agents are used, it is recommended that the appropriate amount of anesthetic be first defined on the basis of results of a similar procedure that used the anesthetic without a blocking agent.

Opin. NMBs prevent voluntary muscle activity so that an animal can no longer respond to painful, or other stimuli, by moving. Since movement in response to a surgical stimulus is used to assess adequacy of anesthesia, the use of NMBs as part of an anesthetic protocol requires special consideration. Before examining what constraints might reasonably be placed on the use of these drugs, the IACUC should first establish why it is thought necessary to use these drugs. In some instances, investigators may simply have taken a human anesthetic protocol and decided to use it in their animal model. This is often inappropriate, since the rationale for NMB use in humans often is not applicable to other animals. It is not necessary, for example, to use a NMB to allow assisted (mechanical) ventilation in animals, nor is it needed to produce sufficient muscle relaxation for the majority of surgical procedures. If, however, NMBs are required, then the following points should be considered:

- The PI should have experience with anesthesia and surgery in the species used and should be using a familiar anesthetic regimen that is effective in the absence of an NMB.
- The NMB should not be administered until after the anesthetic has reached a stable level and surgery has commenced. If practical, the relaxant should be allowed to reversed periodically, so that somatic reflex responses can be assessed before additional doses of relaxant are administered.
- The heart rate and blood pressure should be monitored and elevations (15 to 20% or more) of either in response to painful stimuli used to indicate the need for additional anesthesia. It is important to note that this monitoring technique is not fully reliable and awareness in humans can occur without major changes in these variables.[24] With some anesthetic regimens (e.g., a volatile anesthetic such as isoflurane), monitoring the electroencephalogram can be of value, but this also can prove unreliable.

16:17 If a neuromuscular blocking agent (NMB) is used along with an anesthetic, what might the IACUC request to help assure that the animal is anesthetized and not simply immobilized?

Opin. One of the most common situations in which NMBs are used is in neurophysiological studies. In many of these studies, a very light plane of anesthesia must be maintained to minimize interference between the anesthetic regimen and the study protocol. It is precisely in these circumstances that inadvertent production of inadequate anesthesia is most likely. Some investigators have suggested that after completion of surgery, the surgical wounds can be infiltrated with local anesthetic, general anesthesia discontinued, and the animal immobilized because of the effects of the neuromuscular blocking drug. This proposal raises two concerns. First, it is difficult to ensure the adequacy of local anesthetic blockade either initially or later when the NMB has been administered. Second, even if the animal is pain free, it may become distressed if exposed to other stimuli and is unable to react because it is paralyzed. General anesthesia is used in animals not only to provide insensibility to pain but also to induce unconsciousness, which helps prevent the distress caused by physical restraint and experimental manipulations. A paralyzed, conscious animal, even if pain free, is likely to experience considerable distress. The monitoring techniques that can be applied to animals receiving NMBs as part of an anesthetic regimen are described in 16:16.

16:18 Is the use of a local anesthetic to perform a minor procedure considered to be alleviation of pain by anesthesia?

Opin. Yes. Local anesthetics offer a valuable alternative to general anesthesia for workers carrying out a range of different procedures. In some species, the use of regional anesthesia produced by local anesthetics can be considered the method of choice (e.g., a paravertebral block to carry out abdominal surgery in cattle).[19,25] Provided the administration is carried out competently, pain can be prevented either by local infiltration of drugs such as bupivacaine, infiltration around nerve trunks, or epidural or intrathecal administration. Experience in both humans and other animals suggest that these techniques are useful, providing any distress caused by physical restraint or other (nonpainful) procedures can be controlled.[26-28]

16:19 What records related to use of anesthetics and analgesics should be maintained and reviewed by the IACUC?

Reg. The AWAR (§2.35,f) require all records and reports to be maintained for at least 3 years. This includes records made during the activity and for at least 3 years after the completion of the activity. PHS Policy (IV,E,2) has the same requirement.

Opin. Records of analgesic use should vary depending upon the type and complexity of the procedure. As a minimum, the anesthetic drugs or analgesic drugs, dosage, time, route of administration and effect should be recorded, together with details of the experimental animal (age, weight, sex, strain, etc.). If surgical procedures are to be carried out, the investigator should note the adequacy of anesthetic depth and how this was assessed. If additional doses of drugs are given, the time and route of administration and the effect should be recorded.

The investigator also should record the duration of anesthesia, the recovery time to sternal recumbency, any morbidity, any unexpected adverse effects (e.g., vomiting), and record when full recovery (normal activity, feeding, drink, etc.) commences, and indicate how this assessment was made. As discussed in 16:16 and 16:17, it may be desirable to keep a continuous recording of heart rate and blood pressure, especially if neuromuscular blocking agents are used. These data permit a critical review of anesthetic practices. It is also data that should be considered an essential part of any scientific protocol, since interactions between anesthetic complications and the study objectives can easily occur.

When using analgesics, the drug, dose, and route of administration should similarly be recorded. Any additional doses given also should be noted. It may be considered helpful to link this with the pre- and post-procedure observation of variables such as body weight, clinical appearance, etc.

16:20 Should the use of anesthetics and analgesics be based on a strict dosage schedule or on varying dosages as determined by sound clinical judgment?

Reg. The AWAR (§2.33,b,4) require the AV to guide the PI and other personnel in the appropriate use of anesthesia, analgesia, and immobilization. The AWAR (§2.33,b,5) give the AV additional authority, stating that adequate pre- and post-procedural care shall be in accordance with current established veterinary medical and nursing procedures. PHS Policy (IV,A,3,b,1) states that a veterinarian will have direct or delegated responsibility for activities involving animals. PHS Policy (IV,A,1) requires institutions to follow the *Guide*.

The *Guide* (page 56) notes that adequate veterinary care includes effective programs for anesthesia and analgesia, and the veterinarian must provide guidance to the researcher. Pages 64 and 65 of the *Guide* discuss pain, analgesia, and anesthesia.

Opin. Anesthetics and analgesics should be given initially according to dose schedules. These are determined from the scientific literature, but are often modified because of variations in response by different strains of animal. Administering a standard dose of anesthetic may produce the desired effect, but it also can result in animals seeming to be too deeply anesthetized or inadequately anesthetized. After assessing the response to an anesthetic in the particular strain, age, and sex of animal to be used in a study, a more appropriate dose schedule can be developed. The variation can be substantial. For example, when using the duration of unconsciousness as an indicator of anesthetic efficacy, doubling of sleep time can be observed in different strains of mice.[29]

It is relatively easy to adjust anesthetic drug doses to suit particular groups of animals, but more difficult to do this with analgesics, because of our limited ability to assess postoperative pain. Although dose rates obtained from the literature are helpful, every attempt should be made to assess the adequacy of analgesia in each individual animal. Even if retrospective measurements such as body weight (see 16:14, 16:15) are used, this permits variation of the "standard" dose in subsequent studies. The general inadequacy of postoperative pain relief in humans, among other factors, was often ascribed to use of rigid dosing schedules.[30] We should avoid the temptation to take the easy route of prescribing fixed dose rates at fixed intervals for all procedures. The only way to improve pain relief is to try to assess the effect of analgesia in each individual animal.

16:21 If the use of anesthesia might impair the survival ability of an animal to be released in the wild as part of a field study, can the IACUC appropriately approve procedures using either no anesthesia or anesthetic regimens not producing complete pain alleviation?

Reg. Neither the AWAR nor the PHS Policy have pain relief exemptions for animals used in field studies. A field study is defined in the AWAR (§1.1) as any study conducted on free-living wild animals in their natural habitat which does not involve an invasive procedure, and which does not harm or materially alter the behavior of the animals under study. The PHS Policy does not have a specific definition of a field study.

Opin. The use of animals in field studies poses particular difficulties, since it is usually intended that the animal will survive the study and resume its normal activities. Anesthesia may impair this. However, the introduction of reversible anesthetic regimens has greatly improved management of wildlife anesthesia. Similarly, in smaller species, use of modern inhalational agents (e.g., isoflurane) can result in very rapid recovery with few significant aftereffects. The use of potent inhalational agents such as halothane and isoflurane in simple induction chambers is not generally recommended, as dangerous concentrations of anesthetic are produced.[13] However, in field conditions at moderate to low environmental temperatures (<15°C), it is possible to use them, provided the animals are observed closely for signs of overdose. Finally, the use of local anesthetics can be considered as means of minimizing an animal from experiencing pain. If all these options have been considered and rejected as impracticable or ineffective, then an ethical judgment must be made as to whether the aims of the project outweigh the (presumably momentary) pain or distress caused by the required manipulations. Alternatively, if the project seeks only to obtain tissue samples from an animal, a judgment must be made as to whether the procedure should be allowed without adequate anesthesia and the animal allowed to recover, or the procedure carried out with anesthesia and the animal euthanized rather than released. (See 15:9.)

16:22 What is the maximal practical length of time that the IACUC should allow a rat to be deprived of food or water? How do these suggestions change for a mouse, dog, nonhuman primate, or other common species?

Reg. The AWAR (§2.38,f,2,ii) state that short-term withholding of food or water from animals is allowed when specified in an IACUC-approved activity that includes a description of monitoring procedures. APHIS/AC Policy #11[31] uses food or water deprivation beyond that necessary for normal presurgical preparation as an example of a procedure that may cause more than momentary or slight distress. PHS Policy (IV,A,1) requires institutions to follow the *Guide*. The *Guide* (pages 10 and 12) addresses food or fluid restrictions and notes the need for relevant objective information regarding the procedures and purpose of the study. The *Guide* specifically notes that restriction for research purposes should be scientifically justified, and a program should be established to monitor physiologic or behavioral indices, including criteria for temporary or permanent removal of an animal from the experimental protocol.

Opin. Food and water deprivation can be carried out for many reasons, and attempting to establish maximum periods of deprivation without reference to the aims of a particular study is undesirable. For example, a project might be judged so important in terms of its potential benefits that very prolonged periods of food deprivation could be sanctioned if this was considered a necessity. At the extreme, an IACUC might consider whether any project could justify withdrawal of food until an animal dies as a result of this procedure.

Often, only relatively minor periods of food and water deprivation are required, for example, to ensure an empty stomach or gastrointestinal tract, or to induce a catabolic state in the animal. In each study, it is important to determine the minimum period of deprivation needed to achieve the desired objective. Often, 16 or 24 hour periods are chosen, as these are a convenient interval for removal of the food or water at the end of a working day, followed by use of an animal the following morning. In the case of small rodents, this period may be excessive and may induce unnecessarily severe effects because of the high metabolic rate of these smaller species. It has been shown that when presented with a limited quantity of food (e.g., half the amount normally consumed overnight), rats eat normally until the food is exhausted. Thus, by reducing the amount of food placed in their food hopper, effective food removal (e.g., from 3:00 a.m.) can be achieved. This is sufficient to produce an empty stomach.[32] Investigators should note, of course, that coprophagy occurs in these and other species, so complete food deprivation can only be achieved by combining fasting with the use of an anal cup to prevent ingestion of feces.[33]

A second aspect of the animal's normal biology also should be considered. Some species (e.g., rats) normally only feed in the dark phase of their photoperiod. If food is withdrawn overnight and a procedure carried out the next day and if that procedure causes adverse effects, the rat may not eat the following night or the next day. As a result, 48 hours of fasting may have inadvertently been produced. This may have consequences for the particular study, as well as for the welfare of the animals concerned.

When dealing with larger species, longer periods of food withdrawal might be needed and, generally, are better tolerated than in smaller animals. If food is being withdrawn solely to reduce the risk of vomiting during induction or recovery after anesthesia, a period of 8 to 16 hours is appropriate for dogs, cats, ferrets, or nonhuman primates, but unnecessary in rabbits and rodents as these latter species do not vomit. Withholding food in pigs can be helpful in reducing the volume of gut contents for abdominal surgery, and in ruminants it may help reduce the incidence of rumenal tympany.[13]

16:23 Is it appropriate to use electric shock to stimulate animals to run or walk on a treadmill?

Reg. PHS Policy (IV,C,1,a) requires that procedures with animals avoid or minimize discomfort, distress, and pain to animals, consistent with sound research design. The AWAR (§2.31,d,1; §2.31,d,1,i) state that unless acceptable justification for a departure is presented in writing, procedures involving animals will avoid or minimize discomfort, distress, and pain to the animals.

Opin. When using any conditioning stimulus, it is clearly desirable to use a reward system rather than a mild or moderate noxious stimulus such as electric shock. When examining such a proposal, an IACUC should require evidence to show that an aversive stimulus is the only technique that can be used to produce the required behavior in the animal. If it is concluded that aversive stimuli are the only practical conditioning stimuli, then a minimal level of stimulation should be used, and this should be considered an additional cost to the animal when weighing the ethical issues surrounding the research application.

16:24 Can a mammalian fetus feel pain? If so, at what age can a rodent fetus be presumed to feel pain? Should it be included as a vertebrate animal when the IACUC considers the number of animals requested for the study?

Opin. The concern relating to fetal and neonatal pain has emerged largely from studies in human infants. Prior to the mid-1980s, many procedures were undertaken on human infants without effective anesthesia or analgesia. A series of studies indicated that human neonates can experience pain[34,35] and investigations in rodents have shown that not only can pain (or at least responses to noxious stimuli) be demonstrated, but these experiences produce long-term changes in the nervous system. The stage at which these abilities become functional during fetal development is still under debate. In rodents and other species, anatomical studies suggest that nociceptive responsiveness is present in the second half of gestation. It is uncertain whether this anatomical development translates into a capacity to experience pain. However, since pain has a large subjective component, we have the same doubts about mature animals. It seems appropriate to assume that the mammalian fetus has some capacity to experience pain, and adapt research protocols accordingly. Therefore, it is illogical to exclude the fetus when considering the number of animals to be approved in an IACUC submission. Of course, practical considerations should be noted in that an investigator cannot reasonably be expected to predict exactly how many fetal animals in a multiparous species may be present in each pregnant female. A useful review of pain in the fetus and neonate is given by Fitzgerald.[36] (See 8:12, 13:11, 14:20.)

16:25 Can unhatched avian embryos be presumed to feel pain? Is IACUC approval needed for the use of avian embryos?

Reg. NIH/OPRR has interpreted "live vertebrate animal" to apply to avians (e.g., chick embryos) only after hatching.[37] Birds are not currently regulated by the AWAR.

Opin. The answer to this question is difficult when dealing with avian species, since we have a very limited knowledge of nociception and pain perception in these animals. Although it seems illogical to conclude that the capacity to experience pain emerges at the instant of hatching, we cannot as yet determine the stage of development at which this capacity is sufficiently well developed to warrant concern. In the United Kingdom, arbitrary limits were set when legislation was enacted in 1986 and this pragmatic approach may be the best way forward for IACUCs. That is, acknowledge that pain may occur, acknowledge our uncertainty as to the stage of development at which this occurs, and set some initial guidelines which can be reviewed as our understanding of avian neurobiology improves. This same approach would allow studies on early embryonic stages to proceed without IACUC approval, but would require approval once a particular developmental stage has passed.

16:26 Is hypothermia an acceptable form of anesthesia for fetal or neonatal homeothermic animals?

Opin. Hypothermia produces immobility and apparent insensibility in neonates and fetuses, and has become a well-established means of "anesthetizing" neonatal rodents. At low temperatures nerve conduction slows and may be blocked, and depression of body systems can produce unconsciousness. Nevertheless, to date no convincing studies of neonatal central nervous system responses to noxious stimuli during hypothermia have been carried out. In addition, it has been suggested that since rewarming from hypothermia is associated with pain in humans,[38] the technique may be undesirable even if it produces a state of insensibility. Perhaps a more constructive approach is to examine the alternative anesthetic techniques available for use in neonates. Because volatile and injectable anesthetics can be used to produce safe and effective anesthesia,[39,40] it seems reasonable for an IACUC to ask why it is necessary to use a questionable technique when more acceptable alternatives are available. Investigators who have used all of these techniques often find the use of conventional anesthesia more convenient, especially for prolonged surgical procedures, since the neonates do not need to be maintained on an ice pack for the duration of surgery.

16:27 Is hypothermia an acceptable form of anesthesia for poikilothermic animals?

Opin. Similar concerns to those described in 16:26 relate to the use of hypothermia as a means of "anesthesia" in poikilotherms. As with mammalian neonates, well-established alternative anesthetic regimens are available and it, therefore, seems unnecessary to use a questionable technique.

16:28 Can mammals with a cerebral cortex feel pain if made decerebrate?

Opin. When addressing this question it is important to differentiate between pain and nociception (see 16:3). Nociception is the response to damaging or potentially damaging stimuli, whereas pain has a subjective component, resulting in its being interpreted by most humans (and other mammals) as an unpleasant experience.[3,8] It is widely accepted that the presence of a functioning forebrain is required for the perception of pain, so decerebration should remove that capacity. Since thalamic structures are believed to play a role in pain perception, it is usually considered that removal of the forebrain, including the thalamic nuclei, ensures that pain cannot be perceived. The position in regard to decorticate animals is less clear. Comparison with humans leads one to presume that such animals have no capacity for pain perception. However, given the higher levels of organized behavior shown by decorticate or decerebrate animals of some species (e.g., rats), some caution is required in making these extrapolations.

16:29 Can nonmammalian vertebrates lacking or having only a primitive cerebral cortex (e.g., frogs) feel pain?

Opin. Since the capacity to experience the subjective, unpleasant, component of pain in humans appears to rely on the presence of a functioning cerebral cortex, it is often assumed that animals with a less well developed cortex have less capacity to feel pain. Comparison of the frog with humans is simply a more extreme comparison than the more frequent extrapolation of a rat or mouse to humans. In both instances, we have little insight into the nature of the experience of pain in the animal, nor of its significance to the individual. We have adopted an approach that presumes an animal with a certain level of central nervous system development can experience pain, and research involving these species should be conducted in a way that reduces the likelihood of causing pain or distress.

It seems unlikely that frogs experience pain in the same manner as humans, but since we can demonstrate nociception in amphibians[41] and since their degree of cerebral development cannot be said to preclude the possibility of pain perception, we should assume that these species can experience pain.

16:30 Does postorbital blood collection in rodents normally require anesthesia, analgesia, or tranquilization?

Opin. Postorbital (or retroorbital) blood sampling in rodents can provide moderate quantities of blood quickly and easily, and when carried out expertly seems to cause a minimum of long-term complications. When carried out less competently, injuries to the globe can occur and these may be severe.[42] The procedure itself is probably no more painful than peripheral venipuncture (e.g., of the tail vein). Nevertheless, many research units only carry out the technique under anesthesia, both to reduce any pain and to minimize inadvertent injury should the animal struggle during the procedure. The procedure is undoubtedly stressful, and perhaps a more appropriate question to ask is: why is the retroorbital plexus to be used, rather than alternative techniques?[43] (See 16:31.)

16:31 Relative to pain or distress, how many episodes of blood collection by the postorbital sinus or postorbital plexus method should the IACUC reasonably allow on a single animal?

Opin. A reasonable response to this question is for an IACUC to require a report on the incidence of complications (corneal abrasions, retrobulbar hemorrhage or abscessation, damage to the globe, etc.) produced by the technique. Similar assessments of sequelae to other methods of venipuncture can lead some IACUCs to conclude that repeated orbital sinus puncture is acceptable, while others may conclude that it should never be used. Very few studies have accurately assessed the incidence of complications associated with this procedure, and all guidelines appear based on a compromise between an investigator's wishes and an IACUC's unease at the effects on the animal. This author urges the IACUC to require follow-up (including postmortem histology of the globe and periorbital tissues) when allowing use of this technique. It then is possible to make an assessment based on data rather than opinion. (See 19:19, 19:20.)

16:32 For performing procedures such as Southern blots or polymerase chain reactions, can the tail of a rat or mouse be clipped without causing more than momentary pain to the animal? If yes, what length of tail?

Opin. Removal of a small portion of the tail has become an established procedure in many institutes that work with transgenic animals. There is little doubt that removing the tip of a sensitive structure can cause pain. However, to date, no studies have been carried out to determine whether pain persists for more than a few seconds, or whether this is related to the length of tail removed. Different laboratories vary in their opinion as to whether the procedure requires anesthesia and whether anesthesia should be local or general. Our own laboratory carries out the procedure under brief general anesthesia (with isoflurane) and provides post-procedure analgesia (with carprofen). Rather than debate whether pain is produced and how much tail can be removed, it would be preferable to adopt the technique described by Irwin et al.[44] and sample saliva, thus replacing a procedure that is likely to cause at least momentary pain, with one that causes none.

16:33 What are the important issues and what are some general guidelines for the IACUC to consider relative to methods of animal identification?

Reg. The *Guide* (page 46) states that toe clipping, as a method of identification of small rodents, should be used only when no other individual identification method is feasible and should be performed only on altricial neonates. Although the *Guide* lists other methods of identification (e.g., tattooing, ear notching) which can potentially be painful or distressful to an animal, only toe clipping is singled out. The AWAR (1:1, Painful procedure) define a painful procedure as one that causes more than momentary pain or distress (e.g., as can occur from a needle prick). PHS Policy (IV,C,1,a; IV,C,1,b) has similar wording and the same intent. Identification requirements for dogs and cats are noted in the AWAR (§2.38,g).

Opin. Accurate and reliable identification of laboratory animals is essential for most research projects. Methods should be

- Reliable.
- Cause minimal pain or distress to an animal.
- Simple to use.
- Standardized, so that the numbering system can be readily interpreted by all concerned.

A further factor is the cost associated with the technique. This assumes greater significance when large numbers of animals must be identified. The IACUC should balance the consequences of failure of an identification system (loss of animals from a study and the need to repeat the investigation using additional animals) against the impact on the animal of the marking technique. These issues are often debated when an investigator wishes to use relatively low cost (in economic terms) techniques such as ear punching or toe clipping, rather than more expensive methods such as microchip implantation. Aside from concerns that the former methods are less reliable (fighting can result in loss of further portions of the ear), both methods of physical marking result in a minor mutilation of the animal which may cause more than momentary pain. Checking the identification of an animal may cause stress from physical restraint. In contrast, the use of microchips appears to cause only momentary pain (although no well-controlled studies appear to have been carried out to support this assertion) and identification can be carried out more easily, often without the need to restrain the animal. An IACUC must weigh these issues and determine whether methods of identification which require removal of a piece of ear or a digit are acceptable, or whether less traumatic alternatives are preferred. As with many issues, little published data is available and IACUC members must use their general biological knowledge to reach a conclusion.

16:34 Are there any circumstances under which the IACUC would allow a toe to be clipped as a means of animal identification or to obtain tissue for analysis by Southern blot or polymerase chain reaction?

Reg. (See 16:33.)

Opin. A similar approach to 16:33 must be taken to this question. Investigators should state why they consider it necessary to remove a piece of tissue when less invasive methods are generally available. These issues may seem relatively trivial when weighed alongside questions of endpoints in carcinogenesis studies or subjecting animals to major survival surgery, but they should be addressed in a way that recognizes our general concern to reduce to a minimum the pain or distress caused to animals used in biomedical research. The pain and distress may be minor, but the numbers of animals involved can be considerable.

16:35 What criteria can be used to determine if an animal should be euthanized (or treated) due to the growth of an external or internal tumor?

Reg. "Protocols should include criteria for initiating euthanasia, such as … tumor size, that will enable a prompt decision to be made by the veterinarian and the investigator to ensure that the endpoint is humane and the objective of the protocol is achieved" (*Guide*, page 65).

Opin. Determining endpoints in studies that involve production of neoplasia in animals is difficult and there are no universal guidelines for all species and all tumors. The issue has been addressed thoroughly on two occasions (summarized in Reference 45) and these reports provide a great deal of useful information and guidance. As with any issue of this type, before attempting to determine guidelines for termination, the PI should be asked to state the degree of tumor development that is required to meet the scientific objectives of his particular research protocol, and whether the extent of tumor growth can be minimized.

As in other types of research, termination at an early stage, resulting in loss of animals from a study, can result in the need to use additional animals. Termination late in a study may result in animals experiencing unnecessary pain or distress. If animals die as a result of excessive tumor growth, this also can result in unnecessary pain and distress and in the loss of animals from the study since tissue or blood samples may not be available. Thus, there are compelling reasons to try to establish criteria for euthanasia.

When formulating guidelines, it is critical that they should be unambiguous, clearly understood by all involved (the animal care staff as well as the investigator or AV), and their application should be practical. Attempts to develop such criteria often rely on estimates of tumor mass, coupled with features such as ulceration or obvious necrosis of superficial tumors. These criteria are helpful, but it must be recognized that they may not relate to the degree of pain or distress experienced by the animal. As with postoperative pain, behavioral criteria may be more useful, but these are difficult and time-consuming to develop, except when extreme changes in behavior (e.g., complete immobility) are used. Nevertheless, staff should be encouraged to refine the broad criteria adopted, and the IACUC should ask for feedback on the success or failure of the method used.

16:36 What justification should an IACUC expect if an investigator insists that the death of an animal or the death of 50% of her animals is the most appropriate endpoint for the study?

Opin. The use of death as an endpoint has been regarded as essential in some investigations, but rapid developments are being made in this field, and it is an area where both investigators and the IACUC should make particular efforts to keep abreast of the current literature. The usual reason for selecting death as an endpoint is the difficulty of reliably differentiating animals that will die from those which will recover, despite them showing severe clinical signs of illness or toxicity. Typically, animals gradually develop progressively more severe abnormalities, commencing with mild depression of normal activity and signs associated with lack of grooming activity (e.g., ruffled fur coat) and finally progressing to coma and death. When the clinical signs include subjectively distressing changes such as convulsions or severe dyspnea, then there is particular pressure to euthanize an animal rather than allow further deterioration of its condition. As mentioned earlier, the difficulty is that some of these animals may recover and if, for example, a vaccine potency test is in progress, early euthanasia can invalidate the test result.

Several constructive suggestions have been put forward to reduce the need to use death as an endpoint in studies. For example, guidelines on management of animals undergoing assessment of novel antibacterial or antifungal agents have been proposed by representatives of organizations involved in this type of investigation.[46] In some circumstances, simple clinical indices such as the development of profound hypothermia can be used to reliably predict death. There remain many studies, however, in which no validated criteria have been devised. In these circumstances, an IACUC should carefully assess whether criteria can be developed during the progress of the project under consideration. In many instances, failure to develop criteria may be due to insufficiently frequent observation of the animals, or critical events may occur at times of the day when personnel are not usually available, thereby precluding detailed observation. There seems no doubt that progress in refining endpoints has only occurred as a result of investigators carefully evaluating their own particular models. It may be that even after careful assessment, no progress is made and animals must be allowed to die if the aims of the study are not to be frustrated. It is essential, however, that an IACUC is proactive and requires genuine attempts to develop nonlethal endpoints and not allow investigators to use death simply because it is an unambiguous and easy criterion to apply. Finally, the adoption of nonlethal endpoints can have beneficial effects on a project, since they allow blood and tissue samples to be taken which may enable added data to be obtained in a study.

16:37 What limits should the IACUC place on chronic restraint?

Reg. The PHS Policy (IV,C,1,a; IV,C,1,b; U.S. Government Principle IV) and the AWAR (§2.31,d,1,i; §2.31,d,1,ii) note that the IACUC should determine that a proposed activity avoid or minimize discomfort and distress, and that alternatives to painful or distressful procedures have been considered. The *Guide* (page 11) is more specific. It states that prolonged restraint (including chairing of nonhuman primates) should be avoided unless it is scientifically essential and is approved by the IACUC. It suggests the use of less restrictive systems, such as a tether system for nonhuman primates and stanchions for farm animals. Specific requirements for the use of restraint devices in nonhuman primates can be found in the AWAR (§3.81,d).

Opin. Chronic restraint can be required for a variety of purposes, but is usually needed when administering compounds or sampling body fluids via implanted catheters, or obtaining continuous recording of physiological variables. The need for prolonged physical restraint has been reduced by the development of harness and swivel devices that allow an animal some degree of movement, and implantable telemetry devices or ambulatory infusion systems that allow complete freedom of movement.

When considering protocols requiring restraint, an IACUC should first determine whether physical restraint is needed or whether an alternate approach can be adopted. This must be balanced by an appreciation that these alternate systems may still cause some pain, distress, or discomfort to an animal, both after implantation and during the earlier conditioning period. If a study can be completed with, for example, physical restraint for 1 to 2 hours, this might be preferable to the use of a tether system.

As the period of restraint increases, the balance between different systems changes. It is important to determine whether the animal can be readily trained to accept physical restraint. Measurement of stress sensitive indices, such as blood glucose concentration, suggest that restraint in slings or restraining boxes can be carried out without causing significant distress. Problems arise, however, when physical restraint is carried out for prolonged periods and in such a way that the animal is unable to eat, drink, or carry out any normal behaviors such as grooming.

The issue of restraint, therefore, is not simply one of "how long," but depends critically on the method used, on the prior experience of the animal, and the availability of alternative, less stressful systems.

16:38 A mammal is born with a genetic defect which leads to an abnormal but nonlethal health condition very soon after birth (e.g., an inability to use its hind limbs). It adapts as well as can be expected to that condition and requires no significant additional levels of husbandry or veterinary care. Should the IACUC consider this as nonpainful and nonstressful (Category C on the APHIS/AC annual report) or as unalleviated pain or distress (Category E on the APHIS/AC annual report)?

Opin. In the example, given animals may require additional levels of husbandry to supplement grooming of the affected limbs and there is also a risk of traumatic damage and self-inflicted wounds. Another example is congenital blindness or deafness. These animals have no apparent abnormality on clinical examination, require no special husbandry, but by human standards they have severe sensory deprivation. If the permissible categories are either nonpainful or nonstressful, or unalleviated pain or distress, then it could be argued that it should be placed in the latter category, since we simply do not know the impact of the deficit in the animals. Alternatively, we could argue that since we are unable to detect any apparent adverse effect, except by neurological examination, we should class this as nonpainful. These authors prefer to consider such animals on a case-by-case basis. The loss of use of hind limbs could cause distress. Loss of vision may well not, particularly if vision has been absent since birth.

16:39 In the example in 16:38, would the APHIS/AC classification change if this abnormality was caused by an experimental manipulation in an adult animal?

Opin. The classification might well change if the defect was produced in an adult animal, since the animal's prior experience has some relevance.

An animal that suddenly could no longer see or hear or move might require some time to adapt to the condition and during this time the animal might be distressed by the defect, in contrast to an animal which had the defect from birth.

16:40 Cats are used to teach tracheal intubation techniques. They are anesthetized, recovered, and used again. What is a reasonable number of times, and at what intervals might a cat be used for this purpose? How does the IACUC determine this?

Opin. Not only are limits on frequency unhelpful, they make an assumption that somehow the repetition of a procedure can be added up to produce a "total score" of pain or distress. It is more appropriate to say that an animal can be used as often as required for such a procedure, provided it shows no resentment at the time of restraint for anesthetic induction, and it continues to feed, drink, grow normally, and does not develop clinical abnormalities. It should always be thoroughly examined for evidence of laryngeal, oral, or other trauma at the end of the procedure. Familiarizing an animal to the procedure, providing rewards for appropriate behavior, and an enriched environment (there can be no reason not to keep such animals in groups and in pens, rather than individual cages) can lead to improved welfare of such groups of animals that could not easily be achieved when animals are used on a few occasions, then euthanized, and a naive animal used for the next series of manipulations.

16:41 A drug to be used in a research project is known to cause tremors or mild seizures which occur once and last for approximately 2 minutes. Should the IACUC classify this as unalleviated distress? In this example, if seizures occur frequently, should the IACUC classify this as unalleviated distress?

Reg. The AWAR (§2.36,b,7) specifically require painful procedures for which appropriate anesthetic, analgesic, or tranquilizing drugs are not used to be placed in Category E on the APHIS/AC Annual Report.

Opin. Many drugs produce mild neurological side effects, and when these include seizures, this often raises particular concern over the welfare of the animals used. It is generally perceived that seizure activity can cause pain particularly when marked muscle spasm occurs. In the case of seizures that cause a loss of consciousness, there is no evidence that the seizure itself is painful, although we must consider that postseizure muscle spasm can cause mild or moderate pain. It is the opinion of APHIS/AC[47] that seizures such as this constitute unalleviated distress.

As with other areas of research using animals, we have no clear evidence regarding the degree of pain or distress caused by seizures. It could be argued that if the seizure is of a type that results in loss of consciousness and is not associated with potentially distressing aftereffects, then it should not be considered to produce unalleviated distress. If, in contrast, there are good reasons to believe that the animal remains aware during the seizure, then it should be classified as unalleviated distress. In summary, then, the IACUC should consider the type of seizure, try to determine the likelihood of consciousness persisting, and consider the likelihood of postseizure effects (e.g., muscle pain following severe muscle spasms).

References

1. Russell, W.M.S. and Burch, R.L., *The Principles of Humane Experimental Technique*, Methuen, London, 1959.
2. Broom, D.M. and Johnson, K.G., *Stress and Animal Welfare*, Chapman and Hall, London, 1993.
3. National Research Council, *Recognition of Pain and Distress in Laboratory Animals*, National Academy Press, Washington, D.C., 1992.
4. Appleby, M.C. and Hughes, B.O., *Animal Welfare*, CAB International, United Kingdom, 1997.
5. Australian Council on the Care of Animals in Research and Teaching, *Animal Pain: Ethical and Scientific Perspectives*, Kuchel, T.R., Rose, M., and Burrell, J., Eds., Australian Council on the Care of Animals in Research and Teaching, Australia, 1992.
6. International Association for the Study of Pain, Report of International Association for the Study of Pain subcommittee on taxonomy, *Pain*, 6 , 249, 1979.
7. Zimmerman, M., Neurological concepts of pain, its assessment and therapy, in *Neurophysiological Correlates of Pain*, Bromm, B., Ed., Elsevier, Amsterdam, 1984, 15.
8. Association of Veterinary Teachers and Research Workers, Guidelines for the recognition and assessment of pain in animals, *Vet. Rec.*,118, 334, 1986.
9. Khamis, H.J., Statistics and the issue of animal numbers in research, *Contemp. Topics Lab. Anim. Sci*, 36, 54, 1997.
10. U.S. Department of Agriculture, Animal and Plant Health Inspection Service, Animal Care Policy #12. *Written Narrative for Alternatives to Painful Research.* Available on the World Wide Web at: *http//www.aphis.usda.gov/ac/policy12.html*
11. Morton, D.B. and Griffiths, P.H.M., Guidelines on the recognition of pain, distress and discomfort in experimental animals and an hypothesis for assessment, *Vet. Rec.*, 116, 431, 1985.
12. Flecknell, P.A., Advances in the assessment and alleviation of pain in laboratory and domestic animals, *J. Vet. Anesth.*, 21, 98, 1994.
13. Flecknell, P.A., *Laboratory Animal Anesthesia*, 2nd ed., Academic Press, London, 1996.

14. Sawyer, D.C., Rech, R.H., and Durham, R.A., Does ketamine provide adequate visceral analgesia when used alone or in combination with acepromazine, diazepam, or butorphanol in cats? *J. Am. Anim. Hosp. Assoc.*, 29, 257, 1993.
15. Banknieder, A.R., Phillips, J.M., Jackson, K.T., and Vinal, S.I., Jr., Comparison of ketamine with the combination of ketamine and xylazine for effective anesthesia in the rhesus monkey, *Lab. Anim. Sci.*, 28, 742, 1987.
16. Boschert, K., Flecknell, P.A., Fosse, R.T., Framstad, T., Ganter, M., Sjostrand, U., Stevens, J., and Thurmon, J., Ketamine and its use in the pig. Recommendations of the Consensus Meeting on Ketamine Anesthesia in Pigs, Bergen 1994, *Lab. Anim.*, 30, 209, 1996.
17. Flecknell, P.A., Medetomidine and atipamezole: potential uses in laboratory animals, *Lab. Anim.*, 26(2), 21, 1997.
18. Green, C.J., Knight, J., Precious, S., and Simpkin, S., Ketamine alone and combined with diazepam or xylazine in laboratory animals: a 10-year experience, *Lab. Anim.*, 15, 163, 1981.
19. Thurmon, J.C., Tranquilli, W.J., and Benson, G.J., *Lumb and Jones' Veterinary Anesthesia*, 3rd ed., Williams and Wilkins, Baltimore, 1996.
20. Rubal, B. and Buchanan, C., Supplemental chloralose anesthesia in morphine premedicated dogs, *Lab. Anim. Sci.*, 36, 59, 1986.
21. Field, K.J. and Lang, C.M., Hazards of urethane (ethyl carbamate): a review of the literature, *Lab. Anim.*, 22, 255, 1988.
22. Liles, J.H. and Flecknell, P.A., A comparison of the effects of buprenorphine, carprofen and flunixin following laparotomy in rats, *J. Vet. Pharm. Therapeut.*, 17, 284, 1993.
23. Liles, J.H. and Flecknell, P.A., The effects of surgical stimulus on the rat and the influence of analgesic treatment, *Br. Vet. J.*, 149, 515, 1993.
24. Whelan, G. and Flecknell, P.A., The assessment of depth of anesthesia in animals and man, *Lab. Anim.*, 26, 153, 1992.
25. Hall, L.W., *Veterinary Anesthesia*, 9th ed., Balliere Tindall, London, 1991.
26. Branson, K.R. and Thurmon, J.C., Performing epidural anesthesia in swine, *Vet. Med.*, 85, 1345, 1990.
27. Kero, P., Thomasson, B., and Soppi, A.M., Spinal anesthesia in the rabbit, *Lab. Anim.*, 15, 347, 1981.
28. Valverde, A., Dyson, D.H., Cockshutt, J.R., McDonell, W.N., and Valliant, A.E., Comparison of the hemodynamic effects of halothane alone and halothane combined with epidurally administered morphine for anesthesia in ventilated dogs, *Am. J. Vet. Res.*, 52, 505, 1991.
29. Lovell, D.P., Variation in pentobarbitone sleeping time in mice 1: strain and sex differences, *Lab. Anim.*, 20, 85, 1986.
30. Smith, G., Postoperative pain, in *Quality of Care in Anesthetic Practice*, Lunn, J.N., Ed., McMillan Press, London, 1984, 164–192.
31. U.S. Department of Agriculture, Animal and Plant Health Inspection Service, Animal Care Policy #11, Painful/Distressful Procedures. Available on the World Wide Web at: *http//www.aphis.usda.gov/ac/policy11.html*
32. Vermeulen, J.K., DeVries, A., Schlingmann, F., and Remie, R., Food deprivation: common sense or nonsense? *Anim. Technol.*, 48, 45, 1997.
33. Waynforth, H.B. and Flecknell, P.A., *Experimental and Surgical Techniques in the Rat*, Academic Press, London, 1992.

34. Anand, K., Sippell, W.G., and Aynsley-Green, A., Randomised trial of fentanyl anesthesia in preterm babies undergoing surgery: effects on the stress response, *Lancet*, 1 (8524), 243, 1987.
35. Anand, K., The biology of pain perception in newborn infants, *Adv. Pain Res. Ther.*, 15, 113, 1990.
36. Fitzgerald, M., Neurobiology of foetal and neonatal pain, in *Textbook of Pain*, 3rd ed., Wall, P. and Melzac, R., Eds., Churchill Livingstone, London, 1994.
37. Division of Animal Welfare, Office for Protection from Research Risks, National Institute of Health, The Public Health Service responds to commonly asked questions, *ILAR News*, 33(4), 68, 1991.
38. Morton, D., personal communication, 1998.
39. Danneman, P.J. and Mandrell, T.D., Evaluation of five agents/methods for anesthesia of neonatal rats, *Lab. Anim. Sci.*, 47, 4, 1997.
40. Park, C.M., Clegg, K.E., Harvey-Clark, C.J., and Hollenberg, M.J., Improved techniques for successful neonatal rat surgery, *Lab. Anim. Sci.*, 42, 508, 1992.
41. Stevens, C.W. (mini-review), Alternatives to the use of mammals for pain research, *Life Sci.*, 50, 901, 1992.
42. Van Herck, H., Baumans, V., Brandt, C.J.W., Hesp, M.A., Sturkenboom, P.M., van Lith, H.A., van Tintelen, G., and Beynen, A.C., Orbital sinus blood sampling in rats as performed by different animal technicians: the influence of technique and expertise, *Lab. Anim.*, 32, 377, 1998.
43. BVA/FRAME/RSPCA/UFAW, Removal of blood from laboratory mammals and birds — First report of the BVA/FRAME/RSPCA/UFAW joint working group on refinement, *Lab. Anim.*, 27, 1, 1993.
44. Irwin, M.H, Moffatt, R.J., and Pinkert, C.A., Identification of transgenic mice by PCR analysis of saliva, *Nat. Biotechnol.*, 14, 1146, 1996.
45. United Kingdom Coordinating Committee on Cancer Research, *UKCCCR Guidelines for the Welfare of Animals in Experimental Neoplasia*, 2nd ed., United Kingdom Coordinating Committee on Cancer Research, London, 1997.
46. Rodent Protection Test Working Party, Guidelines for the welfare of animals in rodent protection tests. A report from the Rodent Protection Test Working Party, *Lab Anim.*, 28(1), 13, 1994.
47. DeHaven, W.R., personal communication, 1999.

17
Euthanasia

Peggy J. Danneman

Introduction

This chapter addresses questions on the topic of euthanasia. The first two questions cover issues to consider when setting institutional policies on euthanasia. The remaining ones address dilemmas related to the use of specific techniques (e.g., decapitation, exsanguination, pithing). There is also a section on euthanasia of fetal and neonatal animals, a necessity that is frequently encountered in practice but is rarely addressed in the literature. To obtain the survey information used in this chapter, questions were sent to 21 veterinarians and IACUC coordinators throughout the country. They included seven affiliated with academic institutions, six affiliated with private industry, five affiliated with government, and three affiliated with private research institutions. The respondents were asked to respond to each question based on the policy of their IACUC. They also were encouraged to indicate situations where their own personal opinion differed from official IACUC policy.

17:1 What are some general guidelines for euthanasia?

Reg. Euthanasia of experimental animals may be performed for many reasons, including procurement of tissues or blood as part of the experimental design. Animals also may be euthanized to alleviate otherwise untreatable pain or distress. Both the AWAR (§2.31,d,1,v) and PHS Policy (IV,C,1,c) require investigators to euthanize animals that would otherwise experience severe or chronic pain or distress that cannot be relieved by other means. In such instances, animals should be euthanized at the end of the procedure or, if appropriate, during the procedure. The *Guide* (pages 65 and 66) further indicates that animal use protocols should include specific criteria for initiating euthanasia to "ensure that the endpoint is humane and the objective of the protocol is achieved."

Opin. The term "euthanasia" is derived from the Greek, meaning "easy death." According to definitions in the AWAR (§1.1, Euthanasia); *Guide* (page 65), and the 1993 Report of the AVMA Panel on Euthanasia,[1] euthanasia involves killing in a humane manner that causes rapid unconsciousness and death with little or no pain or distress. These goals are met by selecting an agent or technique that is appropriate for the situation, taking into account the species of animal, its age and temperament, the availability of appropriate equipment for restraint, and the environment in which the euthanasia will be performed. It also is important that the person performing the euthanasia be compassionate, gentle, and technically proficient. Because other animals may be distressed by visual, auditory, or olfactory signals from the animal being euthanized, it is generally preferable to perform the procedure in an area where other animals are not present.[1] (See the *Guide* (page 65) and the AVMA Panel on Euthanasia.[1])

Aside from the all-important humane issues, three broad, overlapping, criteria should be considered when selecting a method of euthanasia.

- *Regulatory*: The method should be in compliance with relevant regulations and guidelines noted above. Deviations from these regulations and guidelines are permitted under certain circumstances, and should be approved via written justification for scientific or medical reasons (*Guide*, page 65; AWAR §2.31,c,1,xi; PHS Policy IV,C,1,g). It should be noted that some states also have specific laws related to euthanasia.
- *Human*: The method should take into account personnel and management issues, including the qualifications and training of the personnel administering euthanasia; the need to minimize emotional distress in human participants and observers by proper attention to the aesthetic implications of the method; and attention to the health and safety of humans and other animals.
- *Scientific:* The method should take into account the potential effect of the method on the scientific objectives of the research or postmortem evaluations.

Other criteria to consider (*Guide,* page 65; References 1 and 3) are the reliability of the method in producing humane death, the relative potential that an apparently dead animal might recover following disposal, the expense and availability of drugs or equipment, the potential for human drug abuse, and the ability to maintain euthanasia equipment in proper working order.

17:2 The AVMA Panel on Euthanasia guidelines[1] are not recognized as law. Must the IACUC use them when determining the appropriateness of the proposed method of euthanasia?

Reg. It is true that the reports of the AVMA Panel on Euthanasia are presented as guidelines, not legal requirements. However, PHS Policy (IV,C,1,g) specifically states that the IACUC must determine that methods of euthanasia in a proposed research project are consistent with the AVMA Panel recommendations "unless a deviation is justified for scientific reasons in writing by the investigator." While this wording allows for deviation from AVMA Panel recommendations under specific circumstances, it clearly indicates that IACUCs are required to give the guidelines careful consideration when evaluating a proposed method of euthanasia. APHIS/AC Policy #3 states that "the method of euthanasia must be consistent with the current Report of the AVMA Panel on Euthanasia."[2]

Opin. The recommendations of the AVMA Panel are supported by reports in the literature, but the Panel's interpretation of those reports and final recommendations are not always consistent with the opinions of other experts. For example, the overwhelming majority of neuroscientists — and many experts in laboratory animal science — are in disagreement with the Panel regarding the humaneness of cervical dislocation or decapitation for small rodents (see 17:8). Similarly, many experts on the pathophysiology of pain, as well as experts in laboratory animal science, are in disagreement with the Panel regarding the humaneness of carbon dioxide asphyxiation (see 17:4). While an IACUC may approve a method of euthanasia that is not consistent with the AVMA Panel guidelines, PHS Policy (IV,C,1,g) states that it may do so only if the investigator has justified use of the method for scientific reasons. There is no leeway for the IACUC to simply disagree with the interpretation and recommendations of the Panel (e.g., to decide, after careful consideration of the existing evidence and literature, that cervical dislocation performed by a skilled individual is a humane and fully acceptable method of euthanasia for small rodents). In summary, the AVMA Panel report is an invaluable resource and there is no question that Panel recommendations should be taken into account by an IACUC in deciding whether to approve a particular method of euthanasia. However, by requiring IACUCs to adhere strictly to the Panel recommendations, PHS Policy imposes rigid constraints on the exercise of professional judgment. Does that mean that IACUCs should feel free to act contrary to the recommendations of the AVMA Panel? The answer is "no," unless the institution is willing to contend with the consequences of running afoul of NIH/OPRR and AAALAC.

17:3 Is the use of dry ice a satisfactory method of generating carbon dioxide to be used for euthanasia?

Opin. Dry ice is generally recognized as an alternative to using compressed carbon dioxide in cylinders for euthanasia.[1,3,4] Nevertheless, dry ice presents certain hazards and technical difficulties that are not associated with the compressed gas. Most important, severe chilling or freezing occurs if animals come into direct contact with dry ice. Also, precise regulation of the inflow of gas into the euthanasia chamber, which is possible with compressed gas cylinders, is not possible with dry ice. Because of these considerations, compressed gas cylinders are often considered preferable to dry ice for euthanasia.[1,4] If dry ice is used, it is imperative that precautions be taken to prevent the animal from coming into direct contact with the dry ice (e.g., by use of a wire mesh grid to elevate the animal or suspension of the dry ice in a compartment above the animal's head).

In the author's opinion, dry ice is a satisfactory agent for euthanasia of small animals (e.g., rats and mice) *provided that* appropriate precautions are taken to prevent the animal from coming into direct contact with the dry ice. It is preferable to suspend the dry ice from the lid of the euthanasia chamber, as this will reduce the tendency of the carbon dioxide gas to pool at the bottom of the chamber, thereby unnecessarily prolonging the onset of narcosis. While time to euthanasia will typically be prolonged with dry ice, the slow buildup of carbon dioxide that results from the use of this agent may actually provide a more humane death than that which would result from the rapid introduction of high concentrations of carbon dioxide from compressed gas cylinders (see 17:4).

Surv. Does your IACUC consider dry ice to be a satisfactory method of generating carbon dioxide for euthanasia?

- Satisfactory alternative (Two of the 14 consider compressed gas cylinders to be preferable, and 13 of the 14 specify that precautions must be taken to prevent direct contact of the animal with the dry ice.)   14/21
- Permissible only with extraordinary justification   1/21
- Not permissible under any circumstances   6/21

17:4 What are the most appropriate conditions (e.g., concentration, flow rate) for euthanasia of adult rodents with carbon dioxide?

Opin. The literature contains numerous recommendations regarding the use of carbon dioxide for euthanasia. Nevertheless, there is no general agreement regarding the optimal concentration or flow rate or whether the euthanasia chamber should be prefilled prior to introducing the animal. The 1993 AVMA Panel[1] recommends prefilling the chamber with ≥70% carbon dioxide. Then, after the animal has been placed in the chamber, the gas flow should be sufficient to displace at least 20% of the chamber volume per minute. These recommendations are controversial. While there is some agreement that prefilling the chamber may reduce anxiety and struggling,[4] other sources suggest that pain and distress can be reduced if the chamber is not prefilled.[5-7] Several sources suggest the use of low gas flow rates.[4,5,7] Two reasons are cited for this recommendation:

- The turbulence and loud hissing associated with the rapid influx of gas into the chamber appears distressing to the animals.
- Slowly rising carbon dioxide concentrations allow the animal to lose consciousness prior to suffocation and without pain.

There is evidence that animals experience significant pain when exposed to high concentrations of carbon dioxide.[6,8,9] This is consistent with recent research indicating that carbon dioxide, which forms carbonic acid in a moist environment (e.g., the nasal mucosa), specifically excites small nerve fibers that subserve sensations of pain.[10,11] For this reason, it has been recommended that conscious animals should not be exposed to carbon dioxide concentrations greater than 70%.[6] On the other hand, it is difficult, if not impossible, to euthanize some animals (e.g., diving animals, such as mink) with carbon dioxide in concentrations less than 80%.[1] For these, and other species, gradual introduction of 100% carbon dioxide into a nonprefilled chamber will result in a painless loss of consciousness at low concentrations followed by death as the carbon dioxide concentration within the chamber builds.

Surv. What conditions does your IACUC recommend for the euthanasia of adult rodents with carbon dioxide?

- Nonprefilled chamber, slow gas flow into chamber, 100% carbon dioxide 10/21
- Various combinations of carbon dioxide, flow rate, and chamber filling (three require nonprefilled, four require prefilled, two have no policy); four recommend low flow rate, five recommend high flow rate; four recommend 100% carbon dioxide, five recommend 60 to 80% carbon dioxide 9/21
- Never addressed chamber filling, concentration, or flow rate issues 1/21
- Recommends dry ice 1/21

17:5 Is carbon dioxide an appropriate agent for euthanasia of neonatal rodents?

Opin. Carbon dioxide is generally viewed as less than optimal for euthanasia of newborn animals of any species. Because during fetal life they have been adapted to the comparatively low oxygen, high carbon dioxide environment of the fetus, neonates are typically more resistant to hypoxia and more capable of coping with high environmental carbon dioxide than are older animals.[1,3,4] For this reason, it has been stated that carbon dioxide should not be used to euthanize newborn animals.[4] Others caution that, while carbon dioxide can be used for euthanizing neonates, the exposure time should be prolonged.[1,3,12] The Canadian Council on Animal Care recommends that neonates euthanized with carbon dioxide be left in the chamber for a full half hour after all movements have ceased.[3]

Surv. Does your IACUC consider carbon dioxide an appropriate agent for euthanasia of neonatal rodents? More than one response is possible.

- Appropriate with prolonged exposure 6/21
- Appropriate when followed by physical method (e.g., decapitation) 7/21
- Permit its use only with justification 5/21
- Not permitted for neonates under any circumstances 4/21

17:6 Is carbon dioxide an appropriate agent for euthanasia of rabbits?

Opin. Depending on the source, carbon dioxide is viewed as either a fully acceptable agent for euthanasia of rabbits[4,12,13] or as a less desirable, but still acceptable, choice for this species.[1,3,12] Indeed, Green[12] (pages 238, 240) notes that it is a useful agent, but does cause apprehension in rabbits. Others also have noted that rabbits are prone to struggle or appear apprehensive or distressed when exposed to this gas.[1]

In the author's opinion, carbon dioxide is a conditionally acceptable, but not desirable agent for rabbit euthanasia. When rabbits are exposed to high concentration of this gas, particularly at first, they tend to struggle and kick, and the potential for injuring themselves or an animal handler is significant. This is a common response of rabbits to any stimulus that causes pain or fear, and is probably not indicative that rabbits experience more discomfort than other animals exposed to carbon dioxide. If scientific or other circumstances necessitate the use of carbon dioxide to euthanize a conscious rabbit, it is essential that the animal is properly restrained (e.g., in a commercial rabbit restrainer) to prevent it from kicking. The rabbit should never be placed loose in a euthanasia chamber. As with any species, the process of euthanasia can be made more humane by techniques that allow for the gradual buildup of carbon dioxide in the chamber (see 17:4).

Surv. Does your IACUC consider carbon dioxide as an appropriate agent for rabbit euthanasia?

- Fully acceptable   2/21
- Only with good justification   5/21
- Only if rabbit anesthetized prior to exposure   1/21
- Not permitted   13/21

17:7 What type of justification should the IACUC request if an investigator states that a rodent cannot receive any anesthetic or tranquilizer prior to decapitation or cervical dislocation?

Reg. With regard to physical methods of euthanasia, the 1993 AVMA report[1] states:

> In general, physical methods are recommended for use only after other acceptable means have been excluded, in sedated or unconscious animal when practical, and when scientifically or clinically justified. Consequently, the panel considers all physical methods, except microwave irradiation, conditionally acceptable.

With regard to cervical dislocations and decapitation, the AVMA report states that until additional information is available to better define the nature of the persistent electroencephalogram activity that has been recorded following these techniques, they should be used in the research setting only when scientifically justified by the user and approved by the IACUC.

PHS Policy (IV,C,1,g) requires the IACUC to assure that proposed methods of euthanasia are consistent with the AVMA Panel Report unless "a deviation is justified for scientific reasons in writing by the investigator." Therefore, to be in compliance with PHS Policy, an investigator who wished to perform decapitation or cervical dislocation on a conscious, unsedated rodent would have to provide written scientific justification. (See opinion provided in 17:8.) The AVMA Panel is currently meeting to revise the 1993 report.

Opin. The requirement for scientific justification implies that the PI must be able to explain how the use of a sedative or anesthetic would interfere with the goals of the research. Such an explanation can vary from a description of how sedation or anesthesia might reasonably be expected to interfere with the research, to a summary of specific data showing how a drug actually interferes with the research. While the latter approach certainly represents the most convincing form of scientific justification, it is neither reasonable nor desirable to ask an investigator to provide this kind of information. To obtain such data, a PI would have to perform multiple small experiments for no purpose other than to document the effect of various drugs on the type of data to be collected. The waste in animal lives — not to mention the enormous expenditure of time and resources on the part of the researcher — hardly justifies any possible benefit to the animals in identifying a drug that does not interfere with the scientific objectives. (See 17:8.)

A more practical approach is to require the PI to explain how sedation or anesthesia might reasonably be expected to interfere with the research. For example, a PI needing to perform hysterectomy derivations on mice might request to withhold sedatives or anesthetics prior to euthanasia of the dams by cervical dislocation. By way of scientific justification, she might explain that those drugs are typically viewed as contraindicated in this instance because, as a class, they depress respiratory drive in the pups. Therefore, they would be expected to diminish the probability of successful resuscitation of the pups.

Surv. What type of justification does your IACUC request if an investigator states that a rodent cannot receive any anesthetic or tranquilizer prior to decapitation or cervical dislocation?

Twenty of the 21 IACUCs surveyed require investigators to sedate or anesthetize animals prior to decapitation or cervical dislocation and to provide some form of written justification if sedation or anesthesia cannot be used. They all specify that the justification must include an explanation of the manner in which the investigator believes a sedative or anesthetic could interfere with the experiment. Twelve of these IACUCs require evidence showing that sedatives or anesthetics would "most likely" have an effect on the research. The remaining eight require proof (in the form of published data or data collected by the laboratory) that the drugs have a specific undesirable effect on the research. One of the latter IACUCs specified that such data has to be provided for each anesthetic or class of anesthetics available.

The remaining IACUC in the survey requires that all people performing decapitation or cervical dislocation undergo training provided by an institutional program. Its position is that decapitation and cervical dislocation are humane procedures when performed by properly trained personnel and that justification for withholding sedation or anesthesia under these circumstances is not necessary. *The reader is cautioned, however, that this institution's position is not in compliance with the PHS Policy, based on the current AVMA report.*

17:8 If an investigator is granted IACUC approval to decapitate a rodent that will not be pretreated with an anesthetic or sedative, is the procedure considered to involve significant pain or distress without the use of drugs to alleviate the same?

Reg. There is no requirement under the PHS Policy or the AWAR to make a determination of "pain without anesthesia" for rodents. (See 17:7.)

Opin. The primary concern here is whether anesthesia or sedation is necessary to prevent pain or distress associated with decapitation. The AVMA Panel on Euthanasia[1] states that decapitation may cause pain or distress as a result of the handling and restraint required to perform the technique or as a result of incorrect performance of the technique. Presumably on the basis of such concerns, that Panel recommends that animals be sedated or unconscious "when practical" before euthanasia by any physical method, including decapitation. Similarly, the Canadian Council on Animal Care recommends that "prior sedation or tranquilization should take place whenever possible" before use of any physical method.[3]

Another source of concern is the brain electrical activity that persists for 13 to 14 seconds following decapitation. This persistent activity has been observed by many investigators, and one study specifically interpreted it as indicative of "conscious awareness of pain and distress."[14] Others have debated this interpretation on the basis that the type of activity recorded from the decapitated brain also may be recorded from normal animals during Rapid Eye Movement sleep and deep anesthesia produced by volatile anesthetics.[15,16] In fact, Vanderwolf et al.[15] concluded that the type of activity recorded from the decapitated brain more closely resembles the response to anesthesia than the response to painful stimulation. They and others observed that the electrical activity persisting following decapitation is actually prolonged by anesthesia, and argue that this is the most convincing evidence that this activity is not a valid indicator of conscious pain perception.[15,16] Even if the decapitated brain was able to perceive pain, hypoxia would result in a complete loss of conscious awareness within 2 to 6 seconds.[16,17] Thus, the preponderance of evidence suggests that sedation or anesthesia is not necessary to prevent pain or distress associated with decapitation, provided that the technique is properly performed by a skilled individual.

Surv. If an investigator is granted IACUC approval to decapitate a rodent that will not be pretreated with an anesthetic or sedative, is the procedure considered to involve significant pain or distress without the use of drugs to alleviate the same?

- Momentary pain or distress   14/21
- Unalleviated significant pain or distress   5/21

- Momentary pain or distress only if personnel performing procedure are skilled and animal accustomed to handling and restraint 2/21

17:9 Under what circumstances might an investigator be permitted to euthanize adult (large) rats or rabbits by cervical dislocation?

Opin. It is generally acknowledged that it is physically difficult to perform manual cervical dislocation on larger, more heavily muscled animals. As a result, there is a high potential for severely injuring, but not killing, such an animal and this method is discouraged as a means of euthanizing larger rats (200 gm or more) and rabbits (1 kg or more).[1,3,4] It is, therefore, especially important that scientific justification be carefully reviewed by the IACUC (see 17:7). Nevertheless, an exception can be made if a person demonstrates proficiency in performing manual cervical dislocation on larger animals or if a mechanical dislocator is used.[1,3] As with cervical dislocation of smaller animals, it is important to assure that the person performing the procedure is properly trained and technically skilled, and that the use of this technique has been justified for scientific reasons by the investigator and approved by the IACUC.[1,3] Prior sedation or anesthesia of the animal (see 17:8) should be given particularly careful consideration with larger animals, as the potential for improper performance of the technique, with resulting pain or distress, is much greater than with a mouse or small rat.

Surv. Under what circumstances does your IACUC permit euthanasia of adult (large) rats or rabbits by cervical dislocation?

- Do not allow this form of euthanasia for large (≥200 gm) rats   9/21
- Do allow this form of euthanasia for large (≥200 gm) rats   12/21
- Do not allow this form of euthanasia for large (≥1 kg) rabbits   10/21
- Allow this form of euthanasia for large (≥1 kg) rabbits   10/21
- Does not have rabbits and has no policy   1/21

Of the 12 IACUCs that permit cervical dislocation of larger rats or rabbits, 10 require scientific justification and demonstration of proficiency. One requires that the animal be anesthetized first, and one requires either anesthesia or justification plus demonstrated proficiency.

17:10 Is sedation (e.g., with diazepam or acepromazine) an acceptable alternative to anesthesia for pretreatment of animals prior to decapitation or cervical dislocation?

Opin. The 1993 AVMA Panel report[1] recommends that, when practical, animals be sedated or unconscious prior to euthanasia by any physical method. However, neither this report nor any other published guidelines suggest that anesthesia is preferable to sedation. In fact, anesthesia is absolutely preferred only if it is imperative that the animal be unconscious during the euthanasia procedure. This would be the case if there was a significant inherent potential for pain, or if it was anticipated that restraint would be a significant problem that might lead to improper performance of the technique. Most of the available data suggest that, if these procedures are performed correctly, neither cervical dislocation nor decapitation is likely to cause significant pain (see 17:8).[3,4,15-17] Still, the potential for pain is high if either procedure is performed incorrectly. This is most likely to occur if the person performing the euthanasia is inadequately trained. This also is the type of situation in which problems related to restraint are most likely to arise. Therefore, anesthesia is preferable to sedation on any occasion where the technical proficiency of the person performing the euthanasia is in question. In most other situations, sedation is an acceptable alternative to anesthesia.

Surv. Is sedation (e.g., with diazepam or acepromazine) an acceptable alternative to anesthesia for pretreatment of animals prior to decapitation or cervical dislocation?

• Sedation acceptable alternative	9/21
• Sedative not acceptable alternative	6/21
• Anesthesia recommended, but would permit sedation with appropriate scientific justification	3/21
• Have not specifically addressed the question, but sedation is probably acceptable	2/21
• Require anesthesia for animals being euthanized by persons undergoing training; once trained, neither sedation nor anesthesia required	1/21

17:11 When reviewing a protocol that involves euthanasia of mice by cervical dislocation or decapitation, how should the IACUC determine that personnel performing these techniques are properly trained?

Reg. PHS Policy (IV,C,1,f), the AWAR (§2.31,d,1,viii), the *Guide* (page 66), and the 1993 AVMA Panel report,[1] emphasize that the personnel performing procedures on animals must be properly trained. In all instances, euthanasia procedures are explicitly or implicitly included in these requirements.

Opin. Although the PHS Policy (IV,C,1,f) and AWAR (§2.31,d,1,viii) specify that the IACUC is responsible for assuring that training requirements are met, neither of these documents define who should provide the training or how the IACUC should determine that a particular individual has been properly trained. The 1993 AVMA Panel on Euthanasia[1] and Canadian Council on Animal Care[3] indicate that inexperienced persons should be trained and closely supervised by experienced individuals. The ILAR Committee on Pain and Distress in Laboratory Animals[4] is more specific in its recommendations, stating that the IACUC and AV should provide for the training and supervision of personnel who will perform euthanasia. This approach provides the IACUC with the highest degree of certainty that personnel are properly trained. Nevertheless, many committees and veterinarians do not have the resources to provide this kind of training for all individuals requiring it. An alternative approach is to have the AV, institutional trainer, or an experienced member of the IACUC directly observe the skills of personnel who will perform euthanasia and make the determination that they are either proficient in the technique or that they require further training.

Surv. When reviewing a protocol that involves euthanasia of mice by cervical dislocation or decapitation, how does your IACUC determine that personnel who will be performing these techniques are properly trained?

- Must complete institutional training program or demonstrate skill to the satisfaction of the AV or designee of the AV   7/21
- AV or other experienced person must observe person with unknown skill level   4/21
- Rely on PI to provide assurance that personnel are trained; one of these committees requires direct observation by the AV if there is any question concerning the investigator's experience or training   9/21

17:12 Must a moribund mouse be anesthetized prior to euthanasia by decapitation or cervical dislocation?

Opin. The 1993 AVMA Panel on Euthanasia[1] recommends that, when practical, an animal should be sedated or unconscious prior to euthanasia by any physical method, including decapitation or cervical dislocation. Presumably, the rationale behind this recommendation is that the potential for pain or distress associated with the procedure will be minimized in a sedated or unconscious animal. In the case of a moribund mouse, the relevant issue is whether this animal might be considered to be at least as insensitive to pain or distress as a healthy mouse that had been sedated. The answer depends on the definition of "moribund." One dictionary defines moribund as "being in the state of dying; approaching death,"[18] while another defines it as "in a dying state."[19] Neither of these definitions implies anything about the ability of the moribund individual to experience pain or distress. In fact, in a clinical setting, the term moribund can appropriately be applied to a fully conscious and responsive dying animal as well as an unconscious and unresponsive dying animal. If the mouse shows severely diminished responsiveness, it is reasonable to consider decapitation or cervical dislocation without prior sedation or anesthesia. However, if there is any question about the capacity of the animal to experience pain or distress, compliance with AVMA Panel recommendations require the use of a sedative or anesthetic. (See 17:10.)

Surv. Does your IACUC require a moribund mouse to be anesthetized prior to euthanasia by decapitation or cervical dislocation?

- Require no anesthesia   15/21
- Require justification for not using anesthesia   1/21
- Require anesthesia   2/21
- Require anesthesia only if animal is conscious   3/21

17:13 Is sedation (e.g., with diazepam or acepromazine) an acceptable alternative to anesthesia for pretreatment of animals prior to exsanguination?

Reg. The 1993 AVMA Panel on Euthanasia[1] states that animals should be sedated, stunned, or anesthetized prior to exsanguination. PHS Policy (IV,c,1,g) requires the IACUC to assure that proposed methods of euthanasia are consistent with the AVMA Panel Report unless "a deviation is justified for scientific reasons in writing by the investigator."

Opin. As with decapitation and cervical dislocation (see 17:10), anesthesia is preferred over sedation only if it is essential for the animal be unconscious prior to the procedure. There is no reason to believe that there is significant pain associated with competently performed exsanguination through a cannula placed in a peripheral vein. However, there is concern that the extreme hypovolemia experienced by the exsanguinating animal prior to death is distressing.[1,4] Furthermore, there is disagreement on whether this distress can be adequately managed using only a sedative. The Canadian Council on Animal Care states that "exsanguination is only acceptable as a euthanasia procedure if the animal is first rendered unconscious."[3] While the AVMA and Canadian guidelines differ on the use of sedation vs. anesthesia prior to exsanguination through a peripheral vein, there is no question that an animal should be anesthetized prior to an exsanguination procedure that involves creation of a surgical incision.

Surv. For your IACUC, is sedation (e.g., with diazepam or acepromazine) an acceptable alternative to anesthesia for pretreatment of animals prior to exsanguination?

- Sedation is not an alternative to anesthesia 14/21
- Never addressed the issue 2/21
- Recommend anesthesia, but permit sedative with justification (heavy sedation required by two of these IACUCs) 3/21
- Sedation fully acceptable 2/21

17:14 Must a moribund rabbit be anesthetized prior to euthanasia by exsanguination?

Opin. In recognition of the distress that would presumably be experienced by a fully conscious animal experiencing extreme hypovolemia,[1,4] the 1993 AVMA Panel on Euthanasia[1] indicates that animals must be sedated, stunned, or anesthetized prior to exsanguination. The Canadian Council on Animal Care[3] is less flexible in this regard, stating that exsanguination is acceptable as a method of euthanasia only if the animal is first rendered unconscious. (See 17:13.) The issue with a moribund rabbit is whether its physical condition would render it sufficiently insensitive to pain or distress that further sedation or anesthesia would be unnecessary. As discussed in 17:12, this cannot be determined only on the basis of the animals' being "moribund." If the dying animal is still fully conscious and responsive, it is not appropriate to exsanguinate it without prior sedation or anesthesia. Conversely, if the rabbit is already unconscious, the usefulness of a sedative or anesthetic is questionable, although some consideration should be given to the possibility of the animal regaining consciousness during the procedure. The decision becomes more difficult if the rabbit is still conscious but exhibits diminished responsiveness. Such an animal can be considered to be in a mental state equivalent to "sedated" but cannot be considered equivalent to "anesthetized." Further, even if an IACUC normally requires anesthesia of animals prior to exsanguination, there is an additional issue to consider in the case of a moribund animal. An animal near death might be unable to tolerate a general anesthetic and could very possibly die before or early in the exsanguination procedure, thereby severely limiting the amount of blood that could be collected. This can have a negative impact on the research if it is necessary to collect blood for use in the

	research project. In such a situation, the need to minimize distress in the rabbit has to be balanced with the requirements of the research.
Surv.	Does your IACUC require that a moribund rabbit be anesthetized prior to euthanasia by exsanguination?

- Require anesthesia 14/21
- Require anesthesia if rabbit is still conscious 4/21
- Require anesthesia only if procedure involves penetrating a body cavity; not for peripheral vein 1/21
- May not require anesthesia if anesthesia might result in animal's death prior to collection of the blood 2/21

17:15 An investigator states that she has many years of experience euthanizing rats by stunning. She performs this by swinging the rat rapidly by the tail and hitting its head against the edge of a table. Should the IACUC accept stunning, if properly performed, as an acceptable form of euthanasia?

Reg.	(See 17:2.)
Opin.	Both the AVMA Panel on Euthanasia[1] and the Canadian Council on Animal Care[3] view stunning as unacceptable as the sole method of euthanasia. The concern is that, while stunning may be an effective means for rendering an animal unconscious, it cannot be relied upon to cause death. It should, therefore, be followed by another technique that will cause death.
	Stunning by swinging the animal and hitting the back of its head on the edge of a table is an old method that is seldom addressed in the more recent literature. Waynforth and Flecknell[7] do, however, recommend this technique as a humane method of performing cervical dislocation. They emphasize that the procedure must be performed correctly and that training should take place using deeply anesthetized animals. The need for training of personnel who stun animals by other methods is similarly emphasized by the AVMA Panel on Euthanasia and the Canadian Council on Animal Care.[1,3] Stunning by hitting the animal's head on the edge of a table requires a high level of concentration in addition to considerable technical proficiency. It takes only a small deviation in the force or angle of the blow to result in severe injury with accompanying pain and distress.
Surv.	Does your IACUC accept stunning, if properly performed, as an acceptable form of euthanasia?

- Not acceptable under any circumstances 14/21
- No experience, but probably not acceptable 2/21

- Acceptable only with excellent scientific justification; of these five, one also requires a demonstration of proficiency in performing the technique; two require both a demonstration of proficiency and immediate followup with another method to ensure death   5/21

17:16 Is thoracic compression an ethical form of euthanasia for small rodents?

Opin. Thoracic compression is not addressed as a method of euthanasia in the 1993 AVMA Panel report[1] or in most other published guidelines. However, the American Society of Mammalogists indicates that this is one of the most commonly used methods for euthanizing small mammals in the field because it is quick and causes little pain.[20] While it is probably true that this technique causes little pain, it is quite likely that it causes considerable distress. As a method of euthanasia, thoracic compression can be compared to the use of neuromuscular blocking agents. With both approaches, the animal is rendered unable to breathe, leading to hypoxia and eventual death. Because of the intense fear that the animal presumably experiences during the comparatively prolonged period between the onset of paralysis and death, the use of neuromuscular blocking agents to euthanize unanesthetized animals is widely and emphatically condemned.[1,3,12,21] Euthanasia of conscious animals by thoracic compression should be viewed as equally unacceptable.

Surv. Does your IACUC consider thoracic compression an ethical form of euthanasia for small rodents?

- Absolutely unacceptable   15/21
- Never received a request and never considered the issue; one IACUC would probably not permit it; another IACUC might consider it acceptable if compelling data could be presented to demonstrate that it was necessary and could be performed humanely   5/21
- Acceptable for small birds, but never considered for rodents   1/21

17:17 Is pithing an acceptable method for euthanasia of conscious frogs? Does it require justification?

Reg. (See 17:2.)
Opin. Pithing of conscious frogs is classified as a "conditionally acceptable" method of euthanasia in the AVMA Panel report[1] and as an "acceptable," but not "most acceptable" method by the Canadian Council on

Animal Care.[3] Although the ILAR Committee on Pain and Distress in Laboratory Animals[4] states that "double pithing is an effective method of killing some poikilotherms," it does not list this procedure among its "General Recommendations for Euthanasia" of frogs. All three sources indicate that pithing should involve destruction of both the spinal cord and brain ("double pithing" or destruction of the spinal cord followed by decapitation). All three also emphasize that pithing requires considerable technical proficiency and that, if the person performing the technique is not skilled, the animal could experience considerable pain and suffering. As for all physical methods of euthanasia, the AVMA Panel[1] recommends that pithing be used only after other acceptable means have been excluded, in sedated or unconscious animals when practical, and when scientifically or clinically justified.

Surv. Does your IACUC consider pithing an acceptable method for euthanasia of conscious frogs? Does it require justification?

- Do not house frogs and have no policy   2/21
- Pithing is acceptable method (Seven of these 17 IACUCs specifically require double pithing, 5 require the frog to be anesthetized, 3 specify that the person performing the procedure must be demonstrably proficient in the technique. Of the 17 committees that consider pithing acceptable, 11 require scientific justification, 3 do not require scientific justification, and 3 require scientific justification only if the frog will not be anesthetized.)   17/21
- Discourage pithing as a method of euthanasia, but would consider it with good scientific justification   2/21

17:18 What agents or techniques are most appropriate for euthanasia of neonatal altricial rodents (e.g., rats and mice)?

Opin. There are few published guidelines that specifically address euthanasia of neonatal animals. The few specific recommendations that do exist are based on the fact that newborn animals are typically more resistant to hypoxia and more capable of coping with high environmental carbon dioxide than are older animals.[1,3,4] For this reason, it is often recommended either that inhalant agents (e.g., inhalant anesthetics, carbon dioxide, carbon monoxide) not be used to euthanize neonates[4] or that these agents be used only when the prolonged exposure times necessary to ensure death can be employed.[1,3,12]

When choosing a method of euthanasia for a neonate, it should be kept in mind that neonates, including neonatal altricial rodents, appear to be at least as capable of perceiving pain as adult animals.[22-26]

In the author's opinion, the most humane method of euthanasia for neonatal mice or rats is rapid decapitation with heavy, sharp scissors. Prior sedation or anesthesia is not necessary (see 17:8). This procedure is esthetically upsetting to many people, so an overdose with an inhalant anesthetic (e.g., halothane, isoflurane) may be a more desirable alternative is some instances. Prolonged exposure of the pups to the anesthetic is essential, as is verification of death prior to disposal. Human safety requires effective scavenging of waste gas from the environment. Less desirable, but still acceptable methods of euthanasia for neonatal altricial rodents are barbiturate overdose or carbon dioxide asphyxiation. Both methods have the potential for causing more pain or distress than does rapid decapitation or inhalant anesthetic overdose. Verification of death prior to disposal is essential with both methods, particularly carbon dioxide asphyxiation.

Surv. What agents or techniques does your IACUC consider acceptable for euthanasia of neonatal altricial rodents (e.g., rats and mice)? More than one response is possible.

- Anesthetic overdose (Three specify an inhalant anesthetic and four specify pentobarbital.)   15/21
- Decapitation (Two of these 11 specify that the animal must first be anesthetized with a volatile anesthetic.)   11/15
- Carbon dioxide (Four of these six specify that this must be followed by prolonged hypothermia or a physical method such as decapitation.)   6/21
- Hypothermia followed by decapitation or cervical dislocation (One specifies that the pup should be chilled for 10 min at 4°C, then decapitated.)   3/21
- Cervical dislocation   3/21

17:19 What agents or techniques are most appropriate for euthanasia of prenatal animals?

Opin. There are no specific guidelines published for euthanasia of prenatal animals. In the absence of such specific guidelines, the few recommendations that have been made for euthanasia of neonates (see 17:18) should be followed.

Surv. What agents or techniques does your IACUC approve for euthanasia of prenatal animals? More than one answer is possible.

Interestingly, the methods recommended by a particular IACUC for prenatal animals often differ from those recommended for neonates.

- Decapitation (Four specify with residual anesthetic from anesthesia of mother.)   12/21
- Intraperitoneal barbiturate (One specifies for larger animals only; use decapitation for rodents.)   8/21
- Hypothermia (One specifies that this must be followed by physical method.)   3/21
- Inhalant anesthetic overdose (One specifies this must be followed by exsanguination.)   2/21
- If the fetus is <15 days gestation, killing the dam is sufficient (For fetal age 15 days/birth, use an intraperitoneal barbiturate, or other drug used to kill the dam, then decapitate. This institution uses only rodents.)   1/21
- Carbon dioxide   1/21
- Whatever is used to kill the dam   1/21
- Never addressed by the IACUC   1/21

17:20 Must neonates or preterm fetuses be anesthetized prior to decapitation or cervical dislocation?

Opin. There are no published guidelines that specifically address the issue of decapitation or cervical dislocation of neonates or preterm fetuses. In the absence of such specific guidelines and in recognition of the considerable data indicating that late-term fetuses and neonates are at least as capable of experiencing pain as adults,[22-26] it is appropriate to follow the guidelines for euthanasia of adult animals by decapitation or cervical dislocation. These include the use of a sedative or anesthetic "when practical," but do not include any specific requirements for anesthesia.[1] (See 17:7–17:12, 17:19.)

Surv. Does your IACUC require that neonates or preterm fetuses be anesthetized prior to decapitation or cervical dislocation?

While 20 IACUCs require anesthesia or sedation of adult animals prior to decapitation or cervical dislocation (17:7), only six have the same requirement for neonates.

- Require anesthesia (One IACUC specifies that only animals more than 3 days of age need to be anesthetized. Five also require anesthesia prior to decapitation or cervical dislocation of late preterm fetuses.)   6/21
- Recommend, but do not require, sedation or anesthesia   3/21
- No policy   12/21

17:21 Twelve mouse pups are taken from their dam by hysterectomy, after which vigorous attempts are made to stimulate the pups to breathe spontaneously. These attempts are successful for eight of the pups, which are subsequently placed with a foster mother. Of the remaining four pups, two are never observed to breathe and two breathe only when stimulated. A decision is made to euthanize these last four pups. Is it necessary to ensure death by some other means (e.g., decapitation) in the pups prior to disposal? Is it necessary to administer some form of anesthesia to the pups prior to followup by a physical method to ensure death (e.g., decapitation)?

Opin. This question involves three issues:

- The need to verify death prior to disposal of an animal.
- The necessity of anesthetizing an animal prior to euthanasia by a physical method.
- The degree to which recommendations for adult animals should apply to neonates.

It is generally recognized as essential that death be verified before disposal of an animal[1,3,4] (*Guide*, page 66). This can be accomplished by following up with another method designed to ensure death, e.g., pharmacologic agent, exsanguination, decapitation, or thoracotomy.[1,3] Alternatively, the animal can be examined by a person who is trained to recognize the cessation of vital signs in that species, e.g., absence of heartbeat, respiration, and reflex movements[1,3,4] (*Guide*, page 66). In the case of a neonatal mouse, followup by a physical method is preferable, since examination of such a tiny creature to verify cessation of vital signs would involve a particularly advanced level of expertise. For an adult animal, compliance with the recommendations of the 1993 AVMA Panel report[1] generally requires sedation or anesthesia prior to euthanasia by a physical method. Anesthesia per sé is not required. In the absence of specific recommendations regarding euthanasia of neonates by physical methods and given the evidence that neonates have much the same capacity to experience pain as adults,[22-26] it is appropriate to consider sedation or anesthesia of newborn mice prior to euthanasia by a physical method. However, in the case of the pups described in this question, the capacity to perceive pain is probably diminished, at least to some extent. The pups that never breathed would almost certainly be at least as insensitive to pain or distress as an adult animal treated with a sedative, and the same conclusion can be made about the pups that breathed only when stimulated.

Surv. A Would physical methods of euthanasia be required by your IACUC to assure the death of the rodent pups in the above scenario?

- Require euthanasia for all pups; all specify use of
  a physical method   14/21
- Encourage euthanasia for all pups using a physical
  method   4/21
- Require euthanasia only of the pups that breathed
  when stimulated   2/21
- Do not require euthanasia of any of the pups   1/21

Surv. B In the question above, is anesthesia required prior to euthanasia by a physical method?

- Do not require anesthesia prior to physical euthanasia
  11/21
- Require anesthesia of all pups prior to physical
  euthanasia   2/21
- Encourage but do not require anesthesia prior to
  euthanasia   3/21
- Require anesthesia only for the pups that breathed   2/21
- No policy regarding anesthesia of neonates prior to
  euthanasia   3/21

17:22 If a rat is deeply anesthetized and then perfused to remove its blood, is this considered euthanasia or nonsurvival surgery?

Opin. Questions 17:22 to 17:24 relate to the classification by the IACUC of terminal procedures as euthanasia or nonsurvival surgery. There are no explicit guidelines for setting IACUC policy on this issue, other than admonitions that euthanasia should involve "rapid unconsciousness" and subsequent death[1,3] (AWAR§1.1, Euthanasia). The maximum interval between unconsciousness and death required to meet these definitions is not specified. In determining whether a particular procedure should be classified as euthanasia or nonsurvival surgery, the following considerations can be taken into account:

- *Length of the procedure.* Longer procedures are more likely to qualify as survival surgery.
- *Invasiveness of the procedure.* More invasive procedures, particularly those that require complex surgical manipulations, are more likely to qualify as nonsurvival surgery.
- *Purpose of the procedure.* If the primary purpose of the procedure is to kill the animal, it is more likely to qualify as euthanasia. It is more appropriately classified as nonsurvival surgery if the death of the animal is a secondary objective or merely a con-

sequence of the primary objective (e.g., removal of a vital organ for *in vitro* or *ex vivo* study).

Surv. If a rat is deeply anesthetized and then perfused to remove its blood, does your IACUC considered this euthanasia or nonsurvival surgery?

- Euthanasia 9/21
- Nonsurvival surgery 4/21
- Euthanasia and nonsurvival surgery (One of these four IACUCs classifies it as a nonsurvival *procedure*, not surgery.) 4/21
- Nonsurvival surgery if thoracotomy is involved, euthanasia if a peripheral vein is used 4/21

17:23 If a rat is anesthetized and exsanguinated or has its heart excised, is this considered euthanasia or nonsurvival surgery?

Opin. (See 17:22.)
Surv. If a rat is anesthetized and exsanguinated or has its heart excised, does your IACUC consider this euthanasia or nonsurvival surgery?

- Both procedures are euthanasia 10/21
- Both are nonsurvival surgery 2/21
- Both euthanasia and nonsurvival surgery (One of these four IACUCs classifies both as nonsurvival *procedures*, not surgery) 4/21
- Exsanguination is euthanasia, heart excision is nonsurvival surgery 3/21
- Heart excision is nonsurvival surgery; classification of exsanguination depends on whether or not the chest is incised (One of these two IACUCs defines nonsurvival surgery as any procedure involving an incision made while the animal is still alive.) 2/21

17:24 Viable rat fetuses are removed from an anesthetized dam, after which they are anesthetized and both they and the dam are decapitated. Is this considered nonsurvival surgery or euthanasia for the dam? Is it considered nonsurvival surgery or euthanasia for the pups?

Opin. (See 17:22.)
Surv. Viable rat fetuses are removed from an anesthetized dam, they themselves anesthetized, and the fetus and dam are decapitated. Does

your IACUC consider this nonsurvival surgery or euthanasia for the dam? Is this nonsurvival surgery or euthanasia for the pups?

- Euthanasia for dam and pups   8/21
- Nonsurvival surgery for dam, euthanasia for pups   11/21
- Both nonsurvival and euthanasia for dam and pups   1/21
- IACUC does not distinguish between euthanasia and nonsurvival surgery   1/21

17:25 Which is preferable for euthanasia of adult mice, carbon dioxide or cervical dislocation?

Reg. The *Guide* (page 66) states that in general, inhalant or noninhalant chemical agents are preferable to physical methods for euthanasia.

Opin. The reasoning behind the above recommendation in the *Guide* is not stated. If a recommendation is to be made on humane grounds, the case can be made that mice can be euthanized humanely or inhumanely using either cervical dislocation or carbon dioxide. Carbon dioxide is generally viewed as an acceptable agent for euthanasia of adult mice[1,3,4] (*Guide,* page 66). There is evidence, however, that under certain conditions (e.g., high concentration or high flow rate), animals may experience pain or distress during carbon dioxide euthanasia.[4-9] Cervical dislocation also is recognized as an acceptable method of euthanasia for adult mice by some,[4] but it is classified as "conditionally acceptable" in the 1993 AVMA Panel report[1] and is not listed as either a "most acceptable" or "acceptable" method by the Canadian Council on Animal Care.[3] There are two concerns associated with cervical dislocation (see also 17:8).

Electroencephalogram activity persists for up to 13 seconds following cervical dislocation,[15] which was interpreted in the 1993 AVMA report[1] as possible evidence that the animal may perceive pain for some period of time before death. However, other evidence, including data presented by the investigators who first documented the persistent brain activity, suggests that the animal feels little or no pain if the technique is performed correctly.[3,4,15,17] A more universal concern is that, if it is performed incorrectly, cervical dislocation may result in painful injury rather than death.[1,3]

In summary, if the technical competence of the individual performing the euthanasia is questionable, the potential for pain or distress is greater with cervical dislocation than with carbon dioxide, and the latter is the preferable choice. However, if technical competence is not an issue, there is no clear basis for choosing one method over the other on purely humane grounds.

Surv. Which is preferred by your IACUC for euthanasia of adult mice, carbon dioxide or cervical dislocation?

- Equally acceptable   3/21
- Carbon dioxide preferred   10/21
- Cervical dislocation preferred   8/21

Three veterinarians sitting on IACUCs that prefer carbon dioxide stated that they disagree with the official IACUC view and personally believe that cervical dislocation is more humane.

17:26 How should death be ensured prior to disposal of euthanized animals killed by carbon dioxide, stunning, or pithing?

Opin. It is generally recognized as essential that death be verified before disposal of an animal.[1,3,4] (*Guide,* page 66). This is particularly important when animals are euthanized by techniques that can cause loss of consciousness or cessation of visible respiration well before death occurs. In some instances, animals that are merely unconscious may regain consciousness following disposal. Stunning, pithing, and carbon dioxide asphyxiation are all recognized as methods of euthanasia that may render an animal unconscious long before it dies.[1,3,4,6] Therefore, it is particularly important when using these techniques that disposal not take place until death has been confirmed. This can be done by following up with another method designed to ensure death, e.g., pharmacologic agent, exsanguination, decapitation, or thoracotomy.[1,3] Alternatively, the animal can be examined by a person who is trained to recognize the cessation of vital signs in that species, e.g., absence of heartbeat, respiration, and reflex movements[1,3,4] (*Guide,* page 66). It is important to note that cessation of respiration alone is not a reliable indicator of death, as continued cardiac function, with the potential for recovery, may persist after visible respiratory movements have stopped.[3,4]

Surv. Prior to disposal of an animal, how does your IACUC ensure that the animal is dead after euthanasia by carbon dioxide, stunning, or pithing?

- Does not require verification of death, but strongly encourages followup with a physical method   1/21
- Requires physical method followup only after carbon dioxide   1/21
- Requires physical method followup only after stunning or pithing   1/21
- Require physical method followup after stunning, pithing, or carbon dioxide; physical methods that these committees recommend include thoracotomy (many

require bilateral thoracotomy), cervical dislocation, decapitation, exsanguination, removal of a vital organ, or necropsy	8/21
• Require followup with careful observation to ensure death; must verify the absence of a heartbeat, absence of respiration, absence of movement, or loss of the corneal reflex	5/21
• Require followup with either physical method or careful observation	5/21

References

1. AVMA Panel on Euthanasia, 1993 Report of the AVMA Panel on Euthanasia, *J. Am. Vet. Med. Assoc.*, 202, 230, 1993.
2. U.S. Department of Agriculture, Animal and Plant Health Inspection Service, Animal Care, Policy #3 — Veterinary Care, April 14, 1997. Available on the World Wide Web at: *http://www.aphis.usda.gov/ac/policy3.html*
3. Canadian Council on Animal Care, Euthanasia, in *Guide to the Care and Use of Experimental Animals*, Vol. 1, Olfert, E.D., Cross, B.M., and McWilliam, A.A., Eds., Canadian Council on Animal Care, Ottawa, 1993, 141–153.
4. Committee on Pain and Distress in Laboratory Animals, National Research Council, National Academy of Sciences, Euthanasia, recognition and alleviation of pain and distress in laboratory animals, in *Recognition and Alleviation of Pain and Distress in Laboratory Animals. A Report of the Institute of Laboratory Animal Resources*, National Academy Press, Washington, D.C., 1992, 102–116.
5. Britt, D.P., The humaneness of carbon dioxide as an agent of euthanasia for laboratory rodents, in *Euthanasia of Unwanted, Injured, or Diseased Animals or for Educational or Scientific Purposes*, Universities Federation for Animal Welfare, Herts, U.K., 1986, 19–31.
6. Danneman, P.J., Stein, S., and Walshaw, S.O., Humane and practical implications of using carbon dioxide mixed with oxygen for anesthesia or euthanasia of rats, *Lab. Anim. Sci.*, 47, 376, 1997.
7. Waynforth, H.B. and Flecknell, P.A., Experimental and surgical technique in the rat, in *Experimental and Surgical Technique in the Rat*, Academic Press, London, 1992, 313–340.
8. Peppel, P. and Anton, F., Responses of rat medullary dorsal horn neurons following intranasal noxious chemical stimulation: effects of stimulus intensity, duration, and interstimulus interval, *J. Neurophysiol.*, 70, 2260, 1993.
9. Thurauf, N., Friedel, I., Hummel, C., and Kobal, G., The mucosal potential elicited by noxious chemical stimuli with CO_2 in rats: is it a peripheral nociceptive event? *Neurosci. Lett.*, 128, 297, 1991.
10. Steen, K.H., Reeh, P.W., Anton, F., and Handwerker, H.O., Protons selectively induce lasting excitation and sensitization to mechanical stimulation of nociceptors in rat skin, *in vitro*, *J. Neurosci.*, 12, 86, 1992.

11. Thurauf, N., Ditterich, W., and Kobal, G., Different sensitivity of pain-related chemosensory potentials evoked by stimulation with CO_2, tooth pulp event-related potentials, and acoustic event-related potentials to the tranquilizer diazepam, *Br. J. Clin. Pharmacol.*, 38, 545, 1994.
12. Green, C.J., *Animal Anaesthesia*, Laboratory Animals Ltd, London, 1979.
13. Bivin, W.S., Basic biomethodology, in *The Biology of the Laboratory Rabbit*, Manning, P.J., Ringler, D.H., and Newcomer, C.E., Eds., Academic Press, San Diego. 1994, 72–86.
14. Mikeska, J.A. and Klemm, W.R., EEG evaluation of humaneness of asphyxia and decapitation euthanasia of the laboratory rat, *Lab. Anim. Sci.*, 25, 175, 1975.
15. Vanderwolf, C.H., Buzaki, G., Cain, D.P., Cooley, R.K., and Robertson, B., Neocortical and hippocampal electrical activity following decapitation in the rat, *Brain Res.* 451, 340, 1988.
16. Holson, R.R., Euthanasia by decapitation: evidence that this technique produces prompt, painless unconsciousness in laboratory rodents, *Neurotoxicol. Teratol.*, 14, 253, 1992.
17. Derr, R.F., Pain perception in decapitated rat brain, *Life Sci.*, 49, 1399, 1991.
18. *Merriam Webster's Collegiate Dictionary*, Merriam-Webster, Inc., Springfield, MA, 1995.
19. *Dorland's Illustrated Medical Dictionary*, W.B. Saunders Co., Philadelphia, 1981.
20. Ad Hoc Committee on Acceptable Field Methods in Mammalogy, Acceptable field methods in mammalogy: preliminary guidelines approved by the American Society of Mammalogists, *J. Mammal.*, 68, 1 (suppl.), 1987.
21. Breazile, J.E. and Kitchell, R.L., Euthanasia for laboratory animals, *Fed. Proc.*, 28, 1577, 1969.
22. Fitzgerald, M., Neurobiology of fetal and neonatal pain, in *Textbook of Pain*, Wall, P.D. and Melzack, R., Eds., Churchill Livingstone, London, 1994, 153–163.
23. Guy, E.R. and Abbott, F.V., The behavioral response to formalin in preweanling rats, *Pain*, 51, 81, 1992.
24. Blass, E.M., Cramer, C.P., and Fanselow, M.S., The development of morphine-induced antinociception in neonatal rats: a comparison of forepaw, hindpaw, and tail retraction from a thermal stimulus, *Pharmacol. Biochem. Behav.*, 44, 643, 1993.
25. McLaughlin, C.R. and Dewey, W.L., A comparison of the antinociceptive effects of opioid agonists in neonatal and adult rats in phasic and tonic nociceptive tests, *Pharmacol. Biochem. Behav.*, 49, 1071, 1994.
26. McLaughlin, C.R., Lichtman, A.H., Fanselow, M.S., and Cramer, C.P., Tonic nociception in neonatal rats, *Pharmacol. Biochem. Behav.*, 36, 859, 1990.

18
Surgery

Elizabeth J. Dawe

Introduction

The IACUC and AV have the responsibility to help assure humane care and use of animals on whom surgery is performed and also to help assure that individuals who perform that surgery are appropriately qualified and trained. This chapter addresses several major considerations of surgery using animals, such as anesthesia, aseptic technique, and the perioperative care associated with the conduct of animal surgery in research and teaching.

Surveys noted in this chapter were based on 12 IACUC protocol forms and institutional guidelines, 8 retrieved from their World Wide Web Internet sites and 4 obtained directly from the institutions. Institutions represented include 11 universities and 1 medical school.

18:1 What information regarding survival surgical procedures should be included by the investigator in the IACUC protocol? How detailed of a description is necessary?

Reg. There are numerous regulatory requirements that relate to survival surgical procedures. In the broadest sense, they revolve around adequate veterinary care (see Chapter 27; AWAR §2.33; PHS Policy IV,C,1,e). Specific areas of concern for the IACUC include:

- Use of appropriate anesthesia and analgesia. (See Chapter 16.)
- Use of aseptic technique. (See 18:10, 18:11, 18:13, 18:14.)
- Occurrence of multiple survival surgeries. (See 18:5.)
- Qualifications of surgical personnel. (See 18:15.)
- Perioperative care. (See 18:16.)

- Whether major or minor surgery will be performed. (See 18:3.)
- A description of the procedures to be performed. AWAR (§2.31,d,2; §2.31,e,3); PHS Policy (IV,C,1; IV,D,1,c). PHS Policy (II; IV,C,1) requires compliance, as applicable, with the AWA.

Opin. An investigator should provide sufficient detail in the IACUC protocol when describing surgical procedures so that the IACUC can confirm that acceptable surgical techniques are proposed and the AV can evaluate the perioperative care program pertaining to adequate veterinary care needed.[1] In this author's opinion, incision location, chronic instrumentation and implants (when applicable), and method of wound closure (including size and type of suture) should be included with the description of surgical procedures. In addition, the description of the method of anesthesia should include preanesthetic, anesthetic and analgesic drugs, doses, routes, and frequency of administration. If neuromuscular blocking agents are to be used, the drug, dose, and route of administration should be listed. (See 18:18, 18:20–18:22.)

Surv. A Seven IACUC protocol forms having only a "general procedures" section were surveyed for instructions to the investigator regarding information pertaining to survival surgical procedures and how detailed the surgical procedure description should be. Only general instructions were given to:

- Describe the surgical procedures in detail   3/7
- Describe the essential elements of the surgical procedures   1/7
- Describe procedures to be performed on the animals   2/7
- Provide a level of detail comparable to that in the methods section of a journal article   1/7

Regarding the method of anesthesia during the surgical procedure:

- List or include preanesthetic and anesthetic drugs, dose, route, frequency of administration   7/7
- What is the duration of anesthesia?   2/7
- Will neuromuscular blocking agents be administered?   6/7
- Provide neuromuscular blocking agent dose and route of administration   4/7

Surv. B Five IACUC protocol forms with a separate surgery or survival surgery section were surveyed regarding information pertaining to survival surgical procedures and how detailed the surgical procedure description should be. Specific information was requested, in addition to a request for a detailed description of the surgical procedure. More than one response from each form is possible.

- Describe the surgical procedures in detail including: 5/5
  - Incision location and length 4/5
  - Organs involved 1/5
  - Instrumentation and implants 2/5
  - Type of suture material to be used 2/5
  - Method of skin or wound closure 4/5
  - Type of suture to be used for skin or wound closure 4/5

Regarding method of anesthesia during the surgical procedure:

- List or include preanesthetic and anesthetic drugs, dose, route, frequency of administration 5/5
- Provide supplemental dose of anesthetic drugs 1/5
- What is the duration of anesthesia? 3/5
- Will neuromuscular blocking agents be administered? 4/5
- Provide neuromuscular blocking agent dose and route of administration 2/5

18:2 What information regarding nonsurvival surgical procedures should be included by the investigator in the IACUC protocol? How detailed of a description is necessary?

Opin. As with survival surgical procedures (see 18:1), an investigator should provide sufficient detail in the IACUC protocol when describing surgical procedures so that the committee can confirm that acceptable surgical techniques are proposed and the AV can evaluate the preoperative and intraoperative management pertaining to adequate veterinary care needed. The method of anesthesia described should include preanesthetic and anesthetic drugs, doses, routes, and frequency of administration. If neuromuscular blocking agents are to be used, the drug, dose, and route of administration should be listed.

Surv. The 12 IACUC protocol forms surveyed used the same instructions for the surgical procedure description and requested the same information as described for survival surgery (see 18:1), except the information relevant to postoperative management was deleted.

18:3 Can the IACUC classify surgical procedures as minor vs. major? By what criteria? How can this classification be useful to the IACUC when reviewing protocols?

Reg. The *Guide* (page 61) states that surgical procedures are categorized as major or minor. Major survival surgery is defined as surgery that "penetrates and exposes a body cavity or produces substantial impairment of physical or physiologic functions (such as laparotomy, thoracotomy, craniotomy, joint replacement, and limb amputation)."

The AWAR (§1.1) similarly define a major operative procedure as "any surgical intervention that penetrates and exposes a body cavity or any procedure that produces permanent impairment of physical or physiological functions."

The *Guide* (page 61) defines minor survival surgery as surgery that "does not expose a body cavity and causes little or no physical impairment (such as wound suturing; peripheral-vessel cannulation; such routine farm animal procedures as castration, dehorning, and repair of prolapses; and most procedures routinely done on an 'outpatient' basis in veterinary clinical practice)." PHS Policy (IV,A,1) requires institutions to use the *Guide* as a basis for their animal care and use program.

Opin. Classification of surgical procedures as major or minor by an IACUC are based on the above regulatory definitions. This classification is useful to an IACUC when reviewing a protocol because it is the basis for the IACUC's determination of the type of surgical facility and the degree of aseptic technique required (see 18:9). Major survival surgery on nonrodents must be performed only in facilities designed, operated, and maintained for that purpose. Minor survival surgery and all surgery on rodents does not require separate dedicated facilities; however, aseptic technique must be used (AWAR §2.31,d,1,ix; *Guide,* pages 62, 78).[2] Further, for animals to be used in multiple major survival surgeries, their use must be scientifically justified in the protocol and approved by an IACUC (AWAR §2.31,d,1,x,A; *Guide,* page 12). There is no regulatory limitation to multiple minor survival surgical procedures. Nevertheless, an IACUC should use professional judgment to limit the number of minor surgical procedures performed on an animal.[2]

18:4 Should laparoscopy be considered a procedure or major surgery?

Reg. (See 18:3.)

Opin. Laparoscopic surgery can be either major or minor surgery. Although a body cavity is penetrated when laparoscopic techniques are used, the abdominal cavity is not significantly opened and exposed. Reduced surgical trauma may result in little or no physical impairment. Using these criteria alone, laparoscopy is minor surgery.[2]

To determine if it is major or minor surgery, the surgical procedure being performed using laparoscopic techniques and the postoperative outcome must be evaluated. Surgery that is expected to have minimal tissue trauma, a rapid postoperative recovery, and return to normal function with minimal complications and minimal postoperative pain would be considered minor surgery. If extensive tissue trauma, postoperative complications, significant postoperative care, and more than minimal pain is expected, the criteria for major surgery are met.

18:5 Under what circumstances can an individual animal be used in multiple survival surgical procedures?

Reg. Although performing multiple major survival surgical procedures on an individual animal is discouraged, the *Guide* (page 12) states that they may be permitted if scientifically justified by the investigator and approved by the IACUC.

The AWA (§13,A,3,D; §13,A,3,E) permits more than one major survival operative procedure for:

- Scientific necessity when specified by research protocol.
- Other special circumstances determined by the Secretary (of Agriculture).

Exceptions to these standards may be made only when specified by the research protocol and are to be detailed and explained in an annual report filed with the IACUC.

The AWAR (§2.31,d,1,x) and NIH/OPRR[3] permit multiple major survival operative procedures when:

- Scientifically justified by the investigator in writing.
- Needed as a routine veterinary procedure or to protect the health and well-being of an animal.
- Other special circumstances authorized by the Administrator, APHIS, USDA on a case-by-case basis.

APHIS/AC Policy #14[4] permits more than one major survival operative procedure on an individual animal when:

- Scientifically justified by the PI within one proposal and approved by the IACUC.
- An animal that has an emergency major operative procedure required for proper veterinary care then may be used in a research proposal requiring a major survival operative procedure.

- Other circumstances authorized by the Deputy Administrator. (Exemption requests are made to the appropriate Animal Care Regional Director who forwards it to the Animal Care Assistant Deputy Administrator for review and recommendation to the Deputy Administrator. Annual IACUC evaluation of the exemption is required and must be included in the Annual Report (APHIS Form 7023) for consideration for renewal or continuation of the exemption.)

Regulations which address multiple survival surgical procedures on an individual animal have similar exemption criteria where it is allowed. According to the AWA (§13,A,3,D), exemptions are authorized by the Secretary of Agriculture. However, as per the AWAR (§2.31,d,1,x) and as noted by the NIH/OPRR,[3] requests are sent to the Administrator, APHIS, USDA. The Secretary of Agriculture has delegated this responsibility to the Administrator of APHIS, who has further delegated this to the Deputy Administrator of Animal Care. As per APHIS/AC Policy #14,[4] requests are made to the appropriate Animal Care Regional Director who forwards it to the Animal Care Assistant Deputy Administrator for review and recommendation to the Deputy Administrator.

Opin. Examples of special circumstances that may potentially justify performing multiple surgical procedures on an animal include procedures that are related components of a research project, involve conservation of scarce animal resources, or are needed for clinical reasons. The *Guide* (page 12), APHIS/AC Policy #14,[4] and interpretation of PHS Policy[3] do not consider cost savings alone as an adequate justification for performing multiple major survival surgeries on a single animal.

18:6 An animal is surgically modified by the vendor to meet a research need (e.g., ovariectomy of a rat). Must that breeder have his own PHS Assurance Statement if the animals are sold to an institution and used in a project supported by PHS funds? Should the IACUC obtain a copy of that assurance statement?

Opin. Purchase of animals that have been surgically modified by a vendor and are available for general sale has not been formally addressed by NIH/OPRR. However, NIH/OPRR has published a response to a related scenario involving a commercial supplier producing standard reagent antibodies using its own resources and offering the antibodies for general sale as a catalog item. When antibodies are readily available and are considered off-the-shelf reagents, the supplier is not required to have an approved Animal Welfare Assurance on file with NIH/OPRR.[5] To apply the same principles, if an institution purchases an animal to be used in a PHS-supported study that has been surgically modified by a breeder in advance of sale and is readily available as an off-the-shelf catalog item, the breeder is not required to have his own Assurance on file with NIH/OPRR. On the other hand, if an investigator subgrants or subcontracts with a supplier to produce antibodies using antigens provided by or at the request of the investigator, the antibodies are considered "customized" and the supplier must file an Assurance with NIH/OPRR.[5] The institution must obtain the supplier's Assurance number by contacting the supplier or NIH/OPRR.[6] Again applying the same principles, if an investigator subcontracts with a breeder to surgically modify an animal specifically for use in a PHS-supported study, the breeder must file an Assurance with NIH/OPRR. In this author's opinion, the institution should obtain the breeder's Assurance number. To obtain a copy of the Assurance statement would be a local decision made at the institutional level.

18:7 Should the IACUC permit a farm animal to recover from major surgery so that it can be sold for slaughter once it has fully recovered?

Opin. The IACUC could permit or withhold permission for recovery from major surgery citing applicable regulations. The AWAR (§2.31,d,1,ix) and the *Guide* (pages 60 to 64) do not prohibit recovery from a first major surgery provided that all related requirements are met. However, PHS Policy (IV,C,1,a) states that "procedures with animals will avoid or minimize discomfort, distress, and pain to the animals, consistent with sound research design." This is similarly stated in the AWAR (§2.31,d,1,i). Assuming that the investigator's research or educational needs can be met with nonsurvival surgery, anesthesia without recovery supports the welfare interests of the animal by fully avoiding potential pain and discomfort rather than *minimizing* potential postoperative pain with analgesics. If meat from the slaughtered animal was destined for human consumption, the IACUC has to assure that the animal received only Food and Drug Administration-approved medications and drug withdrawal periods were met. Ultimately, the IACUC's awareness of the public's expectations and perceptions regarding the use of animals in research and education should take precedence in making this decision. Use of animals in research is held to a higher ethical standard than many other uses. The potential negative public perception of returning farm animals to the food chain following use in research or education might be a basis for the IACUC's decision to not permit this practice.[7] (See 14:31.)

18:8 What is aseptic technique?

Reg. The *Guide* (page 62) states that aseptic technique includes:

- Preparation of the patient, such as hair removal and disinfection of the operative site.
- Preparation of the surgeon, such as the provision of decontaminated surgical attire, surgical scrub, and sterile surgical gloves.
- Sterilization of instruments, supplies, and implanted materials.
- The use of operative techniques to reduce the likelihood of infection.

The *Guide* (page 62) also states that "aseptic technique is used to reduce microbial contamination to the lowest possible practical level. No procedure, piece of equipment, or germicide alone can achieve that objective. Aseptic technique requires the input and cooperation of everyone who enters the operative suite. The contribution and importance of each practice varies with the procedure [references omitted]."

Opin. By definition, aseptic technique is the performance of a surgical procedure in a manner to prevent exposure of the patient to pathogenic organisms.[8] (See 26:20.)

18:9 Is aseptic technique necessary for performing survival surgery in animals?

Reg. The AWAR (§2.31,d,1,ix) require that all survival surgery will be performed on all regulated animals using aseptic procedures which includes surgical gloves, masks, sterile instruments, and aseptic technique. The *Guide* (page 62) states that minor survival procedures may be "performed under less-stringent conditions than major procedures, but still require aseptic technique and instruments."

Opin. An IACUC has a regulatory responsibility to assure use of acceptable standards of aseptic technique as stated above. Performing survival surgery using aseptic technique also is necessary to achieve a satisfactory surgical outcome with reduced risk of infection. Aseptic technique is used to reduce microbial contamination of a surgical wound and exposed tissues to the lowest possible practical level (*Guide*, page 62).[9] To use less than optimal aseptic technique potentially increases bacterial contamination and subsequent risk of infection which can compromise an animal's health postoperatively. In addition, lack of clinical evidence of postoperative infection does not rule out clinically inapparent infection. A study done in rats showed that infection can be clinically inapparent and yet cause adverse physiologic and behavioral responses that can affect research results which may go unrecognized.[10]

18:10 Are the standards for aseptic technique different for rodents compared to other mammals?

Reg. Standards for aseptic technique for rodents differ somewhat from nonrodents. The AWAR (§2.31,d,1,ix) and the *Guide* (page 78) do not

	require a dedicated surgical facility for major survival rodent surgery as compared to nonrodents. The AWAR (§2.31,d,1,ix) state that all survival surgery must be performed using aseptic procedures, including surgical gloves, masks, sterile instruments, and aseptic technique. The *Guide* (page 63) identifies characteristics of rodent surgery, such as small incision sites, a one-person "surgical team," surgery performed on multiple animals at one sitting, and procedures of shorter duration, which make changes in standard aseptic techniques used in larger species necessary or desirable.[9,11]
Opin.	Modifications in standard aseptic techniques which are suitable for rodent surgery include use of one sterile instrument pack for up to five rodents incorporating techniques to maintain sterility between animals, and wearing a mask with cap and sterile gown optional.[9,11] Although modifications in standard aseptic techniques may be necessary or desirable for rodents, the performance standards to prevent or minimize exposure of the patient to pathogenic organisms to reduce the likelihood of infection must be met and the well-being of the animals should not be compromised[11] (*Guide,* page 61). It would seem that for APHIS/AC-regulated rodents a mask is required. A mask also seems required by the *Guide* as part of aseptic technique. (See 26:21.)

18:11 Does aseptic technique need to be followed for nonsurvival surgery?

Reg.	APHIS/AC Policy #3[12] does not require aseptic technique or dedicated surgical facilities when performing nonsurvival surgery if the animal is not anesthetized long enough to show evidence of infection. The area to be used should be clean, free of clutter, and prepared using acceptable veterinary sanitation practices as would be used in a standard examination/treatment room. The *Guide* (page 62) does not require using aseptic technique for nonsurvival surgery; however, "the surgical site should be clipped, the surgeon should wear gloves, and the instruments and surrounding area should be clean." If nonsurvival surgery is conducted in a dedicated "survival" surgery area, then aseptic technique should be used or the surgical room be returned to an appropriate level of cleanliness before it is used for major survival surgery (*Guide,* pages 62 and 63).
Opin.	The extent to which an investigator should exceed the minimum regulatory requirements for nonsurvival surgery and apply any or all of the components of aseptic technique as described in 18:8 depends on the experimental protocol and the surgery being performed. Professional judgment is necessary to evaluate the probability of virulent bacterial contamination and subsequent host responses that would invalidate research results. Slattum et al.[13] reported a study demonstrating a need for aseptic technique when performing nonsurvival

surgery. They found that gram-negative bacteremia and septic shock developed in dogs during nonsurvival cardiopulmonary studies when performed without using aseptic technique. Laboratory-prepared nonsterile intravenous solutions were found to be contaminated with Gram-negative bacteria. Bacteremia and septic shock ceased to occur after initiating some components of aseptic technique such as using sterile commercial saline and other sterile intravenous injectables, disinfection of equipment and instruments, and the use of sterile gloves.

18:12 What level of aseptic technique should the IACUC require for field surgery involving wildlife?

Reg. The AWAR (§2.31,d,1,ix) state that survival surgery conducted at field sites does not require dedicated facilities, but must be performed using aseptic procedures, including surgical gloves, masks, sterile instruments, and aseptic technique. The *Institutional Animal Care and Use Committee Guidebook*[14] states that aseptic practices should be used when performing survival surgery on wildlife in field studies. The *Guide* (page 61) recognizes that modification of standard aseptic and surgical techniques might be necessary when performing field surgery; however, animal well-being should not be compromised. When modifications are implemented, thorough assessment of surgical outcomes should be done to ensure that appropriate procedures are followed. Surgical outcome assessment may require other criteria in addition to clinical morbidity and mortality.

Opin. An experimental animal surgical facility environment may be unnecessary when performing survival surgery on wildlife based on it not improving animal well-being or surgical outcome as measured by lack of postoperative complications, improved surgical survival, or minimized pain and stress. Bringing free-living wildlife to a dedicated facility to perform surgery could, in fact, be detrimental to the well-being and survival of the animals. Settings used in clinical veterinary practice may be suitable for field surgery involving wildlife.[2] However, aseptic technique to prevent or minimize exposure of the animal to pathogenic organisms must be used. This includes preparation of the operative site (including hair removal and disinfection), mask, sterile surgical gloves, instruments, supplies and implanted materials, and use of operative techniques to reduce the likelihood of infection (*Guide*, page 62). Steam sterilization or autoclaving is the preferred method of surgical instrument sterilization. However, chemical sterilization using liquid chemicals with appropriate contact time may be necessary in some field surgery settings. Ultimately, professional judgment must be used to optimize the circumstances for the environment where the surgery is performed and the aseptic techniques used.[2]

18:13 Would less than optimal aseptic technique be acceptable if a PI's records indicated excellent surgical success with a lack of postoperative infections?

Reg. The AWAR (§2.31,d,1,ix) and the *Guide* (page 62) require that all survival surgery on regulated species be performed using aseptic technique in accordance with professionally acceptable standards as described in 18:9 and defined in 18:8.

Opin. No. In this author's opinion and that of another,[15] the performance of survival surgery using less than optimal aseptic technique indicates that although the IACUC and AV reviewed and approved these procedures, the review was not thorough enough or the investigator was inadequately trained in aseptic technique. The IACUC did not fulfill its responsibility to assure use of acceptable standards of aseptic technique.

Aseptic technique is used to reduce microbial contamination of a surgical wound and exposed tissues to the lowest possible practical level[9] (*Guide*, page 62). To use less than optimal aseptic technique potentially increases bacterial contamination and subsequent risk of infection which would compromise the animal's health. Lack of clinical evidence of postoperative infection does not rule out clinically inapparent infection. As noted in 18:9, a study performed in rats showed that infection can be clinically inapparent and cause adverse physiologic and behavioral responses that can affect research results which may go unrecognized.[10]

Of course, it is hard to argue with success. Under some circumstances, less than optimal aseptic technique may be acceptable as long as the animal's health is not compromised and the procedures are approved by the IACUC. Although the IACUC should reconsider their approved aseptic techniques to meet current acceptable standards of proper veterinary care, the IACUC has chosen to focus on the product rather than the process.[15]

18:14 Should the IACUC demand that aseptic technique be used when surgery is performed on farm animals in the field (e.g., standing rumenotomy performed in a barn)?

Reg. The AWAR (§2.31,d,1,ix), which apply to farm animals used in biomedical research, state that survival surgery conducted at field sites does not require dedicated facilities, but must be performed using aseptic procedures, including surgical gloves, masks, sterile instruments, and aseptic technique. PHS Policy (IV,A,1) requires compliance with standards for survival surgery as outlined in the *Guide* when using farm animals in biomedical research. The *Guide* (page 61) recognizes that modification of standard aseptic and surgical techniques might be necessary when performing field surgery; however, animal well-being should not be compromised. When modifications are implemented, thorough assessment of surgical outcomes should be done to ensure that appropriate procedures are followed. Surgical outcome assessment may require other criteria in addition to clinical morbidity and mortality.

When using farm animals in agricultural research, standards for survival surgery as outlined in the *Guide for the Care and Use of Agricultural Animals in Agricultural Research and Teaching (Ag Guide)* should be applied.[16] The *Ag Guide* (page 21) states that "major survival surgeries should be performed in facilities designed and prepared to accommodate surgery, and standard aseptic surgical procedures should be employed." Aseptic surgical procedures include use of cap, mask, gown, gloves, and sterile instruments as well as appropriate operative site preparation and draping. Minor survival surgical procedures that do not expose a body cavity and cause little or no physical impairment (e.g., wound suturing and peripheral vessel cannulation) may be performed under less stringent conditions if performed in accordance with standard veterinary practices. Therapeutic and emergency surgeries (e.g., Caesarean section, bloat treatment, and displaced abomasum repair) are sometimes required in agricultural situations that are not conducive to rigid asepsis. However, every effort should be made to conduct minor and emergency survival surgeries in a sanitary and aseptic manner.

Opin. When farm animals in the "field" require elective major survival surgery, it should be performed in facilities designed and maintained for surgery using appropriate aseptic procedures including cap, mask, sterile gown and surgical gloves, sterile instruments, and aseptic technique (see 18:8). Therefore, an elective rumenotomy should not be performed in a barn. Minor survival procedures may be performed under less stringent conditions than major procedures, but require sterile instruments and aseptic technique. When therapeutic and emergency surgeries do not allow transport to dedicated surgical facilities, the facilities used should be clean and methods of aseptic technique used to prevent or minimize exposure of the animal to pathogenic organisms to reduce the likelihood of infection.

18:15 How can the IACUC assure that personnel are qualified to perform surgical procedures?

Reg. The PHS Policy (IV,C,1,f) and AWAR (§2.31,d,1,viii) place responsibility with the IACUC to determine that personnel performing surgical procedures are qualified and trained in those procedures.

The *Guide* (page 61) recognizes that personnel performing surgery on research animals have a wide range of educational backgrounds and might require various levels and kinds of training to ensure that good surgical technique is used. To assist the AV and IACUC in developing appropriate training programs, the Academy of Surgical Research (ASR) developed and published training guidelines for research surgery commensurate with a person's formal education and training background.[17] A research institution should also perform a continuing and thorough assessment of surgical outcomes to ensure that appropriate procedures are followed.

Opin. The goal is to assess surgical competence before IACUC approval of a protocol. The IACUC should review an individual's education, training, certification,[18] and experience for assessment of general surgical competence and qualifications to perform the specific surgical procedure. The ASR training guidelines noted previously[17] can be used in evaluating educational background to determine what an individual may already know. For example, a physician trained in a surgical specialty is expected to be competent to perform surgery on animals within his area of surgical expertise. However, he may require training in interspecies variations of anatomy, anesthesia, analgesia, and postoperative care methods. The need for formal training can be waived if the surgeon participates in a multidisciplinary team approach to perform the specific surgical protocol and includes additional experienced personnel qualified to work with animals.[2,17] To assure a surgeon's competence, the IACUC may require that a laboratory animal veterinarian observe or assist with at least the first surgery. Assistance could continue with subsequent procedures until competency is achieved and a specific surgical procedure is predictable to the satisfaction of the veterinarian. Although there is no single credential to assure competency, the best credential is a documented record of previous successful performance of the proposed surgical procedure on the specified species, demonstrating minimal operative and postoperative complications.[19] With such documentation, there should be no need for additional training. Following IACUC approval of a surgical protocol, continuing assessment of surgical outcomes should be performed. Participation and input from a laboratory animal veterinarian, animal care staff, surgical technicians, and the investigator are needed to assure ongoing use of appropriate procedures, to address complications and evaluate their rate of occurrence, and to initiate necessary corrective changes.

18:16 What is meant by perioperative care?

Opin. Perioperative care encompasses all events associated with a surgical procedure.[2] A perioperative care program is comprised of three overlapping components:[2,20]

- Preoperative planning and management.

- Intraoperative care.
- Postoperative care.

Detailed descriptions of the perioperative care program components have been published.[2,9,20] *Preoperative* planning should include:

- Identifying members of the multidisciplinary surgical team, all of whom should provide input into presurgical planning.
- Roles and training needs of personnel.
- Equipment and supplies required.
- Facilities for conducting procedures.

An anesthetic protocol should be developed including anesthetic agents, techniques, and methods of anesthetic monitoring to be used. Planning of the surgical procedure should include aseptic techniques to be used (see 18:8) and assessment of indications for perioperative antibiotics. A postoperative care plan should be outlined and the location and facilities for postoperative recovery identified.[2,9,20] (*Guide*, page 61). *Preoperative* management should include: [2,20] (*Guide*, page 61).

- A preoperative animal-health assessment with a physical examination and laboratory examination if indicated.
- A period of stabilization to a new environment for animals before undergoing surgical procedures.
- Preoperative fasting of a specified duration if indicated for the species to be used
- Administration of preoperative medications or antibiotics.

Components of *intraoperative* care include:[2]

- Monitoring of anesthetic level and vital organ function.
- Providing vital organ support such as parenteral fluid administration, supplemental oxygen, and maintenance of body temperature.
- Proper surgical technique which is comprised of:
    - Gentle tissue handling
    - Effective hemostasis
    - Maintenance of sufficient blood supply to tissues
    - Asepsis
    - Accurate tissue apposition
    - Proper use of surgical instruments

- Appropriate use of monitoring equipment
- Expeditious performance of the surgical procedure

The *postoperative* period can be divided into three overlapping phases: recovery from anesthesia, acute postoperative care, and long-term postoperative care.[2] Frequent assessment of thermoregulation and cardiovascular and respiratory function is required during anesthetic recovery and acute postoperative care. Additional care may include:

- Monitoring of the surgical incision.
- Thermal support to combat hypothermia.
- Parenteral fluid administration to maintain hydration.
- Administration of analgesics for postoperative pain.
- Administration of prophylactic antibiotics and other drugs.

Long-term postoperative care following anesthetic recovery and adequate physiologic stabilization requires at least once a day monitoring until suture removal (usually at 10 to 14 days postoperatively) and any postoperative complications are resolved. Monitoring of vital signs, hydration, feed and fluid intake, feces and urine output, attitude and activity, surgical wound condition, body weight, and signs of postoperative pain and infection should be performed. Special diets, analgesics, bandaging, antibiotics, and other medications may be indicated. Following suture removal, postoperative care required depends upon the species and surgical procedure. For example, chronic catheters or other partially exteriorized implants require ongoing monitoring and care. Monitoring of body weight should be scheduled throughout the postoperative period.[2]

18:17 What level of perioperative care is appropriate for rodents when compared to nonrodents?

Opin. The general components of the perioperative care programs (see 18:16) are the same for rodents and nonrodents. Nevertheless, some elements of the programs differ. For rodent surgery, the surgical "team" may be reduced to one person who serves as surgeon, anesthetist, surgical technician, and scrub nurse. One person performing surgery on multiple animals at one sitting, as frequently occurs in rodent surgery, results in the need for careful presurgical planning to assure availability of all supplies and equipment required to perform surgery and support necessary modifications in standard aseptic technique[9,11] (*Guide*, page 63). The preoperative animal-health assessment should include vendor supplied colony health testing (e.g., serology) and a visual examination rather than a physical examination of each animal when received.[2] Serologic testing of sentinel animals in the facility or other health surveillance results also may be useful.

Although sophisticated methods for intraoperative monitoring of anesthetized rodents are available when scientifically required, such methods might not be practical or possible in many research situations. Simply observing chest wall movement to determine respiratory rate and palpating the apical pulse through the chest wall may be sufficient to assess cardiovascular and respiratory stability.[21] Procedures used in larger animal species for intraoperative vital organ support such as intravenous fluid therapy can be difficult to use in rodents and may require other routes of administration of fluids or other strategies for supporting circulating blood volume.[2]

The same intensity of monitoring and supportive care commonly provided for larger animals during recovery from anesthesia and acute postoperative care is not practical or feasible for rodents. In addition, performing surgery on multiple animals at one sitting which frequently occurs requires developing methods of supportive care for the anesthetic recovery of multiple animals simultaneously.[21]

18:18 How much detail should be provided in the IACUC protocol relative to perioperative care?

Opin. In this author's opinion, the perioperative care detail that should be provided in the IACUC protocol includes:

- Names, qualifications, and responsibilities of participating personnel including the surgeon, the person who will be administering and monitoring anesthesia, and the person who will perform postoperative care.
- Location where surgery will be done.
- Duration of fasting with rationale if greater than 24 hours.
- Perioperative medications and antibiotics, dose, volume, route and frequency of administration, duration of treatment.
- Intraoperative monitoring and methods to be used to assess adequate anesthesia level.
- How surgical anesthesia will be monitored if neuromuscular blocking agents are administered.
- Aseptic techniques to be used.
- Frequency of animal monitoring during anesthetic recovery and postoperative period.
- Postoperative monitoring and care.

- Indication of compliance with postoperative monitoring and care guidelines in the institution's IACUC handbook.
- Indicate expected health changes or possible postoperative complications and describe methods of monitoring and care.
- Postoperative analgesics to be given; provide agent, dose, route and frequency of administration; and provide criteria for determining need for analgesics.

(See 18:21 to 18:23.)

Surv. Twelve IACUC protocol forms were surveyed. Perioperative care detail requested on those protocol forms include:

• Surgeon's name and qualifications and/or names and qualifications of participating personnel	12/12
• Name of person administering and monitoring anesthesia	5/12
• Location where surgery will be done	12/12
• Describe preoperative preparation of the animals	6/12
– Preoperative medications and antibiotics, dose, volume, route and frequency of administration, duration of treatment	
• Duration of fasting	5/12
• Describe intraoperative monitoring and methods to be used to assess adequate anesthesia	8/12
• How surgical anesthesia will be monitored if neuromuscular blocking agents are administered	9/12
• Describe intraoperative monitoring and supportive care	2/12
– How is body temperature and fluid balance to be maintained and monitored, and heart and respiratory rate monitored?	1/12
• Aseptic techniques to be used for survival surgery	7/12
– IACUC form has guidelines requiring and/or describing aseptic technique	6/12
• Where animals will be recovered from anesthesia	4/12
• Frequency of animal monitoring during anesthetic recovery	4/12
• Frequency of monitoring during postoperative care	4/12
• Who will perform postoperative care?	8/12
• Describe postoperative monitoring and care	10/12

- List antibiotics to be given, dose and route, frequency and duration of administration   4/12
- Specify other medications to be given, dose and route, frequency and duration of administration   5/12
- Requests indication of compliance with guidelines in institution's IACUC handbook   3/12
• Indicate expected health changes or possible complications and describe methods of care   8/12
• List postoperative analgesics to be used, dose, route, and frequency of administration   9/12
• Will postoperative analgesics be given routinely?   3/12
  - If given as needed, provide criteria for determining need   3/12

18:19 Can expired pharmaceuticals be used for survival surgery? For nonsurvival surgery?

Reg. APHIS/AC Policy #3[12] states that the use of expired medical materials such as pharmaceuticals "on regulated animals is not considered to be acceptable veterinary practice and does not constitute adequate veterinary care." Policy #3 specifies that expired anesthetics, analgesics, and emergency drugs cannot be used on any regulated animals regardless of the procedure intended. Regarding animals undergoing survival surgery, expired drugs and fluids cannot be used. As to animals undergoing *nonsurvival* surgery, expired pharmaceuticals, other than anesthetics, analgesics, and emergency drugs, can be used if "their use does not adversely affect the animal's well-being or compromise the validity of the scientific study."[12] Facilities should have a written policy addressing the use of expired drugs in nonsurvival surgery or require that their intended use be described in the investigator's IACUC protocol submitted for approval.[12]

Opin. In this author's opinion, expired drugs and fluids of any kind should not be used in animals undergoing survival surgery. The use of expired anesthetics, analgesics, euthanasia, and emergency drugs should not be allowed regardless of the procedure intended. Without written documentation of safety and efficacy from the manufacturer and IACUC approval of an investigator's written request for their use in nonsurvival surgery, use of expired drugs and fluids should not be allowed in animals undergoing nonsurvival surgery.

Surv. Two institutions' written policies on the use of expired medical materials were retrieved from their Web Internet sites. Neither institution allowed the use of expired drugs of any kind on an animal undergoing survival surgery. The use of expired anesthetics, analgesics, and emergency drugs also was banned, as this applies to nonsurvival surgery as well. Both institutions allowed the use of expired intravenous fluids in nonsurvival surgery; however, one institution required that the investigator determine that such use would not jeopardize research results or have undesirable effects on the animal's physiology. This institution also requires IACUC approval of an investigator's written request for use of expired drugs in nonsurvival surgery. The other institution allows the use of "nonemergency drugs" in nonsurvival surgery. Although the issue of expired drug use is explicitly addressed by APHIS/AC only, the policy of both institutions applies to all research animals including birds, rats, and mice.

Postings on the Compmed Internet discussion group regarding the use of expired drugs were retrieved. The procedures followed by the six IACUCs represented included not allowing any use of expired drugs (five IACUCs) and, in one, a 6-month extension was allowed on expired drugs for nonsurvival surgery. In the latter instance the investigator was required to obtain written documentation of safety and efficacy from the manufacturer.

18:20 Should the IACUC request information on the type of suture materials that will be used?

Opin. The PI should describe the type of suture material and method of wound closure on the protocol form to assure that they are appropriate for the species and the surgical incision. To facilitate wound healing, the type of suture material used to close the skin should be chosen to minimize tissue reaction and potential for wound infection. For example, to produce a lesion on healthy skin, an inoculum of 10^6 or more *Staphylococcus aureus* organisms is required, but in the presence of a silk suture, the required inoculum is reduced to less than 10^3 bacteria.[22] The capillary action of braided suture material and the inherent difficulty in keeping a wound clean can combine to increase the probability of infection at the surgical site.[20] Appropriate wound closure methods such as subcuticular sutures should be considered when animals may bite or pick at exposed skin sutures.

Surv. Twelve IACUC protocol forms were surveyed. Four IACUCs requested the method of skin or wound closure and the type of suture to be used. Two IACUCs requested inclusion of the type of suture material to be used in the description of the surgical procedures.

18:21 Should the IACUC request information about when sutures (or clips) will be removed, if that is necessary?

Opin. The appropriate time for suture removal should be provided to investigators in an institutional IACUC guidebook which should contain guidelines on postoperative care. Rather than request that the PI restate when sutures or clips will be removed, a written indication on the protocol form by the PI that she has read and will comply with the postoperative care guidelines provides adequate assurance that timely suture removal will be done. If written IACUC guidelines are not available to the PI, the IACUC should request that the PI state on the protocol form when sutures or clips will be removed.

18:22 Should the IACUC request information on how rodents or other small animals will be kept warm during and after surgery?

Opin. Due to the large surface area to body mass ratio of rodents and other small animals, heat loss resulting in hypothermia is likely to occur without adequate thermal support during surgery and recovery from anesthesia. Maintenance of normal body temperature significantly reduces anesthesia-related cardiovascular and respiratory disturbances (*Guide*, page 63) and is often critical to the successful recovery of rodents from anesthesia.[21] Methods to combat hypothermia should be provided to investigators in an institutional IACUC guidebook which should contain guidelines for rodent surgery and postoperative care. A written indication on the protocol form by the PI that he has read and will comply with these guidelines provides adequate assurance that appropriate thermal support will be provided. If written IACUC guidelines describing methods to combat hypothermia are not available to the PI, the IACUC should request that the PI state on the protocol form how rodents or other small animals will be kept warm during and after surgery.

18:23 Should the IACUC request information about the frequency of postoperative observations?

Opin. The frequency of postoperative observation is determined by the nature of the surgical procedure and the stage of recovery.[2] Guidelines for minimum frequency of postoperative observation should be provided to investigators in an institutional IACUC guidebook included with guidelines for postoperative care. A written indication on the protocol form by the investigator that she has read and will comply with these guidelines provides adequate assurance that appropriate postoperative observation frequency will be carried out. However, if a surgical procedure necessitates more frequent observations than the guidelines recommend, the customized monitoring frequency and circumstances requiring monitoring should be described in the protocol. If there are no written IACUC guidelines available to the PI, the IACUC should request that the PI provide information about the frequency of postoperative observations on the protocol form.

18:24 When is an animal considered to be sufficiently recovered to return it to its home cage?

Opin. In this author's opinion and that of others,[2,23,24] an animal should remain in a recovery area until recovered from anesthesia, physiological parameters are adequately stabilized, and it has regained normal ambulatory and protective behaviors before being moved to its home cage.

18:25 Is there a need for a separate recovery area or can animals be recovered in their home cage?

Reg. The *Guide* (page 79) describes that "a postoperative-recovery area should provide the physical environment to support the needs of the animal during the period of anesthetic and immediate postsurgical recovery and should be so placed as to allow adequate observation of the animal during this period." The species and types of surgical procedures will dictate the type of caging and support equipment required, but should be designed to support physiologic functions, such as thermoregulation and respiration.

Opin. During anesthetic recovery, an animal should be in an area that is warm, safe, quiet, comfortable, and appropriate for the needs of the individual animal and species.[25] Dedicated recovery rooms are recommended for species such as rabbits, cats, dogs, and swine.[24] A separate recovery area provides the necessary physical environment, ambient temperature control, and equipment for monitoring and supportive care needed to manage complications that may occur during recovery. Animals can be housed individually in appropriately sized cages designed to avoid injury to occupants. Individual housing also prevents potential trauma from cagemates. In addition, recovery areas can be located to facilitate the appropriate frequency of monitoring an animal by personnel.

Rodents usually recover from anesthesia in the same laboratory where surgery is performed.[24] Provisions to support body temperature are frequently necessary as thermal support is often critical to the successful recovery of rodents from anesthesia.[21] Recovering rodents should be housed individually to prevent injury by cagemates. Although extensive monitoring may not be possible, the animals should be frequently observed until recovered from anesthesia and adequately stabilized before being returned to their home cages.

In this author's opinion and that of others,[2,24] small ruminants too large to fit into recovery cages and large farm animals can recover in their own pen or stall if provisions are made for adequate observation, warmth, and animal safety.

18:26 What records should investigators keep relative to surgery and the perioperative period?

Reg. APHIS/AC Policy #3[12] requires appropriate postoperative recordkeeping in accordance with accepted veterinary procedures for species covered. The *Institutional Animal Care and Use Committee Guidebook*[14] states that recordkeeping for major surgical procedures on nonrodent mammals should include an anesthetic monitoring record and surgeon's report. Records kept during the anesthetic recovery period should document that the animal was observed until it was extubated and able to stand. Postoperative care records should reflect a minimum of daily observations until skin suture removal or surgical incisions have healed which is generally considered the endpoint of the postoperative period. The *Guide* (pages 62 and 63) states that appropriate medical records should be maintained during and after the anesthetic recovery period.

Opin. There are records appropriate to each phase of the perioperative period which include documentation of an animal's preoperative health and physiological status, the anesthetic record kept primarily during the intraoperative phase, the surgeon's report written after completion of a surgical procedure, and medical records documenting monitoring and care during recovery from anesthesia and the postoperative phase.

Preoperative records should include a health profile, and medical and vaccination history, when applicable, provided by the vendor. A physical exam to assess an animal's health status should be performed. A preanesthetic evaluation should include records of body weight, body temperature, heart rate, and respiratory rate to establish baseline data. The need for diagnostic radiographic tests and laboratory evaluation of blood, urine, and feces will depend on the animal species, research protocol, and health history of the animal or colony.[2] Preoperative administration of medications or antibiotics should be documented.

An anesthetic record should document anesthetic administration, monitoring of anesthetic depth including reflexes, and frequent assessment of the physiologic status of an animal including body temperature, and cardiovascular and respiratory function.[26] The extent of recordkeeping and monitoring sophistication depends upon the potential complications associated with the anesthetic regimen, the surgical procedure, and the animal species.[2] Minimal documentation should include vital signs, core body temperature, heart rate and respiratory rate, recorded at an average interval of every 15 minutes, as well as anesthetic administered, dose, route, and time of administration.

The postoperative period can be divided into three overlapping phases which are recovery from anesthesia, acute postoperative care, and long-term postoperative care.[2] The extent of recordkeeping and intensity of monitoring during the postoperative period depends upon the species and the surgical procedure. Generally, the most intensive and frequent monitoring is required during recovery from anesthesia "when the animal is most vulnerable to the potentially adverse effects of anesthetic agents, hypothermia, and physiologic disturbances secondary to surgery."[24] As occurs during anesthesia, frequent assessment of thermoregulation and cardiovascular and respiratory function should continue during anesthesia recovery and acute postoperative care until adequately stabilized. Vital signs, including body temperature, pulse rate and rhythm, and respiratory rate and rhythm, should be recorded approximately every 15 minutes during anesthesia recovery with less frequent monitoring possible as an animal stabilizes. Documentation of clinical observations including mucous membrane color, capillary refill time, pain assessment, and surgical wound condition are recommended. Administration of analgesics, antibiotics, parenteral fluids, and other medications should be recorded including the drug name, dosage, route, and time given. A brief description of the surgery should be recorded with the postoperative care records or as a separate surgeon's report.

During long-term postoperative care following anesthetic recovery and adequate physiologic stabilization, monitoring should continue at least once a day until suture removal (usually at 10 to 14 days postoperatively) and any postoperative complications are resolved. Vital signs, hydration, feed and fluid intake, feces and urine output, attitude and activity assessment, pain assessment, surgical wound condition and care, and monitoring for postoperative infection should be recorded. All postoperative care and medications administered as described above with anesthesia recovery should be documented. Following suture removal, frequency and parameters monitored depend on the species and surgical procedure. For example, chronic catheters or other partially exteriorized implants require documentation of ongoing monitoring and care. Monitoring of body weight should be scheduled throughout the postoperative period.

References

1. Silverman, J., Protocol review: whose responsibility is it? *Lab. Anim.*, 23(2), 22, 1994.
2. Brown, M.J., Pearson, P.T., and Tomson, F.N., Guidelines for animal surgery in research and teaching, *Am. J. Vet. Res.*, 54, 1544, 1993.
3. Potkay, S., Garnett, N., Miller, J.G., Pond, C.L., and Doyle, D.J., Frequently asked questions about the public health service policy on humane care and use of laboratory animals, *Contemp. Topics Lab. Anim. Sci.*, 36(2), 47, 1997.
4. U.S. Department of Agriculture, Animal and Plant Health Inspection Service, Policy #14, Major survival surgery single vs. multiple procedures, April 14, 1997. Available on the World Wide Web at: http://www.aphis.usda.gov/ac/policy14.html
5. Potkay, S., Garnett, N., Miller, J.G., Pond, C.L., and Doyle, D.J., Frequently asked questions about the Public Health Service policy on humane care and use of laboratory animals, *Lab. Anim.*, 24(9), 24, 1995.
6. Office for Protection from Research Risks, Dear Colleague Letters: Sources of Custom Antibody Production, Number 95-02, Animal Welfare, March 8, 1995. Available on the World Wide Web at: http://www.nih.gov:80/grants/oprr/dc95-3.htm
7. Silverman, J., Protocol review: no recovery? *Lab. Anim.*, 23(4), 20, 1994.
8. National Research Council, Survival surgery and postsurgical care, in *Education and Training in the Care and Use of Laboratory Animals: A Guide for Developing Institutional Programs*, Report of the Institute of Laboratory Animal Resources Committee on Educational Programs in Laboratory Animal Science, National Academy Press, Washington, D.C., 1991, 61.
9. Cunliffe-Beamer, T.L., Applying principles of aseptic surgery to rodents, *AWIC Newsl.*, 4(2), 3, 1993.
10. Bradfield, J.F., Schachtman, T.R., McLaughlin, R.M., and Steffen, E.K., Behavioral and physiologic effects of inapparent wound infection in rats, *Lab. Anim. Sci.*, 42, 572, 1992.
11. Brown, M.J., Aseptic surgery for rodents, in *Rodents and Rabbits: Current Research Issues*, Niemi, S.M., Venable, J.S., and Guttman, H.N., Eds., Scientists Center for Animal Welfare, Bethesda, MD, 1994, 67.
12. U.S. Department of Agriculture, Animal and Plant Health Inspection Service, Policy #3, Veterinary care, April 14, 1997. Available on the World Wide Web at: http://www.aphis.usda.gov/ac/policy3.html
13. Slattum, M.M., Maggio-Price, L., DiGiacomo, R.F., and Russell, R.G., Infusion-related sepsis in dogs undergoing acute cardiopulmonary surgery, *Lab. Anim. Sci.*, 41, 146, 1991.
14. U.S. Department of Health and Human Services, Public Health Service, Institutional Animal Care and Use Committee Guidebook, National Institutes of Health Publication No. 92-3415, Bethesda, MD, 1992, Chap. B-6.
15. Banks, R., Protocol review: a question of technique, *Lab. Anim.*, 23(5), 22, 1994.
16. Federation of Animal Science Societies. *Guide for the Care and Use of Agricultural Animals in Agricultural Research and Teaching*, 1st revised ed., Federation of Animal Science Societies, Savoy, IL, 1999, 21.

17. Academy of Surgical Research, Guidelines for training in surgical research in animals, *J. Invest. Surg.*, 2, 263, 1989.
18. Dennis, M.B., Surgical research specialist certification, in *Proc. Laboratory Animal Welfare Training Exchange 1998 Meeting*, St. Louis, MO, August 20–21, 1998, 53.
19. Dennis, M.B., Surgical training and personnel qualifications, in *Research Animal Anesthesia, Analgesia and Surgery*, Smith, A.C. and Swindle, M.M., Eds., Scientists Center for Animal Welfare, Greenbelt, MD, 1994, 11.
20. Brown, M.J. and Schofield, J.C., Perioperative care, in *Essentials for Animal Research: A Primer for Research Personnel*, Bennett, B.T., Brown, M.J., and Schofield, J.C., Eds., National Agricultural Library, Washington, D.C., 1994, 79.
21. Wixson, S.K. and Smiler, K.L., Anesthesia and analgesia in rodents, in *Anesthesia and Analgesia in Laboratory Animals*, Kohn, D.F., Wixson, S.K., White, W.J., and Benson, G.J., Eds., Academic Press, New York, 1997, 165.
22. McCurnin, D.M. and Jones, R.L., Principles of surgical asepsis, in *Textbook of Small Animal Surgery*, 2nd ed., Slatter, D.H., Ed., W.B. Saunders Co., Philadelphia, 1993, 114.
23. White, W.J. and Blum, J.R., Design of surgical suites and postsurgical care units, in *Anesthesia and Analgesia in Laboratory Animals*, Kohn, D.F., Wixson, S.K., White, W.J., and Benson, G.J., Eds., Academic Press, New York, 1997, 149.
24. Smith, A.C. and Swindle, M.M., Post surgical care, in *Research Animal Anesthesia, Analgesia and Surgery*, Smith, A.C. and Swindle, M.M., Eds., Scientists Center for Animal Welfare, Greenbelt, MD, 1994, 167.
25. Haskins, S.C. and Eisele, P.H., Postoperative support and intensive care, in *Anesthesia and Analgesia in Laboratory Animals*, Kohn, D.F., Wixson, S.K., White, W.J., and Benson, G.J., Eds., Academic Press, New York, 1997, 379.
26. Brown, M.J., Principles of anesthesia and analgesia, in *Essentials for Animal Research: A Primer for Research Personnel*, Bennett, B.T., Brown, M.J., and Schofield, J.C., Eds., National Agricultural Library, Washington, D.C., 1994, 39.

19

Antigens, Antibodies, and Blood Collection

Harold F. Stills, Jr.

Introduction

Modern biologic research techniques often require the production of specific polyclonal and monoclonal antibodies as an essential component of the research protocol. Production of both polyclonal and monoclonal antibodies may require the immunization of animals, therein presenting the IACUC with the dilemma and duty of evaluating the immunization procedures and schedules with respect to potential animal pain and distress. Published guidelines from a number of sources are available to assist the IACUC in developing policies and procedures for animal use in antibody production.[1-4] Although *in vitro* alternatives to the use of live animals in the production of antibodies is often appropriate (see 19:16), this chapter primarily focuses on those circumstances in which the IACUC approves the use of live animals for antibody production.

Production of a quality antibody requires an appropriate stimulation of the immune system. The initial immune activating event involves processing of the antigen by antigen-presenting cells (APCs) which are primarily the macrophages and the dendritic cells, such as the Langerhans cells in the epidermis of the skin. Activation of these cells along with the recruitment of additional cells for antibody production is often enhanced by the addition of various biologically active compounds (primarily bacterial products) in the adjuvant. Protecting the antigen from rapid degradation in the body and permitting the slow release of the antigen to the APCs is another function often performed by the adjuvant. The end result of the use of an adjuvant is both an increased antibody response along with the production of a substantial inflammatory response.

Of the available adjuvants in use today, none has proved more effective overall than Freund's complete adjuvant (FCA).[5,6] First described in 1942,[7,8] it is composed of paraffin oil, mannide monooleate, and killed mycobacteria. Injected into animals, FCA produces a chronic granulomatous inflammatory reaction at the injection site.[9] Granulomas often may be detected in draining lymph nodes, spleen, kidney, and other organs where microdroplets of the emulsion have been distributed by the vascular or lymphatic systems. Intradermal injections routinely produce ulcerations as early as 12 to 14 days postinjection and persisting for 8 weeks or longer.[9] The microscopic lesions produced by injection of FCA are primarily those of granulomatous inflammation and focal necrosis.[9-11]

The surveys in this chapter resulted from a questionnaire sent to 37 academic and 9 commercial institutions. There was somewhat less than a 50% return rate.

19:1 Should special justifications be required for the use of Freund's complete adjuvant (FCA) in antibody production?

Reg. (See 19:2.)

Opin. The dramatic inflammatory reaction generated by the injection of FCA has restricted its use to research antibody production and also resulted in efforts to encourage the use of other adjuvants. However, in many instances the alternative adjuvants chosen also produce severe inflammatory reactions.[12,13] All adjuvants, by nature of their mode of action, induce an inflammatory reaction. Since many of the compounds for which antibodies are required are poorly immunogenic, adjuvants are essential. The choice of adjuvant should be based upon sound scientific rationale which considers the specific immunogen, the species being immunized, and the desired antibody.

Surv. Does your institution require special justifications for the use of FCA?

- Approve use without special justifications   17/21
- Approve use only with special justifications   4/21

Institutions requiring special justification for the use of FCA also required justifications for any adjuvant.

19:2 Is the use of Freund's complete adjuvant (FCA) a painful or distressful procedure?

Reg. APHIS/AC Policy #11 lists the use of FCA under "examples of procedures that can be expected to cause more than momentary or slight pain."[14] That same policy goes on to state that FCA used for antibody production may cause perturbations ranging from momentary or slight pain to severe pain, depending on the product, procedure, and species. PHS Policy (U.S. Government Principle IV) and the *Guide* (page 64) require the minimization of discomfort, distress, and pain in concert with good science.

Opin. Although APHIS/AC considers the use of FCA as likely to be painful, other reports have failed to document any pain or distress in animals injected with FCA[10,11,15] or in tuberculin-negative humans.[16,17] In our experience, uncomplicated FCA-antigen emulsion injections are not characterized by tenderness, irritation, or any other indication of distress in the animal. Feed consumption and body weights are unaffected by FCA-antigen injections.

The choice of immunogen, regardless of the choice of adjuvant, may induce an autoimmune condition that is associated with pain and distress. The adjuvant-induced model of arthritis is a prime example and has even been recommended as a model of chronic clinical pain.[18] (See 20:23.)

Surv. What APHIS/AC pain classification does your institution assign to projects using FCA in antibody production?

- Category D (alleviated pain or distress)   7/21
- Category A (no pain or distress)   14/21

19:3 Should the IACUC require different routes of antigen-adjuvant injection to have different APHIS/AC pain level classifications?

Opin. The type of inflammatory reaction produced by any adjuvant, including Freund's complete adjuvant, is not dependent upon the route of injection. Therefore, any differences in pain level classification results from the effect of the inflammatory reaction upon the tissues involved. For the common routes of adjuvant injection (intradermal, subcutaneous, and intramuscular), there is no evidence of any differences in pain or distress to the animal. (See 19:11.)

Surv. Does your institution require different pain classifications for different routes of antigen-adjuvant injection?

- No differences in classification   21/21

19:4 Should the amount of killed *Mycobacteria* in the Freund's complete adjuvant (FCA) be limited?

Opin. Formulations of FCA containing no more than 0.1 mg/ml of dry mycobacterial cell mass have been recommended as a method to reduce the associated inflammatory response.[9] Nevertheless, commercially available formulations of FCA generally range from 0.5 to 1.0 mg/ml of dry mycobacterial cell mass.

Surv. Does your institution limit the amount of *Mycobacteria* in FCA, and if so, what is the limit?

- Limit 0.5 to 1.0 mg/ml   2/21
- No limit set   19/21

19:5 What is the maximum volume of Freund's complete adjuvant (FCA)-antigen emulsion that the IACUC should permit for injection in any single site? What is the maximum total volume that should be permitted?

Opin. Recommended total injection volumes and injection volumes per site vary greatly. For the intradermal route of injection in the rabbit, recommendations vary from maximum volumes of 0.05 ml/site[1,2,19] to 0.1 ml/site in some laboratory immunology texts[17] and other reports.[20] Recommended subcutaneous injection volumes per site are generally higher, ranging from 0.1 ml/site[1] to 0.5 ml/site in the rabbit,[20,21] 0.4 ml/site in the guinea pig, and 0.1 ml/site in the mouse.[2] Intramuscular injection volumes are generally the highest with recommended maximums in the rabbit varying from 0.25 ml/site[21] to 0.5 ml/site[1,2] with a recommended maximum of 0.1 ml/site in larger rodents.[2] In farm animals the recommendations are 1 ml/site intramuscularly.[2]

Surv. What is the maximum volume of FCA-antigen that your institution approves for injection in any single site (intradermally, subcutaneously, and intramuscularly)? What is the maximum total volume of FCA-antigen that your institution approves for injection?

TABLE 19.1

Survey of Maximum Volume Permitted for FCA Injections (Intradermal, Subcutaneous, and Intramuscular)

Species	Maximum Volume (ml) per Site		
	Mean	Range	Number of Respondents
Rabbit	0.22	0.1–0.4	17
Mouse	0.11	0.025–0.25	14
Rat	0.10	0.05–0.2	10
Goat/sheep	0.18	0.1–0.25	5
Guinea pig	0.13	0.05–0.25	8
Species	Maximum Total Volume (ml)		
	Mean	Range	Number of Respondents
Rabbit	1.39	0.4–4.0	14
Mouse	0.40	0.125–0.80	11
Rat	0.54	0.25–0.80	7
Goat/sheep	1.10	0.5–2.0	5
Guinea pig	0.42	0.25–1.0	6

19:6 Should the intradermal (ID) route of injecting Freund's complete adjuvant (FCA)-antigen emulsions be permitted by an IACUC?

Opin. The dramatic inflammatory reaction created by the injection of FCA has led to a number of recommendations regarding injection routes. ID injection of FCA leads to skin ulcerations and the formation of large granulomas.[9-11] The Canadian Council on Animal Care Guidelines on Acceptable Immunological Procedures[2] limits the ID use of FCA in rabbits to only those instances where the induction of cell-mediated immunity is the purpose. Other authors have indicated a preference for the ID route.[4,10,19] Our preference is for the ID route of injection. While the ulcerative granulomas are typical with ID injections, granuloma formation is typical with FCA injection by any route. The localization of the granuloma lesion that is seen with ID injections tends to minimize the total tissue destruction. The ID route also permits monitoring of the lesion's development. These advantages, coupled with the potentially increased immune response due to exposure of the dendritic Langerhans cells of the dermis, favor the ID route for both antibody production and animal well-being.

Surv. Does your institution permit the intradermal injection of FCA-antigen emulsions?

- Intradermal injection permitted   19/20
- Intradermal injection not permitted   1/20

19:7 Should the subcutaneous (SC) route of injecting Freund's complete adjuvant (FCA)-antigen emulsions be permitted by an IACUC?

Opin. The SC route of FCA-adjuvant emulsion injection has been recommended in the literature due to the ease of the injection technique[11] and the usual lack of ulcerations as commonly seen following intradermal injections. The extension of the granulomatous inflammation and formation of fistulous tracts is a drawback of this route of injection.[9-11] The SC injection of FCA-adjuvant emulsions is, in our opinion, an acceptable route of injection, especially in smaller animals where the intradermal route is technically more difficult. Even when the SC route is used, multiple injection sites are recommended to maximize the presentation of the antigen to the immune system.

Surv. Does your institution permit the subcutaneous injection of FCA-antigen emulsions?

- Yes, subcutaneous injection is permitted   20/20
- No, subcutaneous injection is not permitted   0/20

19:8 Should the intramuscular (IM) route of injecting Freund's complete adjuvant (FCA)-antigen emulsions be permitted by an IACUC?

Opin. The IM route is the third most common route of injection in larger animals, after subcutaneous and intradermal injections. IM injections are generally effective for antibody production and have been recommended.[2,19] Granuloma formation, muscle necrosis, and the formation of fistulous tracts have, however, all been reported[9-11] and the difficulty in evaluating lesions has led to cautions.[11,19] Because of the extensive tissue destruction and the lack of any specific advantages of the IM route over either the intradermal or subcutaneous routes, we do not recommend the IM route of injection.

Surv. Does your institution permit the intramuscular injection of FCA-antigen emulsions?

- Permit intramuscular injection   17/20
- Do not permit intramuscular injection   3/20

19:9 Should the intraperitoneal (IP) route of injecting Freund's complete adjuvant (FCA)-antigen emulsions be permitted by an IACUC?

Opin. The IP route of injection is often recommended for use in rodents.[2] The formation of fibrous adhesions, abdominal fluid distension, and the potential associate clinical signs have been described.[19,22,23] The potential complications of IP injection of FCA should be closely evaluated by all IACUCs prior to approving a project which proposes to use this route of injection. Scientific justification for the necessity of IP injection of FCA should be required. For immunization, there is little justification for the IP route over the intradermal or subcutaneous routes of injection.

Surv. Does your institution permit the intraperitoneal injection of FCA-antigen emulsions?

- Permit intraperitoneal injections   11/16
- Do not permit intraperitoneal injections 5/16

19:10 Should the intravenous (IV) route of injecting Freund's complete adjuvant (FCA)-antigen emulsions be permitted by an IACUC?

Opin. The IV injection of FCA-antigen emulsions has been associated with adjuvant emboli in the lungs of rabbits and multiple pulmonary granulomas.[24] Several publications have stated that the IV route for FCA-antigen emulsion injections should not be permitted.[2,20] The IV injection of any nonwater-miscible compound may result in the formation of emboli and have serious consequences. We do not believe

	that there is any acceptable justification for the IV injection of FCA-antigen emulsions. If IV injection is desired (e.g., as in hybridoma formation), then the antigen should be injected in an aqueous form, with or without an aqueous soluble adjuvant.
Surv.	Does your institution permit the intravenous injection of FCA-antigen emulsions?

- Permit intravenous injection — 5/20
- Do not permit intravenous injection — 15/20

19:11 Should footpad injections of Freund's complete adjuvant (FCA)-antigen emulsions be permitted by an IACUC?

Opin.	As shown below, the footpad injection route for FCA-antigen emulsions in rabbits was prohibited by all the institutions responding to the survey, corresponding to the recommendations in the literature.[2,4,19] The use of the footpad injection route in rodents is generally restricted to cases where scientifically justified and then only in one foot. Even in instances where a localized immune reaction is scientifically justified (e.g., lymphocyte isolation from the popliteal lymph node), justification for not using the dorsum of the foot instead of the footpad should be given. In this author's opinion, there are few, if any, acceptable scientific justifications for footpad injections in rodents.
Surv.	Does your institution permit the footpad injection of FCA-antigen emulsions?

- Approve footpad injections in rabbits — 0/20
- Approve footpad injections in rodents — 3/20
- Do not approve footpad injections in rodents — 17/20

19:12 What procedures should be required by the IACUC to minimize the potential for contamination of injection sites during the injection of adjuvant-antigens?

Opin.	Using aseptic procedures for antigen-adjuvant injections, including sterile preparation of the antigen-adjuvant mixture, clipping the injection site and sterile scrubbing of the injection site with an antiseptic has been recommended in the literature.[4,10,11] The injection of viable contaminating bacteria with a Freund's complete adjuvant-antigen emulsion often results in the formation of an abscess with the influx of neutrophils, exudation, hyperemia, and pain.
Surv.	What procedures does your institution require for minimizing contamination of injection sites? More than one response per institution is possible.

- Require sterile preparation of antigen-adjuvant mixture   17/20
- Require clipping of fur prior to injection   13/20
- Require surgical scrub to prepare injection site   12/20

19:13 What type of justification or information should be required by the IACUC for that committee to approve a schedule of antigen-adjuvant injections?

Opin. The administration of booster injections prior to the optimum time not only increases the potential of pain and distress but also may be detrimental to the ultimate antibody level and affinity.[10,25-28] Animals with a robust developing immune response may develop an arthus reaction (an acute inflammatory and painful reaction) at the booster injection site The antigen, quantity of antigen, adjuvant used, and injection route interact in determining the optimum schedule for immunization and, optimally, the individual animal's antibody titer should be followed to determine when the antibody level has plateaued and booster injections are indicated.[29]

Surv. A Does your institution require investigators to justify the proposed time interval between antigen-adjuvant injections?

- Require a justification   3/21
- Do not require a justification   18/21

Surv. B Does your institution require investigators to monitor serum titers before booster injections are given?

- Require monitoring   1/21
- Do not require monitoring   20/21

19:14 Should investigators be required to search for commercially available, acceptable antibodies?

Reg. The AWAR (§2.31,d,1,ii; §2.31,d,1,iii) and the PHS Policy (IV,D,1,a) require that PIs provide assurances that, in the case of painful animal procedures, alternatives have been searched for and found unacceptable and that the proposed activities are not duplicative.

Opin. The availability of a commercially available and acceptable antibody is a clear and definite example of an alternative to additional animal use and duplicative procedures. In fulfilling its responsibilities, IACUCs often require literature searches for alternative methods of antibody production and fail to require literature searches for the

	more obvious and more likely possibility, i.e., that an acceptable antibody is commercially available. The availability of both print[30,31] and Internet resources[32] for locating commercially available antibodies makes the task reasonable.
Surv.	Does your institution require the investigator to search for commercially available antibodies and, if available, justify why these would not be acceptable?

- Require assurance that commercial antibodies are
  not available   3/21
- Require PI to search a standard reference   1/21
- No search required   17/21

19:15 What oversight actions should an IACUC consider if an investigator has contracted with a commercial laboratory for the production of antibodies in animals?

Reg.	PHS Policy II requires that any institution using animals with PHS funds, even if the funds are awarded by subcontracting or subgranting, must either have an approved Animal Welfare Assurance or be included as a component under the primary institution's Assurance. If both institutions have approved assurances, it is only necessary that one of the institutions' IACUC review and approve the activity. The subcontracting or subgranting institution, however, retains partial accountability for "... providing effective oversight mechanisms to ensure compliance with the PHS Policy."[33]
Opin.	The contracting of antibody production is a difficult problem for many institutions. The availability of numerous commercial contracting companies along with the many alliances between peptide-generating and antibody-producing companies makes institutional oversight a nightmare. Institutional IACUCs must be diligent in their attempts to identify potential situations and be proactive in the education of PIs as to the necessity of notifying and working with the IACUC when contracting antibody production from a commercial firm.
Surv.	If an investigator proposes to contract with a commercial laboratory for the production of either a polyclonal or monoclonal antibody, what type of review, if any, does your IACUC perform?

- No review   4/17
- Check for PHS Assurance and USDA license   8/17
- Perform some level of protocol review and approval
  in addition to checking for PHS Assurance and
  USDA license   5/17

Several of the institutions responding also stated that they felt uncomfortable about whether or not they would be aware of such subcontracting if and when it occurred.

19:16 Should additional justifications be requested by the IACUC for the production of monoclonal antibodies using the mouse ascites method?

Reg. NIH/OPRR issued a report directing IACUCs to "... critically evaluate the proposed use of the mouse ascites method. ... IACUCs must determine that (1) the proposed use is scientifically justified; (2) methods that avoid or minimize discomfort, distress, and pain (including *in vitro* methods) have been considered; and (3) the latter have been found unsuitable."[34]

Opin. The development of practical and reasonably priced *in vitro* methods for hybridoma growth coupled with evidence that the mouse ascites method causes discomfort has led many IACUCs to require that *in vitro* methods be utilized. Published comparisons between the *in vitro* methods and the ascites method have shown the results to be highly dependent upon the specific hybridoma.[35] In our opinion, there are few justifications for the use of the ascites method for monoclonal antibody production. All hybridomas can be propagated *in vitro* and the levels of monoclonal antibody production can be increased by the use of specifically designed growth chambers, hollow-fiber bioreactor systems, etc. In all instances, the burden of proof should be upon the PI to show that numerous *in vitro* methods have been attempted and were unsuccessful prior to the IACUC approving the ascites method of monoclonal antibody production. (See 12:17 to 12:19.)

A recent report[36] commissioned by the NIH summarizes the findings of the National Academy of Sciences as to the scientific necessity for producing monoclonal antibodies by the mouse method, including ways to minimize any pain or distress that might be associated with that method. The report also discusses regulatory considerations for the mouse method and summarizes the current stage of development of tissue culture methods.

Surv. Does your institution require additional justifications for using the mouse ascites method for monoclonal antibody production?

- Additional justifications needed   10/16
- No additional justifications needed   6/16

19:17 Should the IACUC limit the number of allowable peritoneal taps for the collection of ascitic fluid?

Opin. The process of removing the excessive peritoneal fluid from mice used in monoclonal antibody production requires the use of a large gauge needle and, often, some anesthesia. Many institutions limit the total number of peritoneal taps (with the last tap being postmortem) while other institutions place limits on the animal's physical condition.[4] Each hybridoma line, based to a large degree upon the plasmacytoma fusion partner, behaves somewhat differently when growing in the animal's peritoneal cavity. Certain hybridomas are extremely invasive, resulting in bloody peritoneal fluid on the first tap. The tumor masses hybridomas which form in the peritoneal cavity may be either disseminated or solid and singular.[37] Likewise, the dynamics of monoclonal antibody production from specific hybridomas also varies.[38] Absolute limitations on the number of taps which fail to recognize the differences between hybridoma lines is inappropriate; however, there is little reason for the maximum number of taps to exceed three.

Surv. Does your institution limit the number of peritoneal taps permitted when using the mouse ascites methods for monoclonal antibody production and, if so, what is the limit?

- Limit to two to three taps   12/19
- No absolute limits   7/19

19:18 Should the IACUC limit the amount of pristane or Freund's complete adjuvant (FCA) used to prime the peritoneal cavity for ascites production?

Opin. Several authors have suggested that lowering the volume of pristane or FCA injected to prime the peritoneal cavity of mice may reduce potential pain and distress.[19,39] Scientific studies, however, have been contradictory with one study indicating that 0.5 ml of pristane was the most effective for ascites production[40] while another study found no significant differences in ascites production following 0.1 or 0.5 ml of pristane.[41] Both FCA and pristane "prime" the peritoneal cavity for hybridoma growth by inducing a granulomatous inflammatory reaction (peritonitis). Since pain is a common sequel to peritonitis, IACUCs should require that the minimum effective quantity of FCA or pristane be used in those instances where the *in vivo* ascites method of monoclonal antibody production is warranted.

Surv. Does your institution limit the volume of pristane or FCA used to prime the peritoneal cavity for mouse ascites production and, if so, what is the limit?

- Limit pristane or FCA from 0.1 to 0.5 ml (mean 0.3 ml)   13/19
- No limit   6/19

19:19 What is the limit for an acceptable volume of blood withdrawal from a single collection?

Opin. Numerous studies have associated severe hemodynamic changes and hemorrhagic shock with blood losses greater than 30% of the total blood volume.[42,43] Smaller volume losses in rats (15 to 20% of the blood volume) reduce cardiac output by nearly 50%.[44,45] Overall, the recommendation of a maximum withdrawal of 15% of blood volume appears reasonable.[46]

Surv. What is the maximum blood volume withdrawal that your institution accepts as reasonable for a single withdrawal?

- Limit of 0.8 to 2.0% of body weight (mean 1.4%)   8/17
- Limit 5.0 to 25.0% of total blood volume (mean 13.9%)   9/17

19:20 What is the limit for an acceptable volume of blood withdrawal over a set period of time?

Opin. One standard recommendation in the literature lists a maximum withdrawal of 15% of blood volume (~1% of body weight) as a limit without additional justification and monitoring procedures.[4] Another reference[46] suggests a maximum weekly withdrawal of 7.5% of blood volume. (See 16:31.)

Surv. What is the maximum blood volume withdrawal and interval that your institution would accept as reasonable for multiple blood withdrawals from a single animal?

- Limit of 0.06% to 1.0% of body weight (mean 0.36%) per day   7/13
- Limit of 0.33% to 1.07% of blood volume (mean 0.63%) per day   6/13

19:21 What clinical tests should be required if repeated blood sampling is a component of the protocol?

Reg. The AWAR (§2.33b,2), the PHS Policy (IV,C,1,e), and the *Guide* (pages 55 to 66) all require the provision of adequate veterinary care for the diagnosis, control, and treatment of diseases and injuries.

Opin. Standard hematology tests are often used to monitor the effects of multiple blood withdrawals upon an animal's well-being. The hematocrit and hemoglobin concentration are commonly used because of the ease of these tests and the availability of rapid results. While these tests are generally excellent indicators of anemia in animals subjected to blood collections, they do have their limitations. The hematocrit is unchanged immediately after blood withdrawal and underestimates blood loss for up to 72 hours after blood removal.[47] Hemoglobin concentration is even slower in responding to blood loss, and in human blood donors is lowest between 1 and 2 weeks after donation.[48]

Surv. Does your institution's IACUC require that clinical hematology tests be performed on animals where repeated blood sampling is a portion of the protocol?

- No routine testing necessary   11/18
- Require some clinical laboratory tests   7/18

Those institutions not requiring routine tests commonly indicated that they relied upon the veterinary staff to monitor and stop the blood collection if anemia should become a problem.

References

1. Grumpstrup-Scott, J. and Greenhouse, D.D., NIH intramural recommendations for the research use of complete Freund's adjuvant, *ILAR News*, 30(2), 9, 1988.
2. Canadian Council on Animal Care, *CCAC Guidelines on Acceptable Immunological Procedures*, Canadian Council on Animal Care, Ottawa, Canada, 1991.
3. Workman, J.P., Balmain, A., Hickman, J.A., McNally, N.J., Rohas, A.M., Mitchison, N.A., Pierrepoint, C.G., Raymond, R., Rowlatt, C., Stephens, T.C., et al., UKCCCR guidelines for the welfare of animals in experimental neoplasia, *Lab. Anim.*, 22, 195, 1988.
4. Jackson, L.R. and Fox, J.G., Institutional policies and guidelines on adjuvants and antibody production, *ILAR J.*, 37(3), 141, 1995.
5. Munoz, J., Effect of bacteria and bacterial products on antibody response, *Adv. Immun.*, 4, 397, 1964.
6. Altman, A. and Dixon, F.J., Immunomodifiers in vaccines, *Adv. Vet. Sci. Comp. Med.*, 33, 301, 1989.
7. Freund, J. and McDermott, D., Sensitization to horse serum by means of adjuvants, *Proc. Soc. Exp. Biol. Med.*, 49, 548, 1942.
8. Freund, J., The mode of action of immunologic adjuvants, *Adv. Tuberc. Res.*, 7, 130, 1956.
9. Broderson, J.R., A retrospective review of lesions associated with the use of Freund's adjuvant, *Lab. Anim. Sci.*, 39, 400, 1989.
10. Stills, H.F., Jr., Polyclonal antibody production, in *The Biology of the Laboratory Rabbit*, Manning, P.J., Ringler, D.H., and Newcomer, C.E. (Eds.), Academic Press, New York, 1994, 435.
11. Stills, H.F., Jr., and Bailey, M.Q., The use of Freund's complete adjuvant, *Lab. Anim.*, 20(4), 25, 1991.

12. Johnson, D.K., Adjuvant comparison in rabbits, in *Rodents and Rabbits. Current Research Issues,* Proceeding of a SCAW and WARDS conference, Niemi, S.M., Venable, J.S., and Guttman, H.N., Eds., Washington, D.C., 1993, 77.
13. Johnson, D.K., Adjuvant comparison in rabbits, *Sci. Anim. Care,* 5(2), 2, 1994.
14. U.S. Department of Agriculture, Animal and Plant Health Inspection Service, Animal Care, Policy #11. Painful/Distressful Procedures, April 14, 1997. Available on the World Wide Web at: *www.aphis.usda.gov/ac/policy11.html*
15. Smith, D.E., O'Brien, M.E., Palmer, V.J., and Sadowski., J.A., The selection of an adjuvant emulsion for polyclonal antibody production using a low-molecular weight antigen in rabbits, *Lab. Anim. Sci.,* 42, 599, 1992.
16. Chapel, H.M. and August, P.J., Report of nine cases of accidental injury to man with Freund's complete adjuvant, *Clin. Exp. Immunol.,* 24, 538, 1995.
17. Hughes, L.E., Kearney, R., and Tully, M., A study in clinical cancer immunotherapy, *Cancer,* 26, 269, 1970.
18. Besson, J.M. and Guilbaud, G., Eds., *The Arthritic Rat as a Model of Clinical Pain?* Excerpta Medica, Amsterdam, 1988.
19. Amyx, H.L., Control of animal pain and distress in antibody production and infectious disease studies, *J. Am. Vet. Med. Assoc.,* 191, 1287, 1987.
20. Harlow, E. and Lane, D., Adjuvants, in *Antibodies: A Laboratory Manual,* Cold Spring Harbor Laboratory, Cold Spring Harbor, NY, 1988, 96.
21. Johnston, B.A., Eisen, H., and Fry, D., An evaluation of several adjuvant emulsion regimens for the production of polyclonal antisera in rabbits, *Lab. Anim. Sci.,* 41, 15, 1991.
22. Toth, L.A., Dunlap, A.W., Olson, G.A., and Hessler, J.R., An evaluation of distress following intraperitoneal immunization with Freund's adjuvant in mice, *Lab. Anim. Sci.,* 39, 122, 1989.
23. Lipman, N.S., Trudel, L.J., Murphy, J.C., and Sahali, Y., Comparison of immune response potentiation and *in vivo* inflammatory effects of Freund's and Ribi adjuvants in mice, *Lab. Anim. Sci.,* 42, 193, 1992.
24. Brooks, R.D., Betz, R.D., and Moore, R.D., Injury and repair of the lung: response to intravenous Freund's adjuvant, *J. Pathol.,* 124, 205, 1978.
25. Herbert, W.J., The mode of action of mineral oil emulsion adjuvants on antibody production in mice, *Immunology,* 14, 301, 1968.
26. Hu, J.-G., Yokoyama, T., and Kitagawa, T., Studies on the optimal immunization schedule of experimental animals. IV. The optimal age and sex of mice, and the influence of booster injections, *Chem. Pharm. Bull.* 382, 448, 1990.
27. Hu, J.-G., Ide, A., Yokoyama, T., and Kitagawa, T., Studies on the optimal immunization schedule of the mouse as an experimental animal. The effect of antigen dose and adjuvant type, *Chem. Pharm. Bull.,* 37, 3042, 1989.
28. Hu, J.-G. and Kitagawa, T., Studies on the optimal immunization schedule of experimental animals. VI. Antigen dose response of aluminum hydroxide-aided immunization and booster effect under low antigen dose, *Chem. Pharm. Bull.,* 38, 2775, 1990.
29. Hanley, W.C., Artwohl, J.E., and Bennett, B.T., Review of polyclonal antibody production procedures in mammals and poultry, *ILAR J.,* 37(3), 93, 1995.
30. Linscott, W.D., *Linscott's Directory of Immunological and Biological Reagents* or the *Manufacturers' Specifications and Reference Synopsis Catalog for Primary Antibodies,* William D. Linscott, 4877 Grange Road, Santa Rosa, CA, 1988.

31. Weimer, R.V., Jr. *Manufacturers' Specifications and Reference Synopsis Catalog: Primary Antibodies*, 3rd ed., Aerie Corp., Birmingham, MI, 1995.
32. The Internet resource for locating certain antibodies and suppliers of antibodies is available on the World Wide Web at: *www.antibodyresource.com*
33. Potkay, S., Garnett, N.L., Miller, J.G., Pond, C.L., and Doyle, D.J., Frequently asked questions about the Public Health Service policy on humane care and use of laboratory animals, *Lab. Anim.,* 24(9), 24, 1995.
34. Ellis, G. and Garnett, N., OPRR Reports, Number 98-01, November 17, 1997. Available on the World Wide Web at: *www.nih.gov/grants/oprr/dc98-01.htm*
35. Peterson, N.C. and Peavey, J.E., Comparison of *in vitro* monoclonal antibody production methods with an *in vivo* ascites production technique, *Contemp. Topics Lab. Anim. Sci.,* 37(5), 61, 1998.
36. Committee on Methods of Producing Monoclonal Antibodies, *Monoclonal Antibody Production,* National Academy Press, Washington, D.C., 1999. Available on the World Wide Web at: *http://www.nap.edu/books/0309064473/html/*
37. Jackson, L.R., Trudel, L.J., Fox, J.G., and Lipman, N.S., Monoclonal antibody production in murine ascites, I. Clinical and pathologic features, *Lab. Anim. Sci.,* 49, 70, 1999.
38. Jackson, L.R., Trudel, L.J., Fox, J.G., and Lipman, N.S., Monoclonal antibody production in murine ascites, II. Production characteristics, *Lab. Anim. Sci.,* 49, 81, 1999.
39. McGuill, M.W. and Rowan, A.N., Refinement of monoclonal antibody production and animal well-being, *ILAR News,* 31(1), 7–11, 1989.
40. Brodeur, B.R., Tsang, P., and Larose, Y., Parameters affecting ascites tumour formation in mice and monoclonal antibody production, *J. Immunol. Meth.,* 71, 265, 1984.
41. Hoogenraad, N.J. and Wraight, C.J., The effect of pristane on ascites tumor formation and monoclonal antibody production, *Meth. Enzymol.,* 121, 381, 1986.
42. Noble, D. and Gregerson, M.I., Blood volume in clinical shock. II. The extent and cause of blood volume reduction in traumatic hemorrhage and burn shock. *J. Clin. Invest.,* 25, 172, 1946.
43. Williams, W.J., Beutler, E., Erslev, A.J., and Lichtman, M.A., *Hematology,* 3rd ed., McGraw-Hill, New York, 1983.
44. Saperstein, L.A., Saperstein, E.H., and Bredemeyer, A., Effect of hemorrhage on the cardiac output and its distribution in the rat, *Circ. Res.,* 8, 135, 1960.
45. Ploucha, J.M. and Fink, G.D., Hemodynamics of hemorrhage in the conscious rat and chicken, *Am. J. Physiol.,* 251, R486, 1986.
46. McGuill, M.W. and Rowan, A.N., Biological effects of blood loss: implications for sampling volumes and techniques, *ILAR News,* 31(4), 5–18, 1989.
47. Lee, G.R., The normocytic anemias, in *Wintobe's Clinical Hematology,* 9th ed., Lee, G.R., Bithell, T.C., Foerster, J., Athens, J.W., and Lukens, J.A., Eds., Lea and Febiger, Philadelphia, 1993, 885.
48. Wadsworth, G.R., Recovery from acute hemorrhage in normal men and women, *J. Physiol.* (London), 129, 583, 1955.

20

Occupational Health and Safety

Stefan Wagener and Susan Stein*

Introduction

The use of animals for research purposes has always been directly or indirectly linked to human health concerns. Animal allergens, zoonotic diseases, and physical injuries caused by bites and scratches are only some of the issues that need to be addressed in a general occupational health program associated with animal research. Although regulations and guidelines dealing with human health and protection are a valuable resource for establishing a comprehensive occupational health program, their scope and level of detail will always be limited. Fortunately, most current health and safety regulations are performance oriented, focusing on the goal (e.g., safety) rather than describing how to get there. This allows institutions to tailor their safety programs to meet individual needs and protect employees in the most effective way. While reviewing the following chapter, the reader should keep an important concept in mind: the usefulness of a safety program is to be measured by its success.

General Occupational Health

20:1 What is the general responsibility of the IACUC toward ensuring a safe work environment for persons working with laboratory animals?

Reg.   The need to protect the health and safety of employees involved in animal-based research is addressed in the *Guide* (page 14) and the

---

* Contributing to the chapter were Robert J. Ceru and Kristin Erickson.

PHS Policy (IV,A,1,f). The PHS Policy (IV,A,1) requires institutions to use the *Guide* as the basis for their institutional program for activities involving animals. The *Guide* states that institutions must establish an Occupational Health and Safety (OHS) program as part of the overall animal research program. The *Guide* (page 14) specifically references the National Research Council publication, *Occupational Health and Safety in the Care and Use of Research Animals*,[1] to be used as a tool for establishing such a comprehensive OHS program. While the main responsibility of the OHS program lies with the institution, the IACUC is specifically charged with the oversight and evaluation of the animal care and use program. Since the OHS program needs to be part of the overall animal care and use program (*Guide*, page 14), the IACUC also has to assume at least partial responsibility for the program.

Opin. None of the referenced guidelines and regulations go into detail in describing or specifically addressing the IACUC's responsibility for occupational health and safety. Therefore, the institution must manage and assign responsibilities covering the various programs required by the *Guide*. The quality of the occupational health and safety program will be significantly impacted by the quality of interaction that the IACUC has with relevant institutional functions like the research program, environmental health and safety, occupational health services, and others.

Institutions are required to address OHS issues and assign responsibilities to various groups or individuals inside the institution for management and oversight of the different components. For example, occupational health requires the input of medical professionals, while physical, chemical, and biological hazards are best addressed by specific committees or safety professionals credentialed in these areas. Ideally, the IACUC assumes the role of a facilitator, interacting with the various programs, committees, and individuals to ensure the existence of a comprehensive OHS program and compliance with the *Guide*. The institution must define and delegate roles and responsibilities.

20:2 What is the general responsibility of the IACUC toward ensuring a safe work environment for persons having access to an animal facility but not working with laboratory animals (e.g., maintenance, clerical personnel, visitors)?

Reg. As stated in 20:1, it is the institution that is primarily responsible for the establishment of an employee health and safety program. The PHS Policy (IV,A,1f) requires a health program for personnel having frequent contact with animals, as well as all persons working in laboratory animal facilities. According to the *Guide* (page 14), the level and extent of personnel participation in the occupational health and safety program should be based on a risk assessment. In addition, the

National Research Council publication *Occupational Health and Safety in the Care and Use of Research Animals*[1] as referenced by the *Guide*, includes not only animal caretakers, technicians, students, volunteers, investigators, and veterinarians, but also facility maintenance personnel, housekeepers, security, and other staff.

Opin. Ideally, the IACUC facilitates the close interaction among institutional groups dealing with employee health and safety as it relates to animal research facilities. For example, the institution's Environmental Health and Safety Office oversees the health and safety of maintenance personnel working in areas with known chemical and biological hazards. The institutional occupational health group and the medical professionals identify any necessary surveillance criteria. The assistance of these and other health and safety groups should be requested by the IACUC to perform risk assessments, project review, and health and safety recommendations.

20:3 Does the IACUC have a responsibility to ensure a safe working environment for research technicians and investigators using animals outside of the animal facility?

Reg. The PHS Policy (IV,A,1,f) is applicable to all institutional programs involving animals, not only in animal facilities. Therefore, the occupational health and safety program must address all risks to investigators and technicians associated with PHS-supported animal activity, regardless of where it occurs.

Opin. The IACUC does not have any specific responsibility as it pertains to a safe work environment for research technicians and investigators using animals outside of the animal facility unless it is specifically charged by the institution with this responsibility. General and specific regulatory health and safety requirements (e.g., Occupational Safety and Health Administration regulations), however, do apply to worker protection in and outside of a facility.

Hazards associated with animal research outside of a facility can be of great concern. One example is "field research" involving wild animals. The incorrect use of equipment to trap or immobilize wild animals can pose a significant physical hazard. Zoonotic diseases, such as rabies or Hantavirus Pulmonary Syndrome, can be life threatening. Prudent practices include the review of all animal use protocols by the relevant institutional safety groups or individuals to assure and maintain worker safety outside of the facility.

20:4 What types of hazards are relevant to animal research settings?

Opin. While animal research settings involve some unique hazards related to working with the animals, such common workplace hazards as sharps injuries, burns, and falls also occur. In addition, the research might involve certain physical, chemical, or biological hazards. An overview of relevant physical, chemical, biological, and protocol-related hazards has been published.[1]

In general, physical hazards in the animal research environment can include bites, scratches, and kicks; sprains and strains caused by moving equipment; operational hazards involving electricity, machinery, and noise; or protocol-induced hazards like radioactivity. Chemical hazards are directly related to the specific agent. Examples include chemicals used for processing tissues, and cleaning or disinfecting research equipment. The accidental inhalation of anesthetics is another concern. Research protocols might require the application of toxins, carcinogens, and other hazardous chemicals. Biological hazards, like infectious pathogens, can be introduced through naturally occurring or experimentally infected research animals or include the application of recombinant DNA technology. Another significant hazard is manifested in the increasing development of animal-related allergies.[1,2]

20:5 What are the relevant federal agencies and institutional committees that the IACUC should interface with relative to personal safety?

Opin. Since the IACUC has primarily an oversight responsibility (see 20:1), issues related to personal safety should be addressed by committees and individuals specifically charged with regulatory compliance, worker safety, and environmental health. These groups and individuals interact directly with agencies like Occupational Health and Safety Administration (OSHA), Nuclear Regulatory Commission (NRC), Centers for Disease Control and Prevention (CDC), etc. It is ultimately the institution's responsibility to establish the necessary oversight and management structure to address personal safety issues. If the IACUC needs to address personal safety issues as part of animal use review, it should seek assistance from the appropriate institutional safety committees and groups. These include, but are not limited to occupational health, biosafety, chemical safety, radiation safety, and public safety.

20:6 Should radiation and biosafety issues be approved by these respective committees before final approval of a protocol is granted by the IACUC?

Reg. All use of radioactive materials in research requires prior approval as outlined in the Nuclear Regulatory Commission (NRC) license for the specific facility or institution. NRC requires the establishment of an onsite radiation safety program.[3] Certain biological agents including recombinant DNA and certain infectious agents and toxins are regulated by agencies such as the Centers for Disease Control and Prevention,[4,5] USDA,[6] NIH,[7] and others. Specific programs and approval of protocols may be necessary depending on the type of agent.

Unless the IACUC is specifically charged with radiation and biosafety responsibility, it will have to rely on safety professionals and safety committee review for assistance. Since the use of certain hazardous materials or agents in animal research requires mandatory approval prior to initiation of the project, the IACUC is well advised to seek the relevant review and approval prior to or simultaneously with their own animal use review. Options are to include safety professionals as members of the IACUC or, more practically, make animal use protocols available to safety committees or safety professionals for review and approval. In this way the IACUC can document that experienced committees and individuals have assured compliance with relevant safety and health standards.

20:7 Must an institution working with any hazardous substance have a safety officer?

Reg. A biological safety officer is required by NIH[7] if the institution is involved in large-scale research or production activities involving viable organisms containing recombinant DNA molecules or performs recombinant DNA research at Biosafety Level 3 (BL3) or Biosafety Level 4 (BL4). (See 20:8.)

A radiation safety officer is required by the Nuclear Regulatory Commission, depending on the type of "Domestic Broad Scope License" for byproduct material issued to the institution.[8]

A chemical hygiene officer (chemical safety officer) is required by the Occupational Safety and Health Administration.[9]

Opin. Most institutions have assigned their safety officers far more responsibility and oversight than initially mandated. The goal should be the establishment of a comprehensive safety program that includes all areas of research and support services.

20:8 Must an institution working with any *bio*hazardous substance have a *bio*safety officer?

Reg. The only agency currently requiring the function or position of a biological safety officer is the NIH as part of the requirements for recombinant DNA research[7] (Section IV-B-3). As outlined in those guidelines: "The institution shall appoint a biological safety officer if it engages in large-scale research or production activities involving viable organisms containing recombinant DNA molecules." In addition, "The institution shall appoint a biological safety officer if it engages in recombinant DNA research at BL3 or BL4."

Opin. Most institutions known to these authors have gone beyond this basic requirement and have appointed a biosafety officer to oversee all aspects of biological safety. Biological safety officers commonly manage institutional programs addressing the Occupational Health and Safety Administration's requirements for blood-borne pathogens (e.g., human immunodeficiency virus, hepatitis B virus), and tuberculosis. They oversee compliance with Centers for Disease Control and Prevention programs, develop biosafety operating procedures, and perform training and education in a variety of other areas. To maintain a high standard of proficiency and education, certification and testing programs for biological safety professionals have been developed by the American Biological Safety Association (ABSA) and the National Registry of Microbiologists (NRM). The IACUC should consult the biological safety officer in all areas related to biosafety.

20:9 Should the IACUC establish safe working rules for an animal facility or laboratory?

Reg. The IACUC has no specific regulatory mandate for establishing safe working rules unless they pertain specifically to animal health and care and are based on the *Guide* or other relevant animal use and care regulations. (See 20:1.)

Opin. Safe working rules are a necessity for all facilities whether or not they are used for research with animals. Most safety rules and policies are hazard specific and based on regulatory requirements addressing those hazards. For example, occupational safety issues in laboratories working with chemical hazards are part of the Occupational Safety and Health Administration's Laboratory Safety Standard.[9] The chemical hygiene officer of the facility is the logical person to establish OSHA-based safe working rules for the facility or laboratory if hazardous chemicals are involved. In the end, it will be up to the institution to clearly define the responsibilities as well as setting the standards for health and safety.

20:10 What roles does the AV have in evaluating the impact of biohazards in the animal facility?

Opin. Biohazards in general or specific infectious agents can have a significant effect on the health of animals. Disease prevention needs to be an essential component of a comprehensive animal use and care program and, as outlined in the *Guide* (pages 12 to 13, 57), is the responsibility of the AV who is certified or has training or experience in laboratory animal science and medicine involving the species being used. The *Guide* (page 13) emphasizes the position and the responsibility of the AV. It is in the IACUC's best interest to support the AV in all aspects of a disease prevention program. There is a special need for close collaboration when a research project submitted to the IACUC involves the use of naturally occurring pathogens, experimentally induced infectious disease, or wild caught domestic and exotic species. The AV and biosafety officer can assess the potential impact and highlight any concerns for the IACUC.

20:11 Should the IACUC require occupational health examinations for those involved in animal care and use? If so, what should be the scope and extent of such a program in terms of personnel and diagnostic procedures?

Reg. The *Guide* (pages 14 to 19) outlines an occupational health program, as required in the PHS Policy (IV,A,1,f). This has been summarized by NIH/OPRR.[10] NIH/OPRR states that "basic elements of any health program, however, should provide: a preemployment medical evaluation and history; immunization against tetanus; detailed training on how to perform required procedures safely; instruction in personal hygiene, zoonoses, and precautions for pregnant women and others at risk; protective clothing and devices; instruction in first aid procedures appropriate to potential hazards; and access to medical attention for the treatment of animal bites, scratches, allergies, and other job-related injuries or illnesses."

Opin. The decision on health examination is usually based on the type of work performed, the occupational risks involved, the species of animal, and type and duration of animal contact required. All these factors are included in a project specific risk assessment that needs to be performed in cooperation with healthcare professionals. Rather than requiring blanket health examinations, the IACUC needs to ensure that occupational health professionals are involved in tailoring the health assessment program for all personnel involved in animal care and use.

20:12 What sources of information can the IACUC access relative to common hazards?

Opin. The *Guide* (pages 85 and 86) provides a listing of excellent references of recommended readings on biohazards and environmental contaminants including the Centers for Diseases Control and Prevention/NIH publication *Biosafety in Microbiological and Biomedical Laboratories*.[11]

The common hazards include chemical, physical, biological, mechanical, and environmental hazards and vary with type of research, geographic location, research species, facility design, and duration of exposure. The IACUC should rely on the safety officer and institutional veterinarian to collect and collate material from local, state, and federal agencies, Material Safety Data Sheet information and recognized experts when defining hazards. (See 20:32.) With the increasing use of the Internet and the World Wide Web, resources on health and safety become more and more readily available. For a comprehensive listing, including Web sites and mailing lists, refer to the 2nd edition of "Safety and Health on the Internet."[12]

Infectious Hazards

20:13 What is the definition of a biohazard?

Reg. Currently there are no regulatory-based definitions for biohazards. A regulatory definition for bloodborne pathogens has been published.[13] Recombinant DNA is defined in the NIH *Guidelines for Recombinant DNA Research*.[7] Certain infectious materials are defined in rules and regulations (e.g., Centers for Disease Control and Prevention,[4,5] Department of Transportation[14]) pertaining to shipment and transportation, import and export, transfer and acquisition.

Opin. Biohazards (or biological hazards) refer commonly to agents or materials of biological origin that are potentially hazardous to humans, animals, or plants. Although there is no standardized definition for biohazards, institutions often include infectious agents, recombinant materials, toxins, tissue cultures, and other biological materials in their definition.

20:14 What should be the goals of the IACUC with respect to use or presence of biohazards in the animal facility?

Opin. The IACUC has a responsibility to address biohazard-related issues. The minimum goal should be to protect animals and personnel from unintentional exposure to hazardous biological materials and agents.

The IACUC should be familiar with the various test, quarantine, and health surveillance procedures instituted by the AV before, during, and after the arrival of animals, tumors, cell lines, and tissue cultures at the institution. These procedures may vary with the species and source of the animals, and source of the biological materials. From a personal safety perspective, the IACUC should seek the input of the biological safety officer for a risk assessment, determination of appropriate biological containment levels, practices, and procedures as well as regulatory requirements.

20:15 Must an institution using biohazardous materials have an independent Institutional Biosafety Committee (IBC), or can the IACUC assume that function for hazards related to animal use?

Reg. As outlined in the *Guide* (page 14), the institution has the responsibility to establish formal safety programs to assess the hazards, determine the safeguards needed for specific safety concerns, and ensure adequate training and facilities. The requirement for an Institutional Biosafety Committee (IBC) is based on NIH recombinant DNA guidelines[7] (Section IV,B,2) and the IBC oversight is primarily for recombinant DNA research. It is up to the institution to assign oversight and management of certain biosafety issues to appropriate committees including the IACUC. However, for all issues pertaining to recombinant DNA, the IBC must be involved in the protocol review and approval process.

Opin. According to NIH[7] (Section IV,B,2), the scope of the IBC is not necessarily limited to recombinant DNA. Institutions, therefore, are free to expand the scope and responsibility of their IBC. In many cases local IBCs are part of the establishment of a comprehensive biological safety program reviewing all aspects of infectious disease work. In other instances, IBCs mainly focus on recombinant DNA while the institutional biosafety officer oversees the biosafety program. IACUCs should take an active role in the process of evaluating biohazardous material use involving research animals. Educating the IACUC members and having the IACUC assume responsibility for certain biosafety issues in agreement with institutional policies and practices is one option. The use of existing safety committees and safety officers for this purpose is highly recommended.

20:16 Should an IACUC member be a voting member of the Institutional Biosafety Committee (IBC)?

Reg. There is currently no requirement to have an IACUC member be a voting member of the IBC. However, NIH[7] recombinant DNA guidelines (Section IV,B,2,a) states that one member of the IBC shall have expertise in animal containment principles if the institution is involved in recombinant DNA research with animals, and those projects require official IBC approval.

Opin. At smaller institutions with limited staff and personnel, it can be a significant advantage to use an individual's expertise in a variety of functions. Institutions, therefore, might select an IACUC member with the appropriate expertise to be part of the local IBC. This double function, however, might involve a significant time commitment on the part of the individual.

20:17 Is approval from the Institutional Biosafety Committee (IBC) necessary before the IACUC can give final approval for biohazardous studies involving animals?

Reg. As discussed in 20:6, no specific requirements as to the timing of approval exist for the IACUC. Depending on the type of biohazard used (e.g., recombinant DNA), projects may need approval prior to initiation.

Opin. Institutional policies and procedures need to clearly establish the approval process for the use of biological agents and recombinant DNA. To meet applicable health and safety requirements and to reduce the workload for the IACUC, it is advantageous to have the project reviewed by the institutional biosafety officer or the IBC prior to IACUC approval.

20:18 What general guidelines can the IACUC follow to assure that biohazardous materials are being used safely in the animal facility and in individual laboratories?

Reg. The use of biohazardous materials including recombinant DNA is guided by a combination of laboratory practices and techniques, safety equipment, and special facilities commonly referred to as biosafety levels. One standard reference[11] contains information on four different biosafety levels specific for laboratories and vertebrate animals. The procedures and practices outlined in this publication have been widely accepted as a de facto standard and are used as a basis for a comprehensive biosafety program. Although not federally mandated, compliance is expected by funding agencies including USDA, NIH, the Food and Drug Administration, and others. Similar information is contained in Appendix G of the NIH guidelines[7] and compliance is mandatory for institutions receiving NIH funding. Local and state requirements should be consulted for all waste-related issues pertaining to biohazardous materials, since specific laws for the treatment and disposal of biohazardous waste have been established. Additionally, the use and disposal of controlled substances requires contact with the Drug Enforcement Administration (DEA) and compliance with the Controlled Substances Act of 1970[15] and any state or local requirements.

Opin. Because of the increasing complexity associated with use and disposal of biohazardous materials, the IACUC needs to rely on biosafety committees and/or biosafety officers for advice, oversight, and compliance. A publication is available[11] that can be used as a guidance document and should be available to all IACUC members. Additional information on biosafety is available through the American Biological Safety Association Website.[16] It is recommended to have the Biological Safety Officer or Institutional Biosafety Committee review and approve animal use protocols involving biohazardous materials prior to final IACUC approval. This process might involve the development of hazard-specific protocols outlining safety procedures, safety equipment, inspection, waste disposal, etc. Protocols requiring specific practices and procedures (e.g., personal protective equipment, immunization) should not only be posted on animal rooms including emergency information, but also read and signed by all personnel involved.

20:19 What generally accepted procedure should an IACUC and Institutional Biosafety Committee require relative to recapping of needles in the animal facility?

Reg. With the exception of the Occupational Safety and Health Administration's *Bloodborne Pathogen Standard*,[13] no regulatory requirements exist that specifically address the recapping of needles. (See 20:8.)

Opin. Accidental sharp injuries involving animal fluids are a significant occupational hazard and have resulted in human infection.[17] Prudent practice should require that needles are not recapped and are disposed of in approved sharp's containers. In certain circumstances, recapping might be necessary, but should only be done when all non-recapping alternatives, including the use of syringes with automated needle retraction function, cannot be employed. If necessary, training in these methods should be provided to investigators and staff.

20:20 If a potential respiratory hazard is present, should the IACUC demand a written respiratory hazard protection program that is approved by the Institutional Biosafety Committee (IBC)?

Reg. The Occupational Safety and Health Administration (OSHA)[18] requires the institution to develop and implement a written respiratory protection program with required worksite-specific procedures and elements for required respirator use. This program and the use of respirators is necessary if the employee is exposed to air contaminated with harmful dusts, fogs, fumes, mists, gases, smokes, sprays, or vapors, and these contaminants cannot be controlled or eliminated by other means (e.g., engineering controls). Typical examples of contaminants and their applications include disinfection of animal rooms with aerosolized chemicals and project-specific feed prepara-

Opin. tion including carcinogens, etc. In addition, OSHA requires that the institution designate a program administrator who is qualified by appropriate training or experience to administer or oversee the respiratory protection program and conduct the required evaluations of program effectiveness.[18]

Opin. The need for a respiratory protection program goes far beyond the limited scope of the IACUC or IBC. Neither of these committees generally has the necessary expertise and training and should not maintain or approve such a program. Institutions should assign this mandatory program to safety professionals (e.g., industrial hygienists with appropriate training and expertise). Program approval needs to be at the institutional level, not the committee level. It is important to note that OSHA's respiratory protection is primarily targeted at chemical and physical inhalation hazards with known concentrations and exposure levels. The only biological agent specifically identified as requiring respiratory protection is the tuberculosis (TB) causing *Mycobacterium* species.[19] Only for TB, have agencies like OSHA, National Institute of Occupational Safety and Health, and the Centers for Disease Control and Prevention established a set of rules and guidelines that clearly defines the type of respirator sufficient to offer the necessary protection.[20]

20:21 What common personal protective outer wear should the IACUC, in conjunction with the veterinary staff, require prior to entering a room housing macaque monkeys?

Reg. As a result of exposures with fatal outcome to *Cercopithecine herpes virus* 1 (B-virus), the Centers for Disease Control and Prevention (CDC) and the National Institute of Occupational Safety and Health (NIOSH) have issued specific guidance documents on personal protective equipment.[21,22] The *Guide* (page 17) specifically states, "Personnel exposed to nonhuman primates should be provided with such protective items as gloves, arm protectors, masks, and face shields."

Opin. The selection of appropriate personal protective equipment (PPE) should be based on the following process:

- Identification of the most likely hazards to be encountered.
- Assessment of the risk and any adverse effects caused by unprotected exposure.
- Identification of all other control measures available and feasible to be used instead of personal protective equipment.
- Performance characteristics for the required protection (e.g., splash or impact protection).
- Need for decontamination (e.g., reuse or disposal).

- Assessment of any constraints that might negatively influence the use of PPE (e.g., vision, dexterity).

PPE such as gowns, aprons, gloves, and eye or face protection is considered the last line of defense. It should only be relied on in cases where the specific hazard cannot be removed or contained with any other control measures. The primary hazard posed by macaques is a potential herpes-B virus exposure through infectious monkey body fluids, or bites and scratches caused by the animal or contaminated objects. The PPE selected should provide protection against the physical hazards as well as fluid exposure. Primary routes of entry are mucous membranes requiring appropriate splash protection in the form of splash goggles and masks (for nose and mouth). The use of face shields will offer additional face protection. Other routes of entry involve parenteral exposure through cuts, breaks in the skin, needle sticks, etc. PPE selection should address these risks and also be based on the task performed. Heavy duty reinforced leather gloves are appropriate for restraining animals, but inappropriate for handling syringes. In the latter case, double gloving is recommended, since it facilitates not only the safe change of gloves (due to contamination) without compromising skin protection, but also adds another level of safety. Depending on the task performed and in accordance with local and state regulations, PPE might have to be disposed of as biohazardous waste. It is highly recommended to use existing CDC and NIOSH guidelines[21,22] as the basis for a comprehensive B-virus protection program.

20:22 Many adventitious agents, such as mouse hepatitis virus, can be transmitted via cell lines or tumors. Is it necessary or advisable for an IACUC to request documentation that cell lines, tumors, or nonsterile biologic fluids are free of such agents before a protocol receives final IACUC approval?

Reg. The *Guide* (page 60) specifically states that "transplantable tumors, hybridomas, cell lines, and other biologic materials can be sources of murine viruses that can contaminate rodents" and advises appropriate testing to detect these agents in biological materials. PHS Policy (IV,A,1) requires institutions to follow the *Guide.*

Opin. Introduction of adventitious agents into an animal colony can alter research results, cause disease and death of animals, and produce a domino effect of waste of animal life and time. Additionally, some agents have zoonotic potential under the right conditions. Therefore, it is strongly advised that the IACUC requests documentation that such agents are not introduced into the animal facility. This role is frequently assigned to the AV as part of the veterinary care program.

20:23 Should the use of Freund's complete adjuvant, which has parts of *Mycobacterium tuberculosis* bacteria in it, be a health concern to the IACUC?

Opin. APHIS/AC Policy #11[23] refers to the use of adjuvants (including complete Freund's) as capable of producing pain, but does not address its role as a human health concern.

There are numerous references in the literature to the delayed hypersensitivity reactions induced by the inoculation of killed mycobacterium — an essential component of Freund's complete adjuvant. There is documentation that an initial, accidental inoculation of Freund's complete adjuvant can produce pain in people and repeat inoculations may produce ulcers and abscesses.[24] A Freund's complete adjuvant "exposure" also may interfere with future tuberculin skin test results. The IACUC should have an awareness of these sequelae and as such stress proper procedures and sharps disposal. (See 19:1, 19:2.)

20:24 Assume an agent such as a *Cercopithecine herpes virus* 1 (B-virus) is known or suspected to be harbored by an animal, but the virus is not part of the proposed research. Should this protocol be reviewed and approved by the Institutional Biosafety Committee (IBC) before receiving final IACUC approval?

Opin. In lieu of specific regulatory requirements, guidelines and recommendations for implementing a safety and health program specific for B-virus and nonhuman primates have been established by the Centers for Disease Control and Prevention[21,22] (CDC) and others.[25] None of these guidelines and recommendations require a separate IBC review or approval of research involving animals considered to be infected with herpes B-virus prior to final IACUC approval.

Herpes B-virus infects primarily primates of the genus *Macaca*. For that reason, CDC, as well as the National Research Council, considers all macaque monkeys in the research environment as being infected with B-virus unless they are specifically known to be free of that virus. It is virtually impossible to determine with certainty that an animal is B-virus free. Both agencies request the establishment of a comprehensive B-virus health and safety program if macaques are used or kept at the institution. The establishment of such a program should be a joint venture between the IACUC, the occupational health group, and the biological safety officer or IBC and the AV.

20:25 In 20:24, would the biosafety committee need to approve a protocol if the animal was a rabbit harboring *Pasteurella multocida*?

Opin. *Pasteurella multocida* is considered an agent of zoonotic disease, resulting in acute inflammation, fever, lymphangitis, and regional lymphadenitis. On rare occasions, septicemia and central nervous system involvement can occur as a result of *P. multocida* infection caused, for example, by an animal bite. If the animal is known or suspected of carrying *P. multocida*, animal handling protocols and precautionary measures need to be established preventing accidental transmission as part of a more comprehensive zoonotic disease transmission prevention program (*Guide*, pages 14 and 18).

Disease prevention is an important aspect of a comprehensive animal use and care program. Preventing the spread of an infectious agent from animal to animal or animal to human is only possible through appropriate awareness, training, and compliance with established safety procedures and practices. A successful disease prevention program relies on the close interaction of the AV, IACUC, occupational health groups, and health and safety professionals/committees. The main focus should be on how we best protect personnel and animals. Nevertheless, review and approval of protocols is important and necessary. Depending on the agent or disease involved, specific review and approval procedures can be established by the institution or required by the IACUC to assure human and animal protection.

20:26 What is meant by "recombinant DNA" and what are some common examples of the same?

Reg. Recombinant DNA is defined by the NIH Guidelines[7] as "molecules that are constructed outside living cells by joining natural or recombinant DNA segments to DNA molecules that can replicate in a living cell, or molecules that result from the replication of those described above."

Opin. The ability to isolate, manipulate, and express genetic material has resulted in a new field of basic and applied research called "genetic engineering." With the help of recombinant DNA, genes can be isolated, expressed, and transferred, resulting in a multitude of possible applications. The development of transgenic and "knockout" animals (e.g., mice) is very valuable for studying a variety of diseases. For example, genetically modified infectious agents are used for vaccine development in animal models. Transgenic technology has resulted in changing livestock by altering their biochemistry, their hormonal balance, and their protein products, to mention but a few applications.

20:27 Does the IACUC have a special responsibility under the NIH or other guidelines relative to reviewing studies which use recombinant DNA in animals?

Reg. No. Studies involving any type of recombinant DNA at the institution are under the oversight of the Institutional Biosafety Committee (IBC) as mandated by NIH.[7]

Opin. Although the IACUC does not have any special responsibility, it reviews studies that use recombinant DNA in animals, since animal project review is part of its function. However, the IACUC has no authority to approve the recombinant DNA part of the project. This responsibility is with the IBC. A close interaction and coordination of these two committees, therefore, is vital for an efficient and responsible review and approval process.

20:28 Do commercially purchased transgenic or knockout mice fall under the NIH guidelines for recombinant DNA?

Reg. As outlined in Appendix C-VI of the NIH Guidelines[7]: "The purchase or transfer of transgenic rodents for experiments that require Biosafety Level 1 (BL-1) containment are exempt from the NIH Guidelines."

Opin. The only aspect of transgenic rodent work that is currently exempt from the NIH Guidelines is the purchase or transfer of those animals that only require BL-1 laboratory containment. Transgenic rodents generated at the institution are still covered by the NIH Guidelines and require at a minimum Institutional Biosafety Committee notification simultaneous with the initiation of the work, even if the rodents require only BL-1 containment. BL-1 is the lowest level of physical containment and appropriate for well-characterized agents not known to cause disease, and of minimal potential hazard to laboratory personnel and the environment. Generally, combinations of laboratory practices, containment equipment, and facility design can be made to achieve different levels of physical containment also referred to as *biosafety level*. Four levels of physical containment (BL-1 to BL-4) are commonly used, with BL-4 being the highest level of physical containment and BL-1 being the lowest. For a comprehensive overview refer to the Centers for Disease Control and Prevention[11] and NIH Guidelines.[7]

20:29 Do transgenic or knockout mice purchased from another university (where they were developed and bred and which has a PHS Assurance statement) fall under the NIH guidelines for recombinant DNA?

Reg. The NIH Guidelines[7] currently only exempt the acquisition or transfer of transgenic rodents that require Biosafety Level-1 (BL-1) containment (see 20:28). All other transgenic rodents (either generated at the research facility or requiring a higher containment level) are still covered, independent of the source.

Opin. The key aspect of determining if transgenic rodents are covered under the NIH Guidelines is to look at the required biological safety containment level (biosafety level) and the source. Transgenic rodents that are acquired from outside of the institution (commercial source, other university, etc.) and which require only BL-1 containment are currently exempt. All others need at a minimum the notification of the Institutional Biosafety Committee (IBC) simultaneously with initiation of the project. The IACUC is well advised to consult with the IBC and the biological safety officer on these issues.

20:30 Can the IACUC approve studies using "in-house" constructed transgenic or knockout mice without the approval of the Institutional Biosafety Committee (IBC)?

Opin. The IACUC has no authority under NIH[7] to approve studies using "in-house" constructed transgenic or knockout mice. There is only one exception: if the IACUC is specifically charged with recombinant DNA oversight by the institution and fulfills (in addition to IACUC responsibilities) the duties of the IBC as outlined by NIH.[7]

No IACUC should approve "in-house" recombinant DNA projects. This responsibility is clearly established for the IBC under the NIH Guidelines.[7] Noncompliance with the NIH Guidelines can result in loss of funds, loss of accreditation, etc.

Chemical Hazards

20:31 What is meant by a chemical hazard?

Reg. A hazardous chemical is defined by the Occupational Safety and Health Administration (OSHA) as any chemical, chemical compound, or mixture of compounds which presents a physical or health hazard.[26] A chemical is a physical hazard by OSHA definition if there is scientifically valid evidence that it is

- A flammable or combustible liquid.
- A compressed gas.
- An organic peroxide.
- An explosive, an oxidizer.
- A pyrophoric.
- An unstable material (reactive).
- A water reactive material.

A chemical is a health hazard by OSHA definition if there is statistically significant evidence based on at least one study conducted in accordance with established scientific principles that acute or chronic health effects may occur in exposed employees. Included are allergens, embryotoxicants, carcinogens, toxic or highly toxic agents, reproductive toxicants, irritants, corrosives, sensitizers, hepatoxins, nephrotoxins, neurotoxins, hematopoietic systems agents, and any agents that damage the lungs, skin, eyes, or mucous membranes.

20:32 What is a Material Safety Data Sheet (MSDS)?

Reg. A MSDS is a document containing chemical hazard identification and safe handling information. It is prepared in accordance with the Occupational Safety and Health Administration's Hazard Communication Standard 29 CFR 1910.1200, also known as "Right-to-Know" law.[26] Chemical manufacturers and distributors must provide the purchasers of hazardous chemicals an appropriate MSDS for each hazardous chemical or product purchased. (See 20:12, 20:18.)

Opin. Copies of the MSDS for hazardous chemicals at a given worksite need to be readily accessible to employees in that area. As a source of detailed information on hazards, they must be located close to workers and readily available during working hours.[26] MSDSs for hazardous chemicals used in the animal facility are a necessity and required by law. Accessibility, however, does not necessarily mean to have copies available in each room or attached to each chemical. Central repositories (e.g., main office) can store all available MSDSs and each area housing hazardous chemicals has a notice posted where MSDSs for this location can be found. More and more facilities use computer-based retrieval systems, loaded with software containing large numbers of MSDSs.

20:33 What are some typical chemical hazards that can be found in a research environment where animals are used?

Opin. Researchers commonly use a wide variety of chemicals. Toxic agents and carcinogens can be used to induce disease or study metabolic processes. Other chemicals are used in analytic techniques. Most laboratories have flammable solvents, corrosive liquids, and a variety of toxic chemicals. Cleaning materials, solvents, acids, disinfectants, and sanitizers, while outside of the animal holding rooms, still can pose an exposure hazard if improperly stored or used.

20:34 Should all approved anesthetic and euthanasia agents be considered hazardous?

Reg. Approved anesthetic and euthanasia agents are considered hazardous if they meet one or more of the conditions identified for hazardous chemicals (see 20:31).

Opin. The degree of hazard varies with each agent. Factors that determine the degree of hazard include the specific agent and its properties, exposure concentrations, work practices, engineering controls, personal protective equipment used, and route of administration.

The IACUC should rely on the AV, the Chemical Hygiene Officer, and Material Safety Data Sheet information to address safety issues related to these agents. It is recommended that institutions have standard operating procedures in place for usage, storage, and disposal of anesthetic and euthanasia agents. Training in proper techniques involving inhalational and noninhalational agents should be provided.

20:35 How can an IACUC evaluate a chemical risk?

Opin. An IACUC could partner with an industrial hygienist within its organization (or hired as a consultant) or any other health and safety professional with specific knowledge and experience (e.g., toxicologist) to review and evaluate chemical risks. Areas to review should include:

- Chemical agents used.
- Engineering controls.
- Special personal protective equipment required.
- Route of excretion.
- Precautions for handling animals.
- Animal disposal.
- Bedding disposal.
- Cage decontamination.
- Special precautions.
- Storage and delivery procedures.

20:36 Is a separate chemical hazard committee necessary?

Reg. The use of a chemical hygiene committee is an option suggested by the Occupational Safety and Health Administration (OSHA) in the laboratory safety standard (29 CFR 1910.1450), to be established by the institution if appropriate. Currently, OSHA does not mandate a chemical hazard committee. However, if hazardous chemicals are used in the laboratory or animal facility, a chemical hygiene plan is required. This plan not only outlines all relevant safety procedures and practices as required by law, but also must designate personnel who are responsible for implementing the plan including the assignment of a chemical hygiene officer.

Opin. A separate chemical hygiene committee is a recommended option in overseeing institutional compliance with all aspects of chemical safety. In addition, review of animal research projects involving hazardous chemicals should be done in cooperation with the chemical hygiene officer.

Physical Hazards

20:37 What is meant by a physical hazard?

Reg. Physical hazards are those generally associated with:

- Noise.
- Temperature.
- Vibration.
- Electricity.
- Nonionizing and ionizing radiation (also considered a radiation hazard).
- Ultraviolet radiation.
- Lasers.
- Illumination.
- Sharp objects.

Occupational exposure limits and protective measures for some of these hazards are covered under specific OSHA standards:

- Occupational Noise Exposure (29 CFR 1910.95).
- Nonionizing Radiation (29 CFR 1910.97).
- Ionizing Radiation (29 CFR 1910.1096).
- Electrical (29 CFR 1910 Subpart S) Electrical (1910.301 to 1910.399).

In addition, a certain hazardous chemical is defined as a physical hazard if there is scientifically valid evidence that it is a combustible liquid, a compressed gas, explosive, flammable, an organic peroxide, an oxidizer, pyrophoric, unstable (reactive), or water-reactive.[26]

Opin. Common hazards like noise are often ignored and accepted as part of the animal facilities environment. Excessive noise can be produced by machinery and animals, especially pigs and dogs. Exposure to intense noise over time will result in hearing loss. If normal talking or phone conversation is not possible due to excessive noise, the noise level should be assessed by a safety professional. Electricity is of concern if electrical equipment is old, not well maintained, wires are loose or exposed, etc. Compliance with electrical code requirements is necessary, as is equipment maintenance.

20:38 What are typical physical hazards that are encountered in the laboratory animal environment?

Opin. Physical hazards typically encountered in the animal research environment are associated with the operation and manipulation of machinery, equipment, or instrumentation. These hazards can range from heat exposure from autoclaves to injuries caused by sharp objects (e.g., cages, scalpels). Animal bites, back injuries, scratches, kicks, and related hazards are also typical for the animal research environment. Project-specific hazards might include potential radiation exposure.

20:39 Should physical methods of euthanasia be considered a physical risk?

Reg. Physical methods of euthanasia as defined by the American Veterinary Medical Association (AVMA)[27] include captive bolt, gunshot, cervical dislocation, decapitation, electrocution, microwave irradiation, exsanguination, stunning, or pithing. The last three are used in conjunction with other methods. According to the AVMA panel, "Given that most physical methods involve trauma, there is inherent risk for animals and human beings; therefore, extreme care and caution should be used."

PHS Policy (IV,C,1,g) requires euthanasia methods to be consistent with the AVMA recommendations unless a deviation is justified for scientific reasons, in writing, by the investigator. APHIS/AC Policy #3[28] requires that the method of euthanasia "must be consistent with the current Report of the AVMA Panel on Euthanasia."

Opin. As with all physical procedures and practices, the skill and expertise of the person performing the procedure determines the degree of risk. Prior to approval of such methods, the IACUC should evaluate the proficiency and level of training of individuals performing these procedures.

20:40 How can the IACUC evaluate physical hazard risk?

Opin. The IACUC has no specific mandate to evaluate physical hazards unless they are related to the animal as part of the research protocol.

Potential physical hazards for animals and research personnel working with them are evaluated through routine inspections and protocol review. Since the IACUC is required to regularly inspect the animal facilities, these inspections should be used to observe practices and procedures involving potential physical hazards. Occupational health professionals, safety officers, and others might be able to assist the IACUC in the evaluation process.

Radiation Hazards

20:41 What is meant by a radiological hazard?

Reg. Radiological hazard refers to the hazards associated with ionizing radiation, generated through licensed radioactive materials or machine produced radiation.[3] Background radiation is not considered a radiological hazard.

According to Title 10, Code of Federal Regulations,[3] the definition is: Radiation (ionizing radiation) means alpha particles, beta particles, gamma rays, x-rays, neutrons, high-speed electrons, high-speed protons, and other particles capable of producing ions. Radiation may be machine produced (x-ray machines, accelerators), byproduct material (radioisotopes produced by a reactor or naturally occurring atoms) that emit radiation, such as uranium, radium, or other naturally radioactive atoms.

Opin. The level of hazard is determined by the amount of radiation, type of radiation, chemical form, method of use (procedures and protocols), protective precautions, etc. Radiological hazards must be assessed carefully by a qualified radiation protection professional. Many radiation safety issues are more public relations and regulatory compliance than real risk problems.

20:42 What are typical sources of radiological hazards in the research animal environment?

Opin. Typical sources of ionizing radiation in the animal research environment are radioactive materials administered as part of a research protocol or machine produced radiation, all of which can be either treatment, diagnostic, or research tools. Some examples are ^{3}H, ^{14}C, ^{111}In, ^{51}Cr, ^{131}I, and Tc_m. These radioisotopes may be administered to animals as radiolabeled antibiotics, chemical toxins, and blood flow tracers for trauma and injury studies. X or neutron radiation may be administered to animals to study brain or other physiology, treat cancers, or other areas of research or treatment.

20:43 Is external beam radiation such as x-irradiation considered a radiation hazard of concern to an IACUC?

Reg. The IACUC should consider external beam radiation effects, as it does with any other hazardous agent. Local and state regulations may require the AV or PI to limit the dose to the animal as low as practical. There also may be specific state or federal regulations governing the use of machine-produced radiation and the effects on animals. Improper use of diagnostic radiation should be of concern to the IACUC as it indicates a hazard as well as an inadequate program of veterinary care (see Chapter 27).

Opin. The use of external beam radiation in the animal research environment is normally limited to therapeutic purposes (diagnostic and treatment). Nevertheless, it should be included in the overall animal care program oversight of the IACUC, for as noted above it can be a hazard and if used improperly indicate an inadequate program of veterinary care. If a *research* protocol requires the use of x-radiation, that usage requires IACUC review and approval. Therapeutic or research uses involving high doses of x-radiation which potentially can cause somatic effects should be considered similar to other procedures causing pain and suffering for animals.

All human health aspects related to external beam radiation must be addressed by institutional safety professionals in the radiation safety and occupational health areas. This is usually beyond the IACUC's expertise and scope.

20:44 How can an IACUC evaluate radiological hazard risk?

Reg. This is regulated by federal, state, and local regulations for radiation. Agencies include the U.S. Nuclear Regulatory Commission, the Department of Energy, and Agreement States. (An Agreement State is a state that has entered into an effective agreement with a federal regulatory agency, becoming the regulatory control for the given area in that state). Machine-produced radiation is regulated by the states.

Opin. The IACUC should only evaluate the effects of animal experiments involving radiation. The institutional Radiation Safety Officer and Radiation Safety Committee must assess radiation risk, determine precautions and management practices, and document the radiation protection evaluation. They also must inspect and assure safety and compliance with the regulations. In certain cases, precautions must be taken by animal care technicians. The instructions for precautions are often posted on the door of the room or on the cages of the animals for which the precautions are necessary.

Animal care staff who will handle or assist with the management of animals that have been administered radioactive materials, or who will handle or manage the radioactive waste or bedding, must be trained in radiation safety at their institution.

20:45 Should the IACUC approve a protocol proposing to use potential radiological hazards prior to approval from the appropriate institutional risk committee?

Reg. The IACUC may approve the animal welfare part of the research, but not the radiation uses. The agencies that regulate the use of radioactive materials for research purposes mandate Radiation Safety Committee approval prior to any use of radioactive materials in animal research. The IACUC has no authority to approve the radiation use aspects of animal research protocols unless it is specifically delegated with Radiation Safety Committee responsibilities.

20:46 What kind of documentation should be provided to the IACUC from a radiation safety office?

Reg. Records are mandated for:

- Radiation licensing and permits.
- Approvals.
- Ordering and receipt.
- Training.
- Inspections (both by researchers and the safety managers).
- Protocols and procedures.
- Radiation monitoring instrument certification.
- Transportation.
- Waste management.
- Bioassay.
- Other aspects of radiation use, including decommissioning of the use locations when done.

Title 10 CFR[3] and Title 49 CFR[14] contain most of the applicable radiation regulations. However, Agreement States (see 20:44) have their own regulations, and all states have regulations for machine-produced radiation. Each institution must have the required licenses and permits which, in turn, mandate further requirements. Records for all of the radiation uses, from "cradle to grave," must be maintained for review by the Radiation Safety Office.

Opin. Use of radiation and radioactive materials entails a very comprehensive and strict program of safety and compliance performed by qualified radiation safety professionals. The IACUC should initiate and maintain a close and friendly working relationship with the radiation safety managers at each institution and defer to their findings.

References

1. National Research Council, *Occupational Health and Safety in the Care and Use of Research Animals*, National Academy Press, Washington, D.C., 1997.
2. National Institute of Occupational Health Alert, Preventing Asthma in Animal Handlers, U.S. Department of Health and Human Services (Publication No. 97-116), 1998.
3. Office of the Federal Register, Code of Federal Regulations, Title 10, Part 20, Standards for Protection against Radiation, Washington, D.C.
4. Office of the Federal Register, Code of Federal Regulations, Title 42, Part 72, Interstate Shipment of Etiological Agents, Washington, D.C.
5. Office of the Federal Register, Code of Federal Regulations, Title 42, Part 72.6, Additional Requirements for Facilities Transferring or Receiving Select Agents, Washington, D.C.
6. Office of the Federal Register, Code of Federal Regulations, Title 9, Animal and Animal Products, Parts 92-95, 122, Washington, D.C.
7. National Institutes of Health, Guidelines for Research Involving Recombinant DNA Molecules, 51 Federal Register 16958, May 7, 1986; as amended 59 FR 34496, July 5, 1994; 63 FR 26018, May 11, 1998.
8. Office of the Federal Register, Code of Federal Regulations, Title 10, Part 33, Specific Domestic Licenses of Broad Scope for Byproduct Material, Washington, D.C.
9. Office of the Federal Register, Code of Federal Regulations, Title 29, Part 1910.1450, Occupational Exposure to Hazardous Chemicals in Laboratories, Washington, D.C.
10. Potkay, S., Garnett, N.L., Miller, J.G., Pond, C.L., and Doyle, D.J., Frequently asked questions about the Public Health Service policy on humane care and use of laboratory animals, *Lab. Anim.*, 24(9), 24, 1995.
11. Centers for Disease Control and Prevention and National Institutes of Health, Biosafety in Microbiological and Biomedical Laboratories, 4th ed., U.S. Government Printing Office, Washington. D.C., 1999.
12. Stuart, R.B. and Moore, C., *Safety and Health on the Internet*, Government Institutes, Inc., Rockville, MD, 1988.
13. Office of the Federal Register, Code of Federal Regulations, Title 29, Part 1910.1030, Bloodborne Pathogens, Washington, D.C.
14. Office of the Federal Register, Code of Federal Regulations, Title 49, Parts 171-180, Hazardous Materials Regulations, Washington, D.C.
15. Office of the Federal Register, Code of Federal Regulations, Title 21, Food and Drugs, Washington, D.C.
16. American Biological Safety Association, 1202 Allanson Rd., Mundelein, IL 60060. Phone: 847-949-1517, Fax: 847-566-4580. Available on the World Wide Web at: *http://www.absa.org*

17. Miller, C.D., Songer, J.R., and Sullivan, J.F., A twenty-five year review of laboratory acquired human infections at the National Animal Disease Center, *AIHA J.*, 48, 271, 1987.
18. Office of the Federal Register, Code of Federal Regulations, Title 29: Part 1910.134, Respiratory Protection, Washington, D.C.
19. Office of Health Compliance Assistance, OSHA Directives CPL 2.106, Enforcement Procedures and Scheduling for Occupational Exposure to Tuberculosis, 1996.
20. Centers for Disease Control and Prevention, Guidelines for Preventing the Transmission of *Mycobacterium tuberculosis* in Health Care Facilities, *MMWR*, October 26, 1994, Vol. 43, No. RR-13.
21. Centers for Disease Control and Prevention, Guidelines for the Prevention of Herpesvirus Simiae (B-Virus) Infection in Monkey Handlers, *MMWR*, October 23, 1987, 36(41), 680, 687.
22. National Institute for Occupational Safety and Health, Health Hazard Evaluation Report: 98-0061-2687, Yerkes Primate Research Center, Lawrenceville, GA, April 1998.
23. U.S. Department of Agriculture, Animal and Plant Health Inspection Service, Policy #11, Painful/Distressful Procedures, April 14, 1997. Available on the World Wide Web at: *http://www.aphis.usda.gov/ac/policy11.html*
24. Jackson, L.R. and Fox, J.G., Institutional policies and guidelines on adjuvants and antibody production, *ILAR J.*, 37(3), 141, 1995.
25. Holmes, G.P. et al., Guidelines for the prevention and treatment of B-virus infections in exposed persons, *CID*, 20, 412, 1995.
26. Office of the Federal Register, Code of Federal Regulations, Title 29, Part 1920.1200, Hazard Communication, Washington, D.C.
27. American Veterinary Medical Association, Report of the AVMA Panel on Euthanasia, *J. Am. Vet. Med. Assoc.*, 202, 229, 1993.
28. U.S. Department of Agriculture, Animal and Plant Health Inspection Service, Policy #3, Veterinary care, April 14, 1997. Available on the World Wide Web at: *http://www.aphis.usda.gov/ac/policy3.html*

21
Personnel Training

Howard G. Rush

Introduction

Personnel training is a subject that has gained increasing attention in the years since revision of the Animal Welfare Act Regulations, the PHS Policy, and the *Guide for the Care and Use of Laboratory Animals*. All three documents specifically require institutions to provide training for personnel engaged in animal research, although the specific recommendations in these three documents vary. The mechanisms whereby institutions provide such training, the identification of individuals to provide the training, and the content of training courses have become the subject of much discussion among professionals in research administration and laboratory animal science and medicine. To provide some insight into the evolution of training in animal care and use, trainers at 12 institutions (10 academic, 1 pharmaceutical company, and 1 private research) were polled by telephone to gather information on the training practices and policies at their institutions.

21:1 Is there any requirement in either the AWAR or the PHS Policy for general training in laboratory animal care and use?

Reg. The AWAR (§2.32,c,2; §2.32,c,4; §2.32,c,5) include requirements for general training in laboratory animal care and use. General training includes information on:

- Research methods that limit the utilization of animals or minimize animal distress.
- Institutional procedures for reporting deficiencies in animal care and treatment.

- Services such as the National Agricultural Library and the National Library of Medicine that can be used to acquire information on appropriate methods of animal care and use, alternatives to the use of live animals in research, the intent and requirements of the AWA, and the use of which can prevent duplication of research involving animals.

The PHS Policy does not specify the content of training programs to the same degree as the AWAR. Nevertheless, the PHS Policy (IV,C,1,f) requires the IACUC, in its review of protocols, to determine that personnel conducting procedures are appropriately qualified and trained. In addition, the PHS Policy (IV,A,1,g) requires the institution to include in its NIH/OPRR Assurance a "synopsis of [the] training or instruction in the humane practice of animal care and use, as well as training or instruction in research or testing methods that minimize the number of animals required to obtain valid results and minimize animal distress, offered to scientists, animal technicians, and other personnel involved in animal care, treatment, or use."

The *Guide* (pages 13 and 14) further specifies that individuals who care for or use animals should be properly trained and that the institution has a responsibility for providing either formal or on-the-job training for personnel. Training of animal care personnel is necessary to implement an effective animal care and use program and to foster humane animal care and use. Investigators and other personnel who perform surgery, administer anesthesia, or perform other manipulations must be appropriately trained to accomplish these tasks in a humane and scientifically acceptable manner.

Opin. The prevailing attitude among personnel engaged in animal research is that animals should not be subjected to unnecessary pain or distress. Universally, animal care and use personnel want to humanely perform research which uses animals. Experience has demonstrated that utilization of animal care and research personnel who are well-trained will ultimately reduce animal pain and distress in research because well-trained personnel can perform animal research techniques with greater skill and fewer adverse outcomes for the animals used. Consequently, it is an institutional imperative to provide training to animal care and research personnel to ensure humane care for animals used in research.

21:2 Is there any requirement in either the AWAR or the PHS Policy for additional training in the care and use of a particular species?

Reg. The AWAR (§2.32,c,1,i; §2.32,c,1,ii; §2.32,c,3) require training on the basic needs of the relevant species; proper handling and care for various species used by the facility; and the proper use of anesthetics, analgesics, and tranquilizers for any animals used by the facility.

The PHS Policy does not specify any type of species specific training, but it does require that Assured institutions use the *Guide* as a basis for their animal care and use program (PHS Policy IV,A,1).

The *Guide* (page 14) states that research personnel who perform experimental manipulations on animals including anesthesia and surgery must have adequate training or experience to perform those procedures humanely. In addition, species-specific training for animal care personnel is implied in the general training statement in the *Guide* (page 13) as noted above.

Opin. There is a vast diversity of species used in biomedical research. Although rats and mice account for more than 90% of the animals used in research, numerous other species have been utilized and will continue to be utilized depending on the suitability of particular models for the research being conducted. It is impossible for any one person to be familiar with all species that are or might be used in research. Therefore, institutions must provide animal care and research personnel with access to training when experiments are planned using new species.

21:3 Do the AWAR or the PHS Policy state whether or not training is required for specific research procedures (e.g., performing an arterial cut-down)?

Reg. Procedure-specific training specified in the AWAR (§2.32,c,1,iii; §2.32,c,1,iv) includes training on preprocedural and postprocedural care of animals and aseptic surgical methods and procedures. The PHS Policy does not require procedure-specific training and, as noted in 21:2, the *Guide* (page 14) implies that procedure-specific training may be necessary to ensure that research personnel who perform experimental manipulations on animals can conduct their research humanely.

Opin. Innumerable animal research techniques are described in the scientific literature for use in the many disciplines in biomedical research. It is impossible for any one person to be skilled in all techniques commonly used in animal research. Training personnel should be prepared to teach common animal research techniques to individuals conducting animal research. In addition, they should identify individuals at their institution who work in research laboratories and have unique skills in performing specific research procedures with animals. These individuals should be cultivated as ancillary training staff who can be called upon to help train personnel when training requests for their particular skills are received. Engaging research personnel to participate in the training effort can be a fruitful and rewarding means to expand the training provided at an institution while conserving fiscal resources.

21:4 In addition to the training topics noted in 21:2 and 21:3, what other information is useful to include as part of a training and education program?

Opin. Training programs for animal care and use personnel can be regarded in a tiered manner for optimal delivery of training. All personnel engaged in research with animals and all personnel who provide animal care services should participate in an introductory training session that includes information on federal laws and national standards, institutional policies and procedures, the institutional animal care and use program, institutional occupational health programs, reporting of concerns about animal care and use, and the politics and ethics of animal experimentation. A second tier or level includes species specific training for personnel. It is directed at increasing their familiarity with the species they will be utilizing or for which they will be caring. Such training can include information on the biology and care of a given species, and basic research techniques such as restraint, injection techniques, and euthanasia. A third tier includes technique specific training for personnel to improve their skill level for more advanced research techniques such as surgery, anesthesia, and even specific surgical procedures (e.g., vascular catheter implantation). Personnel not directly engaged in animal care and use need not participate in the second and third tiers, but all personnel participate in the first tier. Additionally, personnel with specific care or research duties also participate in training that is directed at improving their knowledge of and skill level with the species with which they work. These types of core modules have been described.[1]

Surv. Beyond the regulations, what other information would be useful to include as part of a training and education program? More than one answer is possible.

- Institutional policies and procedures — 12/12
- Organizational structure of the institutional animal care and use program — 12/12
- The politics and ethics of animal experimentation — 3/12
- Alternatives to animal use — 1/12
- Pain and distress and its alleviation — 4/12
- Acceptable euthanasia methods — 8/12
- An introduction to surgery — 5/12
- Occupational health programs — 6/12

- Recordkeeping requirements 3/12
- Reporting of concerns about animal care and use 2/12
- Basic care and use of common laboratory animals 12/12
- Informational resources on animal care and use 3/12

The reader is cautioned that many of the above responses include required training under the AWAR. In addition to the results above, the survey revealed that at many institutions research personnel are provided with species specific and technique specific training that is in addition to the introductory training provided to everyone. These sessions may include lectures and laboratories on the species and procedures identified in the investigator's IACUC protocol. Often, the experience and qualifications of the research personnel are taken into account in determining the second level of training needed.

21:5 Who is responsible for assuring that research and animal care personnel working with animals are adequately trained?

Reg. According to the AWAR (§2.32,a), the research facility (see definition in 15:1) is responsible for ensuring that research personnel are qualified to perform their duties. To accomplish this, the research facility must make training available to personnel engaged in animal care and use. The HREA (Sec. 495,c,1,b) and the PHS Policy (IV,A,1,g) require the awardee institution to assure that all personnel involved in animal care, treatment, or use have training available to them. Both the AWAR (§2.31,d,1,viii) and the PHS Policy (IV,C,1,f) specify that the IACUC, in its review of protocols, should determine that personnel are appropriately qualified and trained. The *Guide* states that the institution is responsible for providing training to animal care personnel (*Guide*, page 13). (See 21:6.)

Opin. Regardless of the letter of the law or the Policy, the responsibility for assuring that personnel are adequately trained must be shared by the institutional officers, the IACUC, the animal care and veterinary staff, and the PI. Overall, it is an institutional responsibility to ensure that animal care and research personnel are adequately trained to care for research animals and conduct animal research procedures. The institution must be willing to commit the personnel and financial resources to accomplish this. Nevertheless, in practical terms, the responsibility for assuring that training is available belongs with the IACUC. The IACUC is able to identify the personnel that should be trained and coordinate the resources and activities necessary to provide the training. The provision for actually providing the training usually rests upon the veterinary staff or the staff that supports the IACUC. These are the individuals who possess the scientific, clinical, and technical skills that the animal care and research staff need to perform their duties. Finally, the PI has a responsibility to convey to her staff the importance of receiving proper training in order to humanely conduct the animal studies in which they will be participating. The PI's attitude toward training sets the tone for laboratory personnel with regard to their participation in the institution's training program.

Surv. At your institution, who is responsible for assuring that research and animal care personnel working with animals are adequately trained?

- Responsibility of the principal investigator and IACUC   6/12
- Responsibility of the principal investigator and veterinarian   1/12
- Responsibility of the principal investigator   3/12
- Responsibility of the IACUC   2/12

21:6 Who is typically involved in training persons for the appropriate care and use of laboratory animals?

Surv. Who is typically involved in training persons for the appropriate care and use of laboratory animals? More than one answer is possible.

- Veterinarians   11/12
- Veterinary technicians   9/12
- Laboratory animal technicians   6/12
- Research personnel   10/12

21:7 What are some effective means of instruction in the proper care and use of laboratory animals?

Opin. The qualifications of trainers can vary considerably between institutions and is likely to be affected by many factors. Among these are the size of the institution, its mission (academic, industrial, etc.), the organization of the IACUC and animal care services, the physical resources, the animal population and species maintained, and others (*Guide,* page 13). Most often, veterinarians, veterinary technicians, and laboratory animal technicians and technologists conduct training activities at institutions. However, at 10 of 12 institutions polled by this author, research personnel with expertise in specific procedures also were called upon to assist the centrally designated trainers in providing procedure specific training to other research personnel (see 21:6). In fact, at many institutions, a wide variety of individuals with differing qualifications are recruited to provide training. The use of personnel outside the central animal care unit to provide training is an effective means to increase the number of personnel involved in the institutional training activity.

By far the most effective means of instruction for use in animal care and use training is the individual or small group hands-on training session, sometimes termed a "wet lab." This setting provides the optimal conditions for student-teacher interaction and is excellent for fostering skill development. This type of training session is ideal for teaching animal research techniques such as handling, restraint, anesthesia, surgery, and others. On the other hand, the wet lab is not an efficient use of time or personnel when it is necessary to present introductory material such as federal laws and national standards, institutional policies and procedures, the institutional animal care and use program, institutional occupational health programs, etc. For this type of material, slide presentations, videos, written material, and even computer-based training are more appropriate and efficient.

Surv. What does your institution do to instruct people in the proper care and use of laboratory animals? More than one answer is possible.

- Individual or small group hands-on training sessions   12/12
- Slide presentations   5/12
- Written material (handouts, newsletters, and institutional manuals on animal research)   2/12
- Videos   6/12
- Computer-based training (either auto-tutorials or Web-based)   4/12

21:8 What are some reference sources that can be used to develop a general training program for animal care and use?

Opin. Individuals charged with developing training programs should have access to the resources listed below.

- Office of the Federal Register, Code of Federal Regulations, Title 9 (Animals and Animal Products), Subchapter A (Animal Welfare), Washington, D.C., 1985.
- U.S. Department of Health and Human Services, Public Health Service, Public Health Service Policy on Humane Care and Use of Laboratory Animals, Washington, D.C., 1996.
- Institute of Laboratory Animal Resources, Committee on Educational Programs in Laboratory Animal Science, *Education and Training in the Care and Use of Laboratory Animals: A Guide for*

Developing Institutional Programs, National Academy Press, Washington, D.C., 1991.

- Institute of Laboratory Animal Resources, Committee to Revise the Guide for the Care and Use of Laboratory Animals, *Guide for the Care and Use of Laboratory Animals*, 7th ed., National Academy Press, Washington, D.C., 1996.
- Interagency Research Animal Committee, U.S. Government Principles for Utilization and Care of Vertebrate Animals Used in Testing, Research, and Training, Federal Register, Washington, D.C., May 20, 1985.
- National Institutes of Health, Office of Protection from Research Risks, Institutional Animal Care and Use Committee Guidebook, Washington, D.C., 1992.
- Bennett, B.T., Brown, M.J., and Schofield, J.C., *Essentials for Animal Research: A Primer for Research Personnel*, National Agricultural Library, Beltsville, MD, 1990.
- Committee on the Use of Laboratory Animals in Biomedical and Behavioral Research, *Use of Laboratory Animals in Biomedical and Behavioral Research*, National Academy Press, Washington, D.C., 1988.
- Lawson, P., *Training Manual Series, Vol. I., Assistant Laboratory Animal Technician*, American Association for Laboratory Animal Science, Cordova, TN, 1998.
- Stark, D.M. and Ostrow, M.E., *Training Manual Series, Vol. II., Laboratory Animal Technician*, American Association for Laboratory Animal Science, Joliet, IL, 1990.
- Stark, D.M. and Ostrow, M.E., *Training Manual Series, Vol. III, Laboratory Animal Technologist*, American Association for Laboratory Animal Science, Joliet, IL, 1991.
- Rollin, B.E. and Kesel, M.L., *The Experimental Animal in Biomedical Research, Volume I: A Survey of Scientific and Ethical Issues for Investigators*, CRC Press LLC, Boca Raton, FL, 1990.
- Russell, W.M.S. and Burch, R.L., *The Principles of Humane Experimental Techniques*, Methuen & Co., London, 1959. (Reprinted as a special edition in 1992 by the Universities Federation for Animal Welfare.)
- Federation of European Laboratory Animal Science Association's Working Group on Education, FELASA Recommendations on the Education and Training of Persons Working with Laboratory Animals: Categories A and C, *Lab. Anim.*, 29, 121, 1995.
- Institute of Laboratory Animal Resources, Committee on Occupational Safety and Health in Research Animal Facilities, *Occu-*

pational Health and Safety in the Care and Use of Research Animals, National Academy Press, Washington, D.C., 1997.
- American Veterinary Medical Association, Report of the AVMA Panel on Euthanasia, *J. Am. Vet. Med. Assoc.*, 202, 229, 1993.
- Laboratory Animal Welfare Training Exchange (available on the World Wide Web at: *http://www.lawte.org*)
- IACUC 101 (available on the World Wide Web at: *http://www.iacuc.org*)

21:9 What are some reference sources for training on the care and handling of a particular species?

Opin. The following references cover a wide variety of species and topics.

- Fox, J.G., Cohen, B.J., and Loew, F.M., *Laboratory Animal Medicine*, Academic Press, New York, 1984.
- Universities Federation for Animal Welfare, *The UFAW Handbook on the Care and Management of Laboratory Animals*, 6th ed., Churchill Livingstone, New York, 1987.
- Olfert, E.D., Cross, B.M., and McWilliam, S.S., *Guide to the Care and Use of Experimental Animals*, Vol. I, 2nd ed., Canadian Council on Animal Care, Ottawa, 1993.
- Canadian Council on Animal Care, *Guide to the Care and Use of Experimental Animals*, Vol. II, Canadian Council on Animal Care, Ottawa, 1984.
- Rollin, B.E. and Kesel, M.L., *The Experimental Animal in Biomedical Research. Volume II: Care, Husbandry, and Well-Being, An Overview by Species*, CRC Press LLC, Boca Raton, FL, 1995.
- Hillyer, E.V. and Quesenberry, K.E., *Ferrets, Rabbits, and Rodents: Clinical Medicine and Surgery*, W.B. Saunders, Philadelphia, 1997.
- Hrapkiewicz, K., Medina, L., and Holmes, D.D., *Clinical Laboratory Animal Medicine: An Introduction*, 2nd ed., Iowa State University Press, Ames, 1998.
- Harkness, J.E. and Wagner, J.E., *The Biology and Medicine of Rabbits and Rodents*, 4th ed., Williams and Wilkins, Baltimore, 1995.
- Wagner, J.E. and Manning, P.J., *The Biology of the Guinea Pig*, Academic Press, New York, 1976.
- Van Hoosier, G.L., Jr., and McPherson, C.W., *Laboratory Hamsters*, Academic Press, New York, 1987.
- Manning, P.J., Ringler, D.H., and Newcomer, C.E., *The Biology of the Laboratory Rabbit*, 2nd ed., Academic Press, New York, 1994.

- Foster, H.L., Small, J.D., and Fox, J.G., *The Mouse in Biomedical Research. Vol. I. History. Genetics. and Wild Mice*, Academic Press, New York, 1981.
- Foster, H.L., Small, J.D., and Fox, J.G., *The Mouse in Biomedical Research. Vol. II. Diseases*, Academic Press, New York, 1982.
- Foster, H.L., Small, J.D., and Fox, J.G., *The Mouse in Biomedical Research. Vol. III. Normative Biology, Immunology, and Husbandry*, Academic Press, New York, 1983.
- Foster, H.L., Small, J.D., and Fox, J.G., *The Mouse in Biomedical Research. Vol. IV. Experimental Biology and Oncology*, Academic Press, New York, 1982.
- Bennett, B.T., Abee, C.R., and Hendrickson, R., *Nonhuman Primates in Biomedical Research: Biology and Management*, Academic Press, San Diego, 1995.
- Baker, H.J., Lindsey, J.R., and Weisbroth, S.H., *The Laboratory Rat, Vol. I. Biology and Diseases*, Academic Press, New York, 1979.
- Baker, H.J., Lindsey, J.R., and Weisbroth, S.H., *The Laboratory Rat, Vol. II. Research Applications*, Academic Press, New York, 1980.
- Svendsen, P. and Hau, H., *Handbook of Laboratory Animal Science, Vol. I. Selection and Handling of Animals in Biomedical Research*, CRC Press LLC, Boca Raton, FL, 1994.
- Svendsen, P. and Hau, H., *Handbook of Laboratory Animal Science, Vol. II. Animal Models*, CRC Press LLC, Boca Raton, FL, 1994.
- Laber-Laird, K., Swindle, M.M., and Flecknell, P., *Handbook of Rodent and Rabbit Medicine*, Pergamon Press, Oxford, 1996.

21:10 What are some reference sources for training personnel in anesthesiology, perioperative care, and aseptic surgery?

Opin. The following sources are helpful:

- Block, S.S., *Disinfection, Sterilization, and Preservation*, 4th ed., Lea and Febiger, Philadelphia, 1991.
- Waynforth, H.B., *Experimental and Surgical Technique in the Rat*, Academic Press, London, 1980.
- Waynforth, H.B. and Flecknell, P.A., *Experimental and Surgical Technique in the Rat*, 2nd ed., Academic Press, London, 1992.
- Bivin, W.S. and Smith, G.D., Techniques of experimentation, in *Laboratory Animal Medicine*, Fox, J.G., Cohen, B.J., and Loew, F.M., Eds., Academic Press, Orlando, 1984, Chap. 19.

- Cunliffe-Beamer, T., Biomethodology and surgical techniques, in *The Mouse in Biomedical Research*, Vol. III, Foster, H.L., Small, J.D., Fox, J.G., Eds., Academic Press, New York, 1983, Chap. 18.
- Kraus, A.L., Research methodology, in *The Laboratory Rat*, Baker, H.J., Lindsey, J.R., Weisbroth, S.H., Eds., Academic Press, New York, 1980, Chap. 1.
- Markowitz, J., Archibald, J., and Downie, H.G., *Experimental Surgery*, Williams and Wilkins, Baltimore, 1964.
- Lumley, J.S.P., Green, C.J., Lear, P., and Angell-James, J.E., *Essentials of Experimental Surgery*, Butterworth and Co., London, 1990.
- Schwartz, A., Experimental surgery, in *The Experimental Animal in Biomedical Research, Vol. I: A Survey of Scientific and Ethical Issues for Investigators*, CRC Press LLC, Boca Raton, FL, 1990.
- Cunliffe-Beamer, T.L., Applying principles of aseptic surgery to rodents, *Anim. Welfare Inform. Cent. Newsl.*, 4, 3, 1993.
- Academy of Surgical Research, Guidelines for training in surgical research in animals, *J. Invest. Surg.*, 2, 263, 1989.
- Lang, C.M., *Animal Physiologic Surgery*, 2nd ed., Springer-Verlag, New York, 1982.
- Berg, J., Sterilization, in *Textbook of Small Animal Surgery*, 2nd ed., Slatter, D., Ed., W.B. Saunders, Philadelphia, 1993, Chap. 11.
- Brown. M.J. and Schofield, J.C., Perioperative care, in *Essentials for Animal Research: A Primer for Research Personnel*, Bennett, B.T., Brown, M.J., and Schofield, J.C., Eds., National Agricultural Library, Washington, D.C., 1994.
- Schofield, J.C., Principles of aseptic technique, in *Essentials for Animal Research: A Primer for Research Personnel*, Bennett, B.T., Brown, M.J., and Schofield, J.C., Eds., National Agricultural Library, Washington, D.C., 1994.
- Brown, M.J., Pearson, P.T., and Tomson, F.N., Guidelines for animal surgery in research and teaching, *Am. J. Vet. Res.*, 54, 1544, 1993.
- Cunliffe-Beamer, T.L., Surgical techniques, in *Guidelines for the Well-Being of Rodents in Research*, Guttman, H.N., Ed., Scientists Center for Animal Welfare, Bethesda, MD, 1990.
- Smith, A.C. and Swindle, M.M., *Research Animal Anesthesia, Analgesia, and Surgery*, Scientists Center for Animal Welfare, Greenbelt, MD, 1994.
- Kohn, D.F., Wixson, S.K., White, W.J., and Benson, G.J., *Anesthesia and Analgesia in Laboratory Animals*, Academic Press, San Diego, 1997.

- Flecknell, P.A., *Laboratory Animal Anesthesia: A Practical Introduction for Research Workers and Technicians*, 2nd ed., Academic Press, London, 1996.
- Institute of Laboratory Animal Resources Committee on Pain and Distress in Laboratory Animals, *Recognition and Alleviation of Pain and Distress in Laboratory Animals*, National Academy Press, Washington, D.C., 1992.

21:11 What are some reference sources that can be used to develop a training program for use of hazardous agents in animal research?

Opin. Some useful reference sources include:

- Centers for Disease Control and Prevention and National Institutes of Health, Biosafety in Microbiological and Biomedical Laboratories, 4th ed., U.S. Government Printing Office, Washington, D.C., 1999.
- Centers for Disease Control and Prevention and National Institutes of Health, Primary Containment for Biohazards: Selection, Installation and Use of Biological Safety Cabinets, U.S. Government Printing Office, Washington, D.C., 1995.
- Office of the Federal Register, Code of Federal Regulations, Title 10; Part 20, Standards for Protection Against Radiation, Washington, D.C., 1984a.
- Office of the Federal Register, Code of Federal Regulations, Title 29; Part 1910, Occupational Safety and Health Standards; Subpart G. Occupation Health and Environmental Control, and Subpart Z. Toxic and Hazardous Substances, Washington, D.C., 1984b.
- Office of the Federal Register, Code of Federal Regulations, Title 29: Part 1910. Occupational Safety and Health Standards; Subpart I. Personal Protective Equipment, Washington, D.C., 1984c.
- Clark, J.M., Planning for safety: biological and chemical hazards. *Lab. Anim.*, 22(7), 33, 1993.
- Department of Health and Human Services, National Institutes of Health, Guidelines for research involving recombinant DNA molecules, *Federal Register*, 59:34496 (amended 59 FR 40170, 60 FR 20762, 61 FR 10004, 62 FR 4782, 62 FR 53335, 62 FR 56196, 62 FR 59032, 63 FR 8052, 63 FR 26018, 64 FR 25361).
- Holmes, G.P., Chapman, L.E., Stewart, J.A., Straus, S.E., Hilliard, J.K., Davenport, D.S., and the B Virus Working Group, Guidelines for the prevention and treatment of B-virus infections in exposed persons, *Clin. Infect. Dis.*, 20, 421, 1995.

- Committee on Hazardous Biological Substances in the Laboratory, *Biosafety in the Laboratory: Prudent Practices for Handling and Disposal of Infectious Materials*, National Academy Press, Washington, D.C., 1989.

21:12 How can training vs. qualifications be effectively evaluated by the IACUC?

Reg. (See 21:1 to 21:3.) The *Guide* (page 14) states that investigators and other research personnel engaged in animal research must be qualified through training or experience in order to conduct the research humanely.

Opin. In order for the IACUC to evaluate personnel training and qualifications, investigators should provide the IACUC with specific information on their prior experience and training with the species and procedures proposed in their protocol as well as the experience and training of their staff members. The IACUC then should determine, in the course of reviewing the investigator's protocol, whether the research personnel has sufficient training or experience to conduct the proposed procedures. For example, an investigator may be trained as a human surgeon but, without specific training or experience in animal surgery, may not be qualified to perform surgical procedures in animals. Such an individual may need additional training on species specific anatomy, physiology, behavior, anesthesia, and analgesia. If the prior training or experience of the research staff is deemed adequate by the IACUC, the protocol, once approved, may be initiated. However, if the IACUC determines that the prior training or experience of research personnel is inadequate, the IACUC should request or require the research staff to receive additional instruction before proceeding with their studies. The survey indicated that at some institutions, the requirement for additional training might be made a condition of approval. That is, the IACUC would place a stipulation on the investigator's protocol approval that the studies could not proceed until the staff had received the necessary training. Another interesting approach, described by some trainers, was for the IACUC to approve the protocol with a stipulation that the procedures be observed by a veterinarian or another staff member the first few times they were attempted. If the observations indicated that additional training was needed, it would be provided by the observer at that time or scheduled in the future before additional animals were used.

When semiannual inspection of animal care and use facilities are performed, the IACUC also can gain some insight into the adequacy of training by interviewing research personnel during these inspections. This approach not only identifies inadequacies in training after protocol approval has taken place, but also can be useful as an audit of the adequacy of the review process in identifying personnel in need of training.

21:13 How, and to what extent, should training and education efforts be documented? What documentation must be provided to the IACUC?

Opin. Neither the PHS Policy nor the AWAR specifically require that training records be maintained. The common view (see 21:13 survey) that it is necessary to maintain training records may stem from the institution's responsibility to make training available and the IACUC's responsibility to ensure that personnel are trained and qualified. Logically, it is difficult to meet these obligations without a mechanism to track the training and experience of individuals engaged in animal research.

Surv. A How and to what extent does your institution document training and education efforts?

At many, but not all institutions, training efforts are routinely documented in some fashion by the training staff. The format of training records vary:

- Files containing the sign-in sheets from training classes — 3/12
- Electronic database of personnel training records — 8/12
- No personnel training records are maintained — 1/12

Commonly, the records contain the following information on each person involved in animal care and use: name, identification number, department or unit, inventory of prior experience, names of classes or training sessions, dates of classes or training sessions, and names of trainers. A few institutions also give a short knowledge test at the end of the training session, the results of which also are kept on record.

Surv. B What training documentation must be provided to your IACUC?

- Do not provide reports on training activities at regular IACUC meetings — 5/12
- Provide the IACUC with a summary report on training activities in the semiannual report to the IO — 6/12
- Provide reports on training activities only upon request from IACUC — 1/12

21:14 Can an investigator offer experience with procedures in another species (including humans) as sufficient evidence of qualification to perform the same procedure in a different species?

Opin. This question is a practical application of the principles stated in 21:12. There are no universal guidelines that can be applied across the board to determine whether an individual has sufficient training or experience to perform a proposed procedure. Thus, the type and duration of experience that is acceptable at one institution may be unacceptable at another. Ideally, the IACUC should take into account a variety of factors in order to make this decision, including prior experience, previous species-specific and procedure-specific training, the difficulty and complexity of the proposed procedures, the possible and probable adverse consequences to the animals, and even the interactions of the investigator and the IACUC in the past. The assessment of this latter issue might include objective and subjective information such as prior problems with investigator compliance, the number of animal health problems encountered with previous animal research protocols, IACUC staff impressions of the competence and skill levels of research personnel, and complaints from other staff members against the investigator or his staff members.

Surv. Can an investigator offer experience with procedures in another species (including humans) as sufficient evidence of qualification to perform the same procedure in a different species?

- Yes, but still must evaluate each person's training and experience on a case-by-case basis   10/12
- No, training is always mandatory   2/12

21:15 Should an IACUC approve a protocol with novel procedures based on the surgical or other procedural experience of the investigator or other personnel involved?

Reg. Although the PHS Policy and the AWAR are mute on this matter, the *Guide* provides some insight into this issue. If novel procedures are encountered in protocols under review, the *Guide* suggests that the IACUC seek pertinent information on the possible effects on the animals from the literature as well as experienced animal care and use personnel (*Guide*, page 10).

Opin. In the absence of sufficient information, the IACUC can require the investigator to perform pilot studies in order to evaluate the effects on the animals (*Guide*, page 10). Similar to what was suggested in 21:12, the IACUC can approve the protocol as a pilot study or might place a stipulation or contingency on the protocol that the procedures be observed by a veterinarian or other qualified staff member to ensure that the novel procedures are performed in a humane and painless manner.

Surv. Should an IACUC approve a protocol with novel procedures based on the surgical or other procedural experience of the investigator or other personnel involved?

- Would likely approve a protocol with novel procedures based on the surgical or other procedural experience of the investigator or other personnel involved   12/12

All respondents further stated that, depending on the specific circumstances, the IACUC at their institution would likely attach a stipulation or condition to the protocol requiring observation of the procedures by the veterinary staff.

21:16 Must students or short-term employees (e.g., summer help) fulfill the same training requirements that the IACUC requires of investigators and their staff?

Reg.   See 21:1, as the AWAR and the PHS Policy do not distinguish between different types of employees or students.

Opin.   The IACUC should determine that all research personnel conducting animal studies are appropriately qualified and trained to perform their duties. Ideally this includes full and part-time employees, students, and visiting scientists. It is generally easier to identify full-time employees at institutions because orientation training for their position would likely include exposure to the IACUC and animal care unit. Individuals falling into the other categories can easily be unaccounted and may only be identified after they have begun their duties. This may be because others in the PI's laboratory remember to send them for the training, or because they are recognized by the animal care staff as a new person in the facility. Regardless, it is important for the trainers to foster good communication with laboratory personnel so that new employees are referred promptly to the IACUC for training.

Surv.   Does your IACUC require that students or short-term employees (e.g., summer help) fulfill the same training requirements as investigators and their staff?

- Receive the same training as full-time research personnel   12/12

Although all trainers polled indicated that students and short-term employees should receive the same training as the regular staff, in one instance the training was only voluntary. Several trainers noted further that it was difficult to identify employees in these categories and that some individuals might not receive training if they did not voluntarily come forward or were not identified to the trainers by other laboratory personnel.

21:17 Can a high school student perform animal research if he works under the guidance of an investigator with an IACUC-approved protocol?

Opin.   It has become increasingly common for high school students to work in animal research settings as volunteers, students, or employees.

	There is no inherent reason that a high school student cannot perform research procedures using animals if he is under the guidance of an investigator with approval to conduct the research. High school students should receive whatever training is necessary for them to perform their duties. Their training should be the same as that provided to college-level students and regular employees of the institution.
Surv.	At your institution, can a high school student perform animal research if he works under the guidance of an investigator with an IACUC-approved protocol?

- Yes   6/12
- Rare or never addressed by IACUC   6/12

21:18 Is there any lower age limit below which the IACUC should disallow student participation in animal care and research?

Opin.	There is no specific reason to place an arbitrary age limit on student participation in animal care and use. However, it is certainly uncommon for students in elementary and middle school to volunteer or seek employment in research settings. IACUCs are advised to treat any such request on a case-by-case basis. Any student, regardless of age, must be adequately supervised by an investigator with approval to conduct the animal studies. Other factors such as labor laws and institutional policies on employment and volunteers potentially may limit the participation of younger students in animal research.
Surv.	At your institution, is there any lower age limit below which the IACUC does not allow student participation in animal care and research?

- No policy   12/12

Most respondents noted that no one below high school age had ever been hired at their institution.

21:19 In terms of animal care and use, who is ultimately responsible for the activities of a student?

Opin.	The PI with approval for animal use bears responsibility for any and all personnel working in her laboratory, including students. In the course of review and approval of her protocol, the PI should provide assurance to the IACUC that all personnel working with animals will be adequately trained to perform their duties. The investigator must understand that this responsibility extends to students working in the laboratory.
Surv.	At your institution, who is ultimately responsible for the activities of a student?

- Principal investigator   12/12

21:20 Is there any regulation or policy requiring periodic retraining or recertification of personnel working with animals?

Opin. There is no statement in the PHS Policy or AWAR that requires periodic retraining or recertification of animal care and use personnel. The requirements for recertification are generally determined by a certifying professional organization or agency that governs each specialty area and should not be required by an IACUC. That is, a state board of veterinary examiners determines whether a veterinarian should be recertified at periodic intervals. Thus, the IACUC cannot impose recertification on an individual where no professional recertification program exists. On the other hand, periodic retraining should be encouraged to improve the understanding and skills of the animal care and use staff but there is no externally imposed requirement to do so. Similarly, institutions can develop in-house certification programs along with recertification requirements, but the author is not aware of any that have been implemented.

Surv. At your institution, is there a policy requiring periodic retraining or recertification of personnel working with animals?

- No policy   12/12

A few respondents noted that their IACUC had discussed retraining requirements (not recertification programs), but none had implemented a retraining program at the time of this writing.

21:21 Should training be continuing and ongoing? What are some effective ways to provide and document continuing education?

Opin. As noted in 21:20, retraining should be encouraged but there is no regulatory mandate that requires personnel to be retrained. However, the biomedical research environment is constantly changing. New methods and techniques are being described, new animal models are being identified, and new management approaches are being implemented at a rapid pace. Ideally then, all personnel engaged in animal care or use should participate in some form of continuing education or training in order to stay abreast of changes in their areas of interest. Such participation can occur through the existing institutional training program or through local, regional, and national meetings, conferences, and workshops.

Surv. A Should training be continuing and ongoing?

- Training should be continuing and ongoing   8/12
- It is not necessary for training to be continuing and ongoing   4/12

Surv. B What are some effective ways to provide and document continuing education?

- At many institutions, animal care and research personnel are notified of and encouraged to attend upcoming training sessions, off-site meetings, and special educational opportunities. If personnel attend any on-site training sessions, their participation is recorded in the institution's regular training records. On the other hand, none of the respondents record participation in off-site training programs. Most indicated that the credentials of all personnel were re-reviewed every 3 years when the protocol was resubmitted to the IACUC. If additional training was deemed necessary, it would be identified at that time.

21:22 Does an investigator who mostly writes grants and protocols, but has little hands-on animal contact, need to be trained and qualified in the same way and to the same extent as an animal caretaker or a research technician?

Opin. Investigators who spend most of their time writing grants and venture into the laboratory less often still have a responsibility to ensure that their personnel are trained and that they have the same level of understanding of current animal care standards. It is because of this responsibility that this type of investigator is obligated to participate in orientation training that reviews federal laws and national standards, institutional policies and procedures, the institutional animal care and use program, institutional occupational health programs, reporting of concerns about animal care and use, and the politics and ethics of animal experimentation. The training should impress upon them their responsibility for the personnel working under them to ensure that the individuals who actually conduct the animal procedures are appropriately trained.

Surv. At your institution, does an investigator who mostly writes grants and protocols, but has little hands-on animal contact, need to be trained and qualified in the same way and to the same extent as an animal caretaker or a research technician?

- Those investigators require the same level of training 1/12
- Those investigators require only basic introductory level training 11/12

Most indicated that these investigators would receive an orientative-type of training, covering the subjects listed in 21:4.

21:23 If a person has successfully completed an animal care and use training course at a different institution, should the IACUC at a new institution require that she retake a similar course at that institution or otherwise demonstrate her capabilities?

Opin. As the IACUC examines the credentials of animal use personnel in the course of conducting its review of proposed activities, any training that has occurred at another institution should be taken into account. In order to do this, the IACUC may request the individual to provide detailed information on the nature of training at the previous institution. The amount of "credit" awarded by the IACUC will likely vary with the institution's policies and the content of the training at the previous institution. Training a new employee on the institutional procedure for reporting animal care and use deficiencies is required. Regardless of the nature of the training at the previous institution, the individual probably should take any orientation training (see 21:4) offered in order to acquire institution-specific information that would not be acquired in any other way.

Surv. If a person has successfully completed an animal care and use training course at a different institution, does your IACUC require that she retake a similar course at your institution or otherwise demonstrate her capabilities?

- Must take at least orientation training at new institution   9/12
- Original institution's training program is sufficient   3/12

Reference

1. Institute of Laboratory Animal Resources, Committee on Educational Programs in Laboratory Animal Science, *Education and Training in the Care and Use of Laboratory Animals: A Guide for Developing Institutional Programs*, National Academy Press, Washington, D.C., 1991.

22
Academic Freedom and Proprietary Information

Sallie Thieme Sanford and Lisa A. Vincler

Introduction

The purpose of this chapter is to provide a basic framework for understanding how the concepts of "academic freedom" and "proprietary information" may factor into arguments supporting the confidentiality of IACUC information. In providing institutional review and approval for research activities using animals, IACUCs review information that may be considered confidential by the researchers. IACUCs at government organizations, such as public (state agency) universities, are subject to federal and state laws for open meetings and Freedom of Information acts for "public records." However, even among such government IACUCs, the application of these access-to-information laws varies. Some IACUCs have open meetings and, therefore, whatever information is discussed by the committee becomes public information. Conversely, other IACUCs have closed meetings and do not have this automatic release of information into the public domain. Similarly, the state laws on access to public records may lead to radically different results, some jurisdictions requiring release and other jurisdictions finding an exemption to protect confidentiality.

In reading the court decisions cited in the following sections, it is important to keep in mind that a court's decision is based on the set of facts before it. Thus, a different factual situation might lead to a different legal decision. This is one reason why one might find courts within the same jurisdiction reaching seemingly contradictory decisions. It also is important to remember that a published appellate-level decision provides binding precedent only within the court's jurisdiction and only as to the issue addressed. Of course, a broadly and well-written decision might provide persuasive authority in other jurisdictions and as to other related issues. Thus, for example, a decision by the Vermont Supreme Court, the state's highest court, on the applicability of that state's public records act to an IACUC is binding only within Vermont. The Vermont decision, however, might provide persuasive reasoning that would guide other states' judges who are confronted with questions about the applicability of their states' public records laws to IACUCs or other entities. Readers are encouraged to check with their institution's general counsel or State's Attorney General for specific laws and interpretations in their particular state.

22:1 What are the sources of law that can apply to an IACUC's activities?

Reg. (See 2:1 and 2:2 for specific laws and regulations requiring the formation of an IACUC.)

Opin. Many sources of law can bear on an IACUC's activities. They include federal and state constitutions, federal and state statutes, federal and state regulations, and federal and state case law.

Constitutions set out the fundamental, broad legal principles that govern the nation or the state which enacted them. The principle of academic freedom, for example, is often considered to be grounded in the First Amendment to the U.S. Constitution. State constitutions are sometimes found to be more protective of certain rights (such as the right of privacy) than is the federal Constitution, even where the wording in the two documents is similar.

Statutes are laws enacted by the federal Congress, a state legislature, or, in some states, by a vote of the people. They are codified, or collected, in books that are divided into titles or chapters for different categories of laws. The AWA, for example, is codified in the U.S. Code at Title 7, Section 2131 et. seq. The Health Research Extension Act of 1985 is codified in various places within the U.S. code, most particularly with respect to animal research at Title 42, Section 289d. Citations to codified law properly include a year in parentheses; this is the year the code collection was published, not the year the statute was enacted. Code collections are generally not published every year, although a supplement with only those sections that have changed might be published following each legislative session. Having the publication date is helpful if you want to make sure that the cited statute includes any recent amendments.

Regulations are rules enacted by a federal or state agency pursuant to authority granted in a statute. As with statutes, these rules are collected in books divided into titles or chapters, typically by topic. Pursuant to the AWA, for example, the USDA has enacted various rules which are found in the Code of Federal Regulations. Various administrative procedures govern the adoption of regulations. Typically, for example, the agency is required to publish a notice of proposed rule-making and to provide an opportunity for comment.

Federal notices of this sort are published in the Federal Register; the states have similar publications for the rule making activities of their agencies.

Case law sets out judges' written interpretations of constitutions, statutes, regulations, and other sources of law. The three primary types of judges in the U.S. are federal judges, state judges, and administrative law judges; each type of judge has jurisdiction over specified types of cases, and sometimes the jurisdictions overlap. Only published decisions of appellate-level judges (those who review trial-level decisions) can be cited as binding precedent in other lawsuits. Although it may seem as though there are innumerable books of appellate-level decisions, comparatively few lawsuits result in this kind of decision. The vast majority of filed lawsuits settle prior to a hearing; only some that go to hearing are appealed and only a few of those that are appealed result in a decision written for publication. The citation to a case will indicate where to find the published version, which court made the ruling, and in what year.

22:2 What sources of authority other than those noted in 22:1 might apply to an IACUC's activities?

Reg. The PHS Policy is applicable to activities supported by the PHS. In order to be eligible to receive a grant or contract award from the PHS, an institution must comply with the PHS Policy. This is sometimes referred to as a "condition of award." Conversely, without the required Animal Welfare Assurance and verification of IACUC review and approval, an institution is ineligible to receive support or funding from the PHS for any activities involving animals.

Opin. In addition to the types of legal authority described above, other sources of authority may apply, with varying degrees of force, to an IACUC's activities. The *Guide* certainly carries a great deal of authority in any legal setting, as does the PHS Policy. In addition, a court is likely to consider as presumptively appropriate standards set out by the AAALAC or a similarly influential professional association. Furthermore, courts often look for evidence of a "community standard" among similar facilities if there is a dispute as to whether a facility acted appropriately.

Other important sources of authority are any internally adopted policies that apply to the IACUC. These can be policies adopted by the animal facility, the university, the department, or any entity that has authority to set policy for the IACUC, including the IACUC itself. If an IACUC is to be subject to an internal policy which goes beyond the law's requirements, it is important that the IACUC be able to comply with that policy. Having a policy whose standards are not met can be worse than not having a policy at all.

22:3 What is proprietary information? How is this concept used in Freedom of Information (public records) laws?

Opin. The concept of "proprietary information" can be defined in different ways. Sources defining this concept can include: standard dictionaries, legal dictionaries, statutes, regulations, case law, and/or institutional policies. "Proprietary" may be defined as exclusive ownership rights. It also may be linked with trade secret or patent law which protects ownership interests in intellectual property.

One current definition of "proprietary information" is "In trade secret law, information in which the owner has a protectable interest." *Black's Law Dictionary* (6th ed., 1990).

Many state Freedom of Information (public records) acts have exemptions for proprietary information or research documents. There is, however, a wide variety in how such statutory exemptions are worded and how they are interpreted by the state courts. This has led to a sort of patchwork quilt in terms of the contrary legal decisions from various states. While the federal Freedom of Information Act (FOIA) and court decisions interpreting its provisions may be persuasive for how a state statute should be interpreted, federal law is generally not controlling as to the state law issue.[1] State public records laws often contain a competitive advantage test in exemptions from disclosure for research-type documents based on confidential commercial or trade secret information.[2] For example, the "research formulae, designs, and data" exemption in Washington's public records law protects from disclosure:

> Valuable formulae, designs, drawings, and research data obtained by any agency within 5 years of the request for disclosure when disclosure would produce private gain and public loss.[3]

When a state agency claims an exemption applies and chooses to withhold information from public release, the requestor may sue the agency to gain access to the information. If such a controversy involves a trial court decision which is appealed by an aggrieved party, then an appellate court decision may be published and become precedent-setting case law. Only appellate-level court decisions create precedential standards for future cases and may be cited as legal authority.

An example of case law involving the issue of compelled release of proprietary information is *Progressive Animal Welfare Society v. University of Washington*.[4] In this case, the University of Washington (UW) argued (citing both federal law and state public records statutes) that unfunded grant proposals were preliminary and proprietary information and, therefore, no portion of such documents be required to be released. The Progressive Animal Welfare Society

(PAWS), seeking public access to the unfunded grant proposal, argued that the information is not private because of the public interest in access to information used or possessed by a government agency. The Washington Supreme Court ultimately held that certain portions of information in an unfunded grant proposal are protected from compelled disclosure, but the entire unfunded grant proposal could not be withheld from public inspection. The court rejected the UW's arguments, which cited federal FOIA, the Bayh-Dole Act, and patent law. In applying state FOIA law, the court reasoned that certain portions of unfunded research grant information fall within the exemptions in Washington State's public records law for protecting "research formulae, designs, and data." In its decision, the Washington Supreme Court acknowledged the importance of appropriately protecting certain types of proprietary information and reasoned that the potential harm associated with premature release may outweigh any public interest for accessing the information. The Court noted:

> ... in science, data and hypotheses are inextricably intertwined. Valuable "research data" include not only raw data but also the guiding hypotheses that structure the data. Accordingly, the trial court properly excised hypothesis and other information from which an informed reader might deduce relevant data or hypotheses. Moreover, the valuable research data implicit in unfunded grant proposals are precisely the kind of information or record envisaged by this exemption. If the data or hypotheses contained in the unfunded grant proposal were prematurely released, the disclosure would produce both the private gain constituted by potential intellectual property piracy and the public loss of patent or other rights.[5]

Subsequent to the PAWS case, the Washington Supreme Court again interpreted the scope of the "research formulae, designs, and data" exemption. The court broadly defined "research data" as:

> ... a body of facts and information collected for a specific purpose and derived from close, careful study or from scholarly or scientific investigation or inquiry.[6]

This broad definition of research data gives deference to the norms of the scientific community.

For other examples of state law interpretations on state FOIA research exemptions, see *ASPCA v. Board of Trustees of the State University of New York at Stonybrook, 556 N.Y.S. 2d 447 (1990)* wherein the court held in part that information relating to procedures to be performed on animals, including treatment and method of disposal, do not explore the underlying hypothesis of the researcher, the researcher's methods, analysis or results, and are not exempt from disclosure as trade secrets. See also Reference 7 of this chapter, in which the court held in part that generally research records cannot be characterized as trade secrets and certain IACUC documents must be disclosed. This court decision overturned a lower court ruling that denied the plaintiffs access to research records for reasons of academic freedom, prohibition against forced disclosure of trade secrets, and protection of researchers from harassment. The court did, however, protect the confidentiality of pending grant applications and certain proprietary information contained in IACUC documents.

The specific language of the governing state public records statute and any case law interpreting the exemption generally determine what information a government agency within that jurisdiction should release. State agencies also may have regulations or institutional policies to provide additional guidance on what information to produce or maintain as confidential. For example, academic research institutions may enter into agreements with companies promising to maintain confidentiality of certain information. For such exchanges of research-related information, proprietary information may be defined as any information relating directly or indirectly to a technology and not known to the public or in the public domain. Such proprietary information may be conveyed in written, graphic, oral, or physical form and may include scientific knowledge, know-how, processes, inventions, techniques, formulae, products, business operations, customer requirements, data, plans, or other records and information. Public research institutions must be careful not to promise a level of confidentiality they may not be able to maintain. Regardless of the contractual promises in any confidentiality agreements or any institutional policies, public research universities are still governed by state FOIA (public records) laws and must adhere to legal requirements to produce information. See Reference 8 of this chapter for examples of public records case law holding that agencies have no authority to create confidentiality for records absent statutory authority.

Readers are advised that proposed changes to Circular A-110 (Office of Management and Budget) may have a significant impact on the above comments, as the proposal would extend the FOIA to grantee research records.[9]

22:4 How specific should IACUC protocols be when proprietary information related to techniques, compounds, or devices is at stake?

Opin. Researchers at public research institutions and others who collaborate with public research institutions should consider the nature of information supplied in IACUC documents. If the information is proprietary, its confidentiality may be jeopardized if the IACUC is in a state jurisdiction which requires IACUC meetings to be held in an open forum or if IACUC documents are accessible under the state Freedom of Information Act (FOIA).

The required amount of information to be supplied to an IACUC related to techniques, compounds, or devices may vary depending on the specific review functions of the committee. Certainly, researchers must supply sufficient information to enable the IACUC to perform its functions. Researchers should consider asking the IACUC for advance clarification about how it would handle any sensitive information which the researchers want to maintain as confidential, and IACUCs should consider developing a policy on handling such information. Once provided to a government agency, however, information may be subject to disclosure under federal or state FOIA, regardless of whether the researcher considers the information proprietary or not.

For an example of how this FOIA release decision may be made, consider the following information supplied in a signed declaration of Joanne Belk, the then-acting FOIA officer of the NIH, in a case involving a public records request for access to information in an unfunded grant proposal.[10] Belk's declaration (dated 10/16/91) outlines the NIH approach (which is still current) to release of any information from unfunded grant proposals (no information released) and how the federal FOIA exemptions are interpreted and applied by the NIH:

Commercial Exemption — Exemption 5 U.S.C. §5.65 protects trade secrets and commercial or financial information that is obtained from a person and is privileged or confidential. In general, records are commercial if the submitter has a commercial interest in them. 45 C.F.R. Part 5.65(b) defines information as commercial if it relates to businesses, commerce, trade, employment, profits, or finances. The regulation mandates that this category shall be broadly interpreted. The regulation defines information as confidential if its disclosure would (1) impair the government's ability to obtain necessary information in the future, (2) harm the competitive position of the person submitting the information, or (3) impair the government's interests such as program effectiveness and compliance (45 C.F.R. Part 5, §5.65,b,4).

The unfunded initial grant proposal is protected under this commercial exemption because the grant proposal often contains proprietary information that directly impacts the livelihood of the investigator. Not only does the research proposal frequently disclose information of a patentable nature, but the premature release of such confidential information can impair the investigator's ability to compete for limited research funds. The competition for funding is fierce and disclosure of an investigator's research ideas can result in plagiarism and scientific pirating if other scientists have access to the design. Release of confidential information can enable other individuals to borrow the research design in competition against the original investigator, thus substantially harming his competitive position.

Release of unfunded initial proposals can impair the government's ability to obtain information in the future because investigators would be reluctant to disclose research ideas to an agency if they knew their ideas would be prematurely disseminated to the public. Also, early disclosure would impair the government's ability to effectively and objectively evaluate a proposal because the confidentiality of the peer review system would be destroyed.

Privacy Exemption — The information in an initial unfunded grant proposal also is protected under the privacy exemption.[11] This exemption protects information that would constitute a clearly unwarranted invasion of personal privacy if disclosed. When deciding to release information under this exemption, the agency must weigh the foreseeable harm of invading an individual's privacy against public benefit.

In the case of an unfunded grant proposal, the risk of foreseeable harm is great. Initial grant proposals are often preliminary in nature, and the investigator initially receives feedback on his proposal through the highly confidential peer review process. If the investigator does not receive funding during the initial agency review, he may then revise and redraft his proposal to incorporate the suggestions of his peers. This process is extremely confidential and reviewers are screened for conflicts of interest. If this information is released before the investigator is ready to make the proposal public in its final form, then the investigator is forced to disclose his proposal in a public forum before he is able to refine and finalize his ideas. The information at that point is not of great public benefit because it is preliminary and subject to revision. Additionally, the public may be harmed by the release of proprietary information in its preliminary stages because the intellectual property rights necessary for commercialization of any research discoveries may be jeopardized.

The Department of Health and Human Services has recognized the need to protect the initial unfunded grant application from public disclosure. See *PHS Grants Policy Statement, Department of Health and Human Services Publication No. (OASH) 90-50,000 (Rev.) October 1, 1990*. This Policy Statement expressly states that only *funded* applications are available under the federal Freedom of Information Act. This policy is reiterated in the current PHS Grant Application which states that *unfunded* grant applications will not be disclosed.

Even if a project is funded, certain types of information may not be released under federal law. If a public records disclosure request is received after a project is funded, the investigator is notified to allow him an opportunity to identify patentable or other valuable or confidential information exempted under the Freedom of Information Act. Also, under the guidelines set forth in *Washington Research Project v. Department of Health, Education, and Welfare 504 F.2d 238 (D.C. Cir. 1974)*, the agency need not release information contained within agency research site visit reports or the agency summary statement discussing the merits of the research proposal.

22:5 Could indicating the sponsor of the research on the IACUC protocol form imperil the confidentiality of proprietary information?

Opin. The identity of the research sponsor may or may not need to be disclosed on IACUC protocols depending upon your institution's policies. There does not appear to be any legal requirement to indicate a sponsor on IACUC protocols. Even if the research sponsor's name is not listed on the IACUC documents, such information may be otherwise obtainable by a public records request to a public research institution. As a practical matter, the identity of the sponsor may be beneficial and may need to be released to help determine whether a research proposal has scientific merit. It also may be important for an IACUC to know the identity of a research sponsor to help ensure a potential conflict of interest, such as a researcher with inappropriate financial interests in a sponsoring company, does not exist. In addition, research institutions using a standard grant and contract proposal routing form may require the sponsor to be identified once the research proposal has been approved by the IACUC. The mere identity of the research sponsor is unlikely to jeopardize the confidentiality of proprietary information.

22:6 Can researchers protect their own identities on IACUC protocol forms?

Opin. Researchers and research sponsors may be able to protect their own identities in the IACUC review process. In *ASPCA v. Board of Trustees of the State University of New York at Stonybrook*, the court ruled that the name, department, location, and telephone number of individual researchers and grant number or application number of funding source were exempt from disclosure. The identity of researchers also was protected from disclosure in the North Carolina case, *SETA UNC-CH v. Huffines, 399 S.E. 2d 340 (N.C. Ct. App. 1991)*. Other state decisions have held just the opposite. See *Thomas v. Ohio State University, 643 N.E.2d 126 (1994)* in which the Ohio Supreme Court held that its public records act required the university to release the names and work addresses of research scientists using animals.

22:7 What procedural mechanisms may an IACUC adopt to flag information considered proprietary by the researchers?

Opin. IACUCs at public research institutions may choose to notify researchers that there may be public access to information they supplied to the IACUC. This can happen through both open meeting and public records laws. As noted in 22:11 and 22:12, there are differences between state jurisdictions concerning whether IACUC activities are subjected to state open meetings acts and about what IACUC documents are accessible under state Freedom of Information Acts (FOIA). IACUCs may choose to provide researchers with a mechanism to flag information they believe to be confidential and proprietary in nature. The mere flagging of this information, however, does not mean that it may not be obtainable under federal or state law. If the IACUC receives a request for the documents, the flagging may give a measure of additional protection only to the extent it helps facilitate a careful review of the scope of information capable of being protected from compelled release under the applicable standards. IACUCs that choose to give researchers the ability to flag information should be certain to have sufficient administrative support to follow through on noting the researchers' concerns. It also is important for the IACUC to tell the researchers how the flagging will be used and the limitations it may have.

22:8 Should IACUC members be directed to turn in or destroy copies of protocols reviewed by the committee at the end of a meeting or once a final action has been taken by the IACUC?

Opin. IACUCs may have policies and procedures which include collecting copies of protocols and other documents reviewed by committee members either at the end of the relevant meeting or when the committee's review is completed. These practices would likely be governed by institutional policy and may vary. Caution is warranted, however, to ensure that the IACUC's practices are not in conflict with other record retention requirements. These requirements may include any other federal laws and regulations pertaining to animal-based research that the IACUC must comply with.

The care of animals housed by universities, medical schools, hospitals, and research centers is monitored by APHIS/AC under the provisions of the AWA. This Act, first passed in 1966, has been amended several times by Congress, to reflect changing knowledge about animal care, as well as to include specific areas of community concern about animal research. Each institution is required under the AWA to establish an IACUC to oversee the animal care program and to review each proposed use of animals (see Chapter 2). APHIS/AC inspectors may review the records of the IACUC to ensure that the committee is properly constituted and functioning and that all animal research studies have been approved by it. IACUCs must, therefore, retain records consistent with the federal requirements, which is for at least 3 years (AWAR §2.35,f; PHS Policy IV,E,2).

Other regulatory agencies also may impact the requirements for retaining IACUC documents; for example, the Good Laboratory Practices Regulations. These regulations cover nonclinical laboratory studies that support applications for research or marketing approval for products regulated by the Food and Drug Administration and the Environmental Protection Agency. The agency representatives inspect laboratories as well as all records and specimens. Facilities also are required to have a sufficient number of well-maintained animal housing areas and appropriately trained veterinary and animal care personnel. These regulations also require animal housing areas and all accessory equipment to be clean and sanitized. Testing facilities risk rejection of their data and possible subsequent rejection of their products if they fail to comply with these regulations. To the extent the IACUC functions and records address these areas of performance, the committee should probably maintain the records as long as relevant for these agencies, too.

Finally, the IACUC should comply with any institutional policies for records retention.

22:9 May an IACUC be sued for breach of proprietary confidence?

Opin. There is always the possibility that an IACUC (like any other entity) could be sued by a plaintiff asserting a legal violation and resulting harm. The better question to ask is: if an IACUC is sued for breach of proprietary confidence, will the IACUC prevail in the case? There are several liability-reduction strategies an IACUC may want to adopt to reduce the potential of claims for breach of proprietary information. First, an IACUC should clearly tell researchers submitting information to it what its functions are and whether it is subject to open meetings or state public records laws. Second, if an IACUC receives a public disclosure request for records, it should notify the researchers and allow them an opportunity to review the request and the proposed release of information.

The input of the researchers, if any, should be considered by the IACUC in making the determination about what, if any, documents or information to release. The researchers' opinions may not be in accord with the applicable laws and regulations, so ultimately the IACUC may still decide to release information that the researchers believe to be proprietary in nature. Third, the IACUC may want to offer the researchers who disagree with an institutional decision to release information an opportunity to seek a legal injunction in court to prohibit the release of their proprietary information. As a final strategy, the IACUC may decide to go into court for a judicial ruling on what, if any, portions of the records requested under state law must be released. An IACUC cannot generally be sued for disclosure of information if ordered to do so by a court; however, if the IACUC misled the researchers by promising more confidentiality than actually permitted by applicable laws, then the researchers may have a cause of action based on breach of contract.

22:10 Must an IACUC at a state institution open its meetings to the public?

Opin. Whether an IACUC at a state institution is required to open its meetings to the public depends on the wording and interpretation of the state's open public meetings statute. Some states' statutes require open meetings and other states' statutes do not.

The reasoning of the Vermont Supreme Court in a 1992 case is typical of those cases which have found a state institution's IACUC to be subject to the state's open public meetings act.[12] Vermont's law applied to a "public body" which was defined to include "any board, council, or commission ... of any instrumentality of the state ... or any committee."[13] The Court first determined that the University of Vermont was a "public body" based on prior case law and the language of the university's charter. The Court then went on to hold that the IACUC is a "committee" of the university because it "exercises authority delegated to it in significant part by the [university's] Board of Trustees," and because it is endowed with "considerable policymaking authority."[14] Therefore, the IACUC was subject to the state's open public meetings act.

On the other hand, those courts which determined that a state institution's IACUC meetings need not be open generally made their determinations on one or more of three interrelated bases. The three bases are that the committee was not a "public body" as defined by the applicable statute, was not performing a "state function," or was generally governed by federal law rather than by state law.

One case of this nature was decided by the New York Court of Appeals (New York's highest court) in 1992, the same year as the Vermont Supreme Court's ruling.[15] The New York Court held that the State University at Stonybrook's IACUC was not a "public body" under the state's Open Meetings Law and, therefore, its meetings could remain closed. The Court reasoned that because the powers and function of the IACUC derived primarily from federal law, the committee was not, as specified in the state statute, "performing a governmental function *for the state.*"[16]

In a 1991 case, the Massachusetts Court of Appeals also found that a state institution's IACUC was not covered by the state's open meetings act, though it relied on a slightly different rationale than did the New York Court.[17] In that case the Massachusetts court held that the IACUC is not a state policymaking body and that the state statute only applies to governmental entities whose duties include policymaking.[18]

22:11 Must an IACUC in a private institution open its meetings up to the public?

Opin. As a general rule, no laws require an IACUC in a private institution to open its meetings up to the public. As described in 22:10, state public meeting laws apply to IACUCs that can be characterized as state entities. Of course, a private institution might have a *policy* of allowing public attendance at its meetings.

It is important to bear in mind that if a private IACUC can be characterized as functioning on behalf of a state institution — either in general or as respects a specific project — that IACUC might be found to be covered by the state's public meetings act. We found no published case directly on this point and know of no specific examples, but this is certainly a theoretical possibility. In a 1979 case, the U.S. Supreme Court noted in passing that a private research institution could be characterized as a federal agency for purposes of the federal Freedom of Information Act if there was proof of "extensive, detailed, and virtually day-to-day [federal] supervision."[17]

22:12 Is an IACUC in a state institution subject to a public records act?

Opin. Whether an IACUC in a state institution is required to make its documents available to the public depends on the wording and interpretation of the state's public records statute. Some states' statutes have been found to apply and other states' statutes have been found not to apply. In general, if a state institution's IACUC is subject to the state's open meetings act, that IACUC also will be subject to the state's public records act. This is because the two statutes are often written with similar language. In addition, those seeking access will often raise both open meetings and open records issues in the same lawsuit, or in separate lawsuits that are filed one after another by the same group, against the same institution. (The federal public records statute — the Freedom of Information Act (FOIA) — generally applies only to requests made of federal entities. Thus, for most IACUCs, the relevant statute to consider will be their state's public records act.)

The Vermont Supreme Court case[12] discussed in 22:10 raised both open meeting and public records issues. The Court's rationale on the record issue mirrored its rationale on the meeting issue. The Court held that the IACUC was subject to the state's Public Records Act because the University of Vermont is a "public agency" and the IACUC is a "committee" of the university as those terms are defined in the statute.[19] The Court recognized that exceptions carved out by the statute might apply to specific documents.[19] Exceptions commonly found in these statutes are discussed below.

The New York decision on open meetings[16] discussed in 22:10 established the rationale for the public records decision that followed it by only a few months. In March 1992, New York's highest court held that Stonybrook's IACUC was not subject to the state's open public meetings act. In June 1992, New York's appellate-level court considered a companion case involving the same parties and the same university IACUC in which applicability of the state's public records act was at issue.[20] The appellate-level court cited the highest court's decision and held that, under the same rationale, the IACUC was not performing a state function and, thus, was not subject to the public records act.

22:13 Is an IACUC in a private institution subject to a public records act?

Opin. In general, laws do not require IACUCs in private institutions to make available to the public the records they maintain. The U.S. Supreme Court ruled in 1979 that private organizations participating in federally funded research are not by that fact alone federal "agencies" for the purposes of the federal Freedom of Information Act (FOIA) and, thus, that law presumptively does not apply to them.[21] The Court did note that it was possible for a research institution to be appropriately characterized as a federal agency for purposes of FOIA if there was proof of "extensive, detailed, and virtually day-to-day (federal) supervision" of the private institution.[22]

Of course, the federal government will maintain some documents about the private research facility (such as inspection reports). Unless an exemption applies, those documents are presumptively available to the public from the federal government under FOIA. In addition, information about a collaborative research project with a public institution might be available to the public from the public institution pursuant to the state's public records act.

22:14 If a state's public records act does apply (see 22:12), what disclosure exemptions might be relevant?

Opin. State public records acts typically include a list of exemptions. If one or more of the enumerated exemptions apply, the otherwise presumptively public document (or portions of it) may be withheld from disclosure. In addition to the enumerated exceptions, federal law or constitutional principles might preclude disclosure in a particular situation. This section discusses some of the more common enumerated exceptions, as well as the other sources of legal authority that might be relevant. Note that more than one exception or other law might apply and that there are significant differences in state laws, making it difficult to generalize. For example, in *PAWS v. University of Washington*, 125 Wn.2d 243, 884 P.2d 592 (1994), the university argued that several different exemptions and other sources of legal authority applied to preclude release of unfunded grant proposals.

Many public records acts include an exception for research-related documents. They differ widely in their wording and interpretation. (See Reference 1 which describes differing research exemptions and includes an appendix with the relevant language from the various state statutes.) In 1995, the Indiana Court of Appeals held that Indiana University records on animal use in research were "information concerning research" and thus exempt under that state's Public Records Act.[23] By contrast, the Washington Supreme Court held in 1994 that unfunded grant proposals, as a whole, were not "valuable formulae" or "research data" as defined in Washington's act, although much of the material in the grant proposals did come within that exemption.[24]

Public records acts typically include an exception aimed at protecting the privacy of state employees. Whether an IACUC can delete ("redact") names and addresses of researchers or others when responding to a public records request has been the subject of litigation, with differing results primarily stemming from statutory language. In 1994, for example, the Ohio Supreme Court required Ohio State University to release the names and work addresses of animal research scientists.[25] The university had redacted that information from research-related documents provided in response to a public records request by the group called Protect Our Earth's Treasures. The university redacted the information for security reasons. The Court acknowledged the likelihood of increased harassment, but noted that there did not appear to be the "high potential ... for victimization" required to establish a constitutionally based exemption.[26]

The North Carolina Court of Appeals, however, ruled in 1991 that under that state's Public Records Act, the identity of researchers and staff members could be withheld.[27] The court reasoned that "public policy *does require* that any information contained in the [research] applications relating to the names of the researcher and staff members, their telephone numbers, addresses, their experience, and the department name be redacted from the IACUC applications."[28]

The North Carolina court[27] also ruled that applications which are not approved need not be made public. Many states have exceptions for "preliminary" or "draft" information, and these exceptions might preclude release of unapproved or unfunded proposals. In addition, intellectual property and public policy arguments often have great force when the research is at this early stage. See Reference 1 (pages 421 to 440) for a discussion on the applicability of Bayh-Dole Act, patent law, copyright law, and practical considerations.

22:15 How does the principle of academic freedom apply to IACUC activities?

Opin. Researchers are often aware of the principle of academic freedom and are interested in whether it protects from public disclosure information related to their research. Academic freedom is a multifaceted principle, invoked in support of allowing scholars to study, discuss, and publish ideas free from inappropriate restraints. This principle is invoked in a variety of contexts, from debates over tenure to challenges over publication rights.

This principle is explicitly reflected in the policies of numerous universities and academic associations. In addition, it is generally acknowledged to be firmly grounded in the free speech protections of the federal Constitution and state constitutions.[29]

One of the facets of the principle of academic freedom is what is often referred to as the "research scholars' privilege." Some have argued that this privilege is rooted in the First Amendment and that it protects documents which would otherwise be available under public records statutes, because to release the information would chill research, the unfettered discussion of ideas, and other activities that underlie academic freedom. This is the context in which IACUCs are most likely to face the potential applicability of the principle of academic freedom.

In this context, the principle has generally not fared well in the courts. In a discrimination lawsuit in which peer review materials were sought, the U.S. Supreme Court rejected the argument that the protection afforded by the academic freedom principle trumped the legal presumption that the subpoenaed documents be made available.[30] In this 1990 case, the federal Equal Employment Opportunity Commission sought to enforce its subpoena after the University of Pennsylvania declined to release tenure review materials with respect to a former faculty member who had alleged racial and sexual discrimination. The Supreme Court held that the First Amendment right of academic freedom would not be expanded to protect the materials from disclosure.

A year later, the North Carolina Court of Appeals considered the applicability of this Supreme Court decision to an argument that research applications submitted to a state university's IACUC were exempt from public disclosure pursuant to the principle of academic freedom.³¹ The North Carolina court ruled that the U.S. Supreme Court decision was conclusive on this point. In this court's analysis, the principle of academic freedom is not a basis for withholding this type of document in the face of a state public records act that presumptively mandates release.³¹ As discussed below, the court did allow withholding of unapproved applications and of researchers' names, phone numbers, and addresses on legal bases *other than* the principle of academic freedom.³²

Confronted with a similar factual situation and a similar argument about the applicability of the principle of academic freedom, the Indiana Appeals Court³³ declined to follow the logic of the North Carolina decision.³² The Indiana court noted that it also was faced with a public records request for documents submitted to a state university's IACUC, and then went on to stress a crucial aspect of the Indiana law. "There is a critical distinction between the present case and [North Carolina's case³²] which dictates against following North Carolina's precedent; North Carolina's Public Records Act does not contain a concerning research exception."³⁴ Indiana's statute did. Thus, based on the law's research exception as well as the principle of academic freedom, the Indiana court held that the documents need not be released.

The Indiana decision³³ may be an aberration. More typical are decisions like that in Washington's PAWS case in which courts acknowledge the importance of the principle of academic freedom, but hold that it does not trump statutes that otherwise mandate disclosure.³⁵

The prevailing view seems to be that academic freedom for research constitutes a qualified privilege. This qualified privilege needs to be balanced against the competing interests favoring disclosure, to determine whether it is reasonable to force disclosure of the research information sought. If the competing interest favoring disclosure is a clear public records act without an applicable exemption, disclosure is likely. If the competing interest is a less clear public records act, or a lawsuit to which the documents might be relevant, predicting the outcome is more difficult. This type of case might be very fact-specific and partial disclosure might be the court's conclusion.³⁶

22:16 What mechanisms are available to protect against harassment and vandalism?

Opin. Where researchers' names and addresses are publicly available or information about a controversial project has been publicly released, facilities may wonder how they can protect against harassment and vandalism. In determining whether and how to invoke preemptive measures, it is important to consider the state's criminal and civil laws regarding trespassing, harassment, and property destruction. If a public facility is involved, the constitutional right to free speech also must be considered in, for example, establishing boundaries for a protest.

In addition, many states have laws specifically penalizing vandalism of laboratories and researchers' homes, as well as specifically penalizing certain forms of harassment of researchers. These laws were passed in response to the belief that general state laws against vandalism and harassment provided insufficient protection or compensation.[37]

Washington State is one of those that has these additional protections, both in civil and criminal statutes.[38] For example, a section of the civil statute allows an individual to apply for an injunction based on the reasonable belief that he or she may be injured by the commission of animal research-related vandalism or harassed because of animal research activities.[39] If an injunction is issued to prevent anticipated vandalism or harassment, the violation of the injunction itself gives rise to legal action. Although this type of statutory provision might be valuable in some circumstances, its practical benefits are limited by the fact that it requires researchers to file court documents identifying themselves, the harassers, and the anticipated unlawful action sufficiently in advance of that action.

Other provisions of Washington's statute note the importance of protecting animal research while recognizing legitimate protest rights. As do many states, Washington provides for increased penalties for those who are found to have vandalized an animal research facility.[40]

In considering how to protect facilities and individuals, it is worthwhile to consider seeking guidance from research associations. They often are excellent resources for practical steps that can be taken to reduce the level of confrontation over animal research. The National Association for Biomedical Research and its state affiliates, for example, have very helpful resources regarding community outreach, relations with animal rights groups, and response to protests.

References

1. Lewis, T. and Vincler, L., Storming the ivory tower: the competing interests of the public's right to know and protecting the integrity of university research, 20 *J.C. and U.L.* 417 (Spring 1994).
2. 37 *Am. Jur.* 2D Freedom of Information Acts, § 138 (1994).
3. RCW 42.17.310(1)(h).

4. *Progressive Animal Welfare Society v. University of Washington*, 125 Wn.2d 243, 884 P.2d 592 (1994). The University of Washington received support for its legal arguments by the following organizations which participated in the case as *amici curiae*: American Council on Education, Association of American Medical Colleges, American Psychological Association, The Johns Hopkins University, Washington State Biotechnology Association, and Washington Association for Biomedical Research. PAWS received support for its legal arguments from the American Civil Liberties Union which participated as *amicus curiae* on its behalf.
5. Id. at 255.
6. *Servais v. Port of Bellingham*, 127 Wn.2d 820, 832, 904 P.2d 1124 (1995).
7. *SETA UNC-CH v. Huffines*, 399 S.E. 2d 340 (N.C. Ct. App. 1991).
8. *Van Buren v. Miller* 22 Wn. App. 836, 592 P.2d 671 (1979) and *Hearst Corp. v. Hoppe* 90 Wn. 2d 123, 580 P.2d 246 (1983).
9. Implications of Notice of Proposed Rule Making, Office of Management and Budget Circular A-110, "Uniform Administrative Requirements for Grants and Agreements With Institutions of Higher Education, Hospitals, and Other Nonprofit Organizations." Available on the World Wide Web at: *http://www.nih.gov/grants/policy/a110/a110implications.htm*
10. *PAWS*, 125 Wn. 2d 243 (1994).
11. The privacy exemption, 55 U.S.C. 552 (6), is also addressed in 45 C.F.R. Part 5, § 5.67.
12. *Animal Legal Defense Fund v. IACUC of the University of Vermont*, 616 A.2d 224 (Vt. 1992).
13. Id. at 136 (quoting statute).
14. Id. at 138, 139.
15. *Association for the Prevention of Cruelty to Animals v. State University of New York*, 582 N.Y.S.2d 983 (N.Y. 1992).
16. Id. at 984 (emphasis to statutory language added).
17. *Forsham v. Harris*, 445 U.S. 169, 180 n.11 (1979).
18. Id. at 391-392.
19. See Reference 11, at 140.
20. *Association for the Prevention of Cruelty to Animals v. State University of New York*, 584 N.Y.S.2d 198 (N.Y. App. Div. 1992).
21. *Forsham v. Harris*, 445 U.S. 169, 171 (1979).
22. Id. at 180, n.11 (citations omitted).
23. *Robinson v. Indiana University*, 659 N.E. 2d 153 (Ind. Ct. App. 1995).
24. *PAWS*, 125 Wn.2d at 254–255.
25. *Thomas v. Ohio State University*, 643 N.E. 126, 130 (Ohio 1994).
26. Id. at 129 (citation omitted).
27. *SETA UNC-CH v. Huffines*, 399 S.E.2d 340, 341-343 (N.C. Ct. App. 1991).
28. Id. at 342 (emphasis added).
29. J. Peter Byrne, *Academic Freedom: A "Special Concern of the First Amendment,"* 99 Yale L.J. 251 (1989).
30. *University of Pennsylvania v. Equal Employment Opportunity Commission*, 493 U.S. 182, 199–200 (1990).
31. *SETA*, 101 N.C. App. at 297.
32. Id. at 296.
33. *Robinson v. Indiana University*, 659 N.E.2d 153 (Ind. Ct. App. 1995).
34. Id. at 156.

35. *PAWS*, 125 Wn.2d at 264-65.
36. *In Re Application of R.J. Reynolds Tobacco Co.*, 880 F.2d 1520, 1529-30 (2nd Cir. 1989).
37. Blumenstyk, G., Ten More States Enact Laws on Vandalizing Animal Laboratories, *Chron. Higher Edu.*, Aug. 7, 1991, A15.
38. RCW 4.24.570 and .580 (civil), 9.08.080 and .090 (criminal) (1996).
39. RCW 4.24.580 (1996).
40. RCW 4.24.570(3); see also Reference 36, p. A15.

23

General Concepts of the Facility Inspection and Program Review

Stephen K. Curtis*

Introduction

This chapter discusses the general aspects of two important responsibilities of the IACUC: program review and facility inspection. As described below in greater detail, both the PHS Policy and the AWAR require that the IACUC conduct facility and program evaluations at least once every 6 months.

The material presented in this chapter reflects the procedures used at the author's institution and those of the contributors. These procedures are commonly used at research facilities regulated under the AWAR and the PHS Policy. State laws may add additional requirements.

23:1 What is meant by "reviewing the program of animal care and use"?

Reg. The PHS Policy (IV,B,1; IV,B,2) and the AWAR (§2.31,c,1; §2.31,c,2) require the IACUC to review the institution's program of animal care and use and inspect the institution's animal facilities. The PHS Policy requires the IACUC to use the *Guide* as a basis for evaluation, whereas the AWAR requires the IACUC to use the standards of the AWAR as a basis for evaluation.

Opin. Institutions and their IACUCs have a great deal of flexibility in determining the method used to conduct the program review. In general, the IACUC should review all aspects of the animal care and use program using the *Guide* as an outline. Program reviews should prima-

---

* Contributing to this chapter were Deborah M. Faryna, Molly Greene, Debbie Hampstead, Richard M. Harrison, Todd A. Jackson, Beverly Keniston, William W. King, Gregory R. Reinhard, and Julie Watson.

rily focus on the administrative policies and procedures for conducting the animal care and use program. Areas to cover include:

- *Institutional policies, procedures, and responsibilities*
  - IACUC organization and procedures
  - Personnel qualifications and training
  - Occupational health and safety of personnel
- *Animal environment, housing, and management*
  - Animal housing and physical environment
  - Behavioral management and environmental enrichment
  - Husbandry practices
  - Population and genetic management
- *Veterinary medical care*
  - Animal procurement and transportation
  - Preventive medicine
  - Surgery and postoperative care
  - Management of pain and distress
  - Methods of euthanasia
- *Physical plant*
  - This last section of the *Guide* entitled "Physical Plant" is generally reviewed as part of the facility inspection, discussed elsewhere in this chapter. (See Chapters 24 to 26.)

23:2 What is the purpose of the facility inspection and the program review?

Opin. Most animal research facilities (depending on the source of funding and the species of animals used) fall under the jurisdiction of the PHS Policy or the AWAR. Both documents mandate semiannual program and facility evaluations. However, aside from fulfilling a regulatory requirement, all research facilities using animals should adopt procedures to monitor their program of animal care and use. This helps ensure that animal health and well-being are optimized and pain and suffering are minimized. Monitoring also helps ensure that all personnel are trained in the care and use of animals and protected from occupational hazards associated with animals. In addition, a well-managed and documented animal care and use program helps improve public confidence in the research conducted at the facility and engenders increased support for government sponsored research in general. Commercial entities may find that clients will seek out laboratories that adhere to the highest standards of animal care. Clients may be willing to pay a premium for high-quality work. In other words, a well-managed animal care and use program is a vital component of good scientific research.

23:3 Can the review of animal care and use programs be based on observations made during the inspection of the facilities?

Opin. In general, more than a visual inspection of the facilities is required to determine if the animal care and use program is operating in accordance with law and public policy. For example, facility inspection cannot identify a problem with the composition of the IACUC, or uncover problems with documenting the occupational health program.

A common practice is to conduct a separate program review using an outline or a specific checklist designed to identify areas of review (see 23:24). Checklists help ensure that no important item is overlooked. The *Guide* is generally used as a basic outline for the program review (see 23:1). The NIH/OPRR Web-based training tutorial[1] also lists the following parts of the animal care program:

- Designation of an Institutional Official.
- Appointment of an IACUC.
- Administrative support for the IACUC.
- Standard IACUC procedures.
- Arrangements for a veterinarian with authority and responsibility for animals.
- Adequate veterinary care.
- Formal or on-the-job training for personnel that care for or use animals.
- An occupational health and safety program for those who have animal contact.
- Maintenance of animal facilities.
- Provisions for animal care.

The specific methods for conducting the program review are discussed in 23:10.

23:4 What laws, guidelines, and policies delineate what to identify when an IACUC inspects an animal facility and reviews the program of animal care and use?

Reg. Two basic documents mandate and govern the administrative aspects of the facility inspection and the program review: PHS Policy (IV,B) and the AWAR (§2.31,c). (See 23:1.)

Opin. Other references also should be used to evaluate an animal care and use program. The primary standard for conducting an animal care and use program is the *Guide*. In fact, the PHS Policy mandates that the *Guide* be used as the basic standard of care (PHS Policy IV,B). It is customary for research facilities to follow the tenets of the *Guide* even if they have no legal obligation to do so. In addition, if agricultural research is conducted, the *Guide for the Care and Use of Agricultural Animals in Agricultural Research and Teaching*[2] (*Ag Guide*) should be consulted. Both the *Guide* and *Ag Guide* contain extensive references; a few of the more frequently used references are

- U.S. Government Principles for the Utilization and Care of Vertebrate Animals Used in Testing, Research, and Training, which is part of the PHS Policy.
- *Occupational Health and Safety in the Care and Use of Research Animals.*[3]
- Biosafety in Microbiological and Biomedical Laboratories.[4]
- Canadian Council on Animal Care — *Guide to the Care and Use of Experimental Animals.*[5]
- Institutional Animal Care and Use Committee Guidebook.[6]

23:5 How frequently should facility inspections and program reviews be conducted?

Reg. PHS Policy (IV,B,1; IV,B,2) and the AWAR (§2:31,c,1; §2.31,c,2) require inspection of the animal facilities and review of the program of animal care and use at least once every 6 months. The *Guide* (page 9) uses similar language.

Opin. It is generally accepted that the interval between inspections and review should not exceed 6 months. To minimize the number of inspections in a given year the inspections should occur on a regular basis approximately 6 months apart. At least one USDA-registered and AAALAC-accredited facility conducts its inspections and review in April and November, a 7/5-month interval, without criticism to date. Nevertheless, if the inspection and review took place at the beginning of April and the end of November, this would essentially be an 8-month interval and likely lead to a citation on an APHIS/AC inspection report. Inspectors typically allow a 1 month leeway.[7]

Another facility conducts their review on an ongoing basis but files a report every 6 months. An individual component of that program may not be reviewed exactly every 6 months. However, conventional wisdom dictates that major departures from the 6-month rule should be avoided.

23:6 Can the twice yearly inspections occur in November and again during the following January?

Reg. (See 23:5.)

Opin. No. (See 23:5.) The IACUC would have to conduct a third inspection in July and could not wait until the subsequent November. It would be more efficient to conduct a review each May and November or each January and July.

23:7 Can an IACUC inspection team inspect areas more often than twice a year?

Opin. Yes, facilities may be inspected as often as the IACUC deems necessary. The PHS Policy and the AWAR do not preclude IACUC inspections more frequently than once every 6 months. An IACUC may "reset the clock" if it wishes to perform an inspection early and then perform the next inspection 6 months later. More frequent inspections may be required to verify that previously identified concerns have been corrected. Some IACUCs conduct inspections on an ongoing basis. For example, one IACUC sets aside a time each month to visit laboratories where animal research is conducted. Under this system, an individual laboratory may be inspected more frequently. It is also a common practice to evaluate new areas before housing animals.

Members of the animal care staff, such as directors and managers, are often members of the IACUC. In addition, at least one veterinarian should be a member. These individuals are usually in the animal facility on a regular basis between formal IACUC inspections and, therefore, have the opportunity to evaluate and report to the IACUC on the day-to-day operation of the facility.

23:8 Should animal facilities be inspected only if animals are present at the time of inspection?

Reg. (See animal facility definitions in 15:1.)

Opin. Areas where animals are currently used, even if they are not in that area at the time of an inspection, should be visited and evaluated relative to its physical structure. If it is likely that animals will never again be housed or used in a particular area, the area can be removed from the list of those to be inspected.

It may not be possible at large, decentralized facilities to evaluate every research laboratory where live animal work is conducted. Practically speaking, those areas where animals are housed in excess of 12 hours or where surgery is conducted should be inspected semiannually. Laboratories where only minor work is performed, such as terminal tissue collection, could be inspected less frequently. Although this practice is not in keeping with the letter of the law, in the author's experience, reduced inspections using these criteria has not created concerns with regulatory agencies. The reader is cautioned, however, that the definition of "minor" work can be open to interpretation. In the example presented, if the animals are euthanized where terminal tissue collection also takes place, it is suggested that this area be inspected every 6 months.[7] NIH/OPRR has suggested that reduced inspections or alternate methods of monitoring might be adopted by the IACUC.[8] (See 23:9.)

23:9 Should the facility inspection team visit research laboratories where animal research (such as rodent surgery) is conducted?

Reg. (See 15:1, 25:1 to 25:3.)

Opin. It is our opinion that all areas where live animals are housed or used (even areas where animals are taken for euthanasia and tissue collection) should be included as part of the semiannual facility inspection. Nevertheless, not all IACUCs strictly follow this practice (see 24:10), and there is leeway within the NIH/OPRR policies. NIH/OPRR[9] states that when considering IACUC responsibilities for semiannual review, it is important to keep in mind that each assured institution, acting through its IACUC or facility veterinarian, is responsible for all animal-related activities at the institution regardless of where the animals are maintained or the duration of their stay. The PHS Policy allows institutions some discretion regarding specific methods to assure compliance, but the institution is clearly responsible for what happens to animals in investigators' laboratories. The degree, frequency, and method of IACUC oversight often depends on the nature of the activity. For example, satellite holding facilities and areas in which surgical manipulations are performed must always be included in semiannual reviews. Other activities, such as routine dosing, weighing, or immunization of animals in laboratories, may be monitored using other methods such as random site visits and evaluation.

Inclusion of these laboratories in the semiannual IACUC review is another way to satisfy the PHS Policy requirements. In any case, the IACUC must have access to all investigators' laboratories for the purpose of verifying that activities involving animals are conducted in accordance with the proposal approved by that committee. This implies that IACUCs at institutions regulated by NIH/OPRR have broad discretion in how they oversee animal work that is conducted in laboratories.

Surv. The following quotations from contributors are offered for further clarification and as examples of what is done:

- "At our facility, much of the animal work is done in procedure rooms within the central animal facilities. These areas are inspected semiannually along with the animal housing rooms. The researchers who take animals back to their laboratories to do procedures only have those laboratories spot-checked. One or two labs are checked with each semiannual inspection of the facilities."
- "We use a system such that all recovery surgery areas are part of the inspection at least once a year, and terminal surgery areas every 2 years."
- "The facility inspection team should visit research laboratories where animal research is conducted. Evaluations should be performed during each semiannual inspection as mandated by the AWAR and PHS Policy. At our facility, the IACUC keeps a list of all animal areas by room number. The site visits are conducted in accordance with this list. Additionally, members of the IACUC are familiar with the animal areas of other investigators. These members are valuable in assuring animals are not being used in areas unless identified on the animal study proposal. If there is no work being performed in a given laboratory, the committee checks for such things as outdated drugs, appropriate scavenging of anesthetic gases, and safety concerns such as certification of downdraft tables. Other items to evaluate include appropriate disposal of sharps, appropriate rodent surgical areas, and cleanliness of equipment that will be used with animals."

23:10 What is an effective method for conducting the program review?

Reg. Footnote 7 of the PHS Policy (March 1996 reprint) states that the IACUC may, at its discretion, determine the best means of conducting an evaluation of the institution's programs and facilities.

Opin. Lacking specific guidance by the AWAR and PHS Policy, there are wide variations in the methods institutions use to conduct the semiannual program review. However, there are two accepted formats in common use. One involves an ongoing review of the program as part of normal (generally monthly) IACUC meetings, while the other method involves a discrete review at a special meeting or designated regular meeting. In the author's opinion, it is easier to document the program review when using a discrete review conducted over a short period of time. Regardless of the method used, all areas outlined in 23:3 should be addressed by the committee during the review process. Common means of addressing areas for review include varied combinations of the following methods. In all instances, the full committee or a designated subcommittee may conduct the review.

- The IACUC can review the program using a checklist based on the *Guide*. At least two checklists are available online (see 23:24). Checklists provide a convenient method of documenting the program review. However, checklists have the disadvantage that with continued use, the review becomes mechanical in nature and without sufficient depth to adequately identify problems that invariably occur in large programs.
- The IACUC can interview key staff members such as the veterinarian, animal care supervisors and technicians, and members of the scientific research team. Alternatively, key individuals may be asked to submit reports to or respond to questionnaires from the IACUC. For example, one IACUC asks facility managers to complete portions of the program review checklist before the program review. Information from these forms is consolidated into the final report.
- The IACUC can review selected standard operating procedures, committee reports, protocols, surgical records, and other animal-related records.
- The IACUC can request that special presentations be made. For example, the veterinarian can report on the disease monitoring program or the nonhuman primate enrichment program. Reports can be scheduled on a routine and rotating basis covering essential aspects of the animal care and use program as outlined previously.
- The IACUC can organize various subcommittees with special emphasis on areas such as surgery, occupational health, education and training, veterinary care, and protocol review. The subcommittees then review the area of interest and report to the entire committee on a regular basis.

23:11 Does the entire IACUC need to participate in the facility inspection and program review?

Reg. To comply with the AWAR (§2.31,c,3), a subcommittee participating in the facility inspection and program review must be composed of at least two Committee members and "no Committee member wishing to participate in any evaluations ... may be excluded." Under PHS Policy (IV,B,3 Footnote 7, 1996 reprint) the IACUC "may, at its discretion, determine the best means of conducting an evaluation of the institution's programs and facilities."

The AWAR require that the majority of the members of the IACUC sign the report (§2.31,c,3).

Opin. The entire IACUC need not participate in the facility inspection and program review. It is common practice, especially at large facilities, to appoint a subcommittee to participate in the facility inspections.

Generally, the full IACUC will review and ratify the final report before submission to the IO. It should be noted that regardless of the method chosen, the IACUC retains responsibility for the animal care and use program.

23:12 Minimally, who should be included on every inspection and review team?

Reg. In accordance with the AWAR (§2.31,c,3), at least two IACUC members may be delegated by the IACUC to conduct the required inspection and program review, and no member wishing to participate may be excluded. The PHS Policy has no mandate regarding team size or composition.

Opin. There is wide variation in how institutions conduct their review. It is common practice to have different persons evaluate different parts of the facility and program. Ideally, a typical team consists of several IACUC members with at least one scientist, one veterinarian, and, where possible, one unaffiliated member. Other useful members (not necessarily from the IACUC) are the animal care supervisor, a person with occupational health experience, and a member of the physical facilities maintenance department. No IACUC member who wishes to participate should be excluded from participation. A useful practice at large facilities is to have one member of the team with prior knowledge of the facility, someone who participated in the last evaluation, and a new member to provide a fresh or unbiased review. It is important to keep the inspection and review process vibrant so that the process does not become routine and just another walk around the facility or a "rubber-stamping" of the program.

23:13 Can outside consultants be used by the IACUC during inspection and review?

Reg. PHS Policy (IV,B,3 Footnote 7, 1996 reprint) states that the IACUC may invite ad hoc consultants to assist in conducting the evaluation. However, the IACUC remains responsible for the evaluation and report. The AWAR (§2.31,c,3) have similar language.

Opin. Yes, however, two members of the IACUC must participate in the inspection and review to satisfy the AWAR (see 23:11, 23:12). In practice, outside consultants (other than veterinary consultants employed by some institutions) are not commonly utilized. Most institutions have ample expertise within their own organization.

23:14 Should facility inspections be announced or unannounced?

Opin. Opinions differ, but most contributors to this chapter favor announced inspections. There are no requirements in the AWAR or

PHS Policy stipulating announced or unannounced inspections. The advantages of both have been briefly discussed.[6]

An unannounced inspection gives the inspection team a clear picture of what is routinely occurring in the facility. Announced inspections allow essential persons to be available to answer questions. As there are benefits for both announced and unannounced inspections, it may be advisable to conduct one unannounced site visit per year. The choice is up to the individual IACUC. In practice, it is difficult to keep routine evaluation schedules confidential since key members of the animal care staff often serve on the committee.

It should be remembered that APHIS/AC inspections are always unannounced while AAALAC evaluations are always scheduled well in advance. PHS visits are generally scheduled with short notice. All three systems have merit.

23:15 Should the IACUC inspectors speak with individuals participating in research using animals?

Opin. Yes, this is a generally accepted practice and is highly desirable. The PHS Policy and the AWAR do not mandate IACUC discussions with specific individuals, nor do they preclude them. There are two groups of people from which the committee can obtain information; the research staff and the animal care staff. For example, through informal interviews at the time of the inspection, it is possible to verify that SOPs are being followed, that persons are appropriately trained, and that anesthesia and euthanasia methods are appropriate and in accordance with approved protocols.

Some institutions require that each laboratory have a representative present to answer questions from the inspection team. Other IACUCs schedule inspections at times when persons are normally working, and question those persons available on a less formal basis. The inspection process can and should be a learning experience for both the IACUC and the staff. Free and open communication facilitates this learning process.

The IACUC should take time at the end of a visit to give feedback to the persons involved. When warranted, positive as well as negative comments should be made. IACUCs should consider having an "exit interview" with interested persons as a means of exchanging information.

23:16 How are results of inspections and reviews recorded and processed?

Reg. After the review is completed, a report signed by a majority of the IACUC is forwarded to the IO (AWAR §2.31,c,3; PHS Policy IV,B,3). IACUC members have the right to file formal minority opinions which must be included with the final report (PHS Policy IV,E,1,d; AWAR §2.31,c,3). (See 6:1, 23:18, 23:23.)

Opin. Three methods are often used to document the facility inspection and program review:

- Completed checklist.
- A written report.
- The IACUC minutes.

The method to use and the precise format depend on the complexity of the institution and the preference of the IACUC. Extensive checklists have been written to assist IACUCs in conducting reviews and inspections (see 23:24). The main pitfall with any system is that with repeated use (every 6 months) the process becomes mechanical and members of the IACUC may become complacent. It is best to vary the method of review from time to time to maintain vitality in the review process.

23:17 How long should records be maintained?

Reg. PHS Policy (IV,E,2) states "all records shall be maintained for at least 3 years; records that relate directly to applications, proposals, and proposed significant changes in ongoing activities reviewed and approved by the IACUC shall be maintained for the duration of the activity and for an additional 3 years after completion of the activity." The AWAR (§2.35,f) uses similar wording.

Opin. In practice, there are two opinions regarding recordkeeping. Many institutions maintain records indefinitely because of the difficulty in segregating records into activities that are either complete or ongoing. At these institutions, records are kept as long as storage is available. However, some institutions have concerns about storing unnecessary records due to freedom of information laws. These institutions generally store minutes and other general records for only 3 years and specific protocols for the duration of the project plus 3 years. It is unusual, in the author's opinion, for regulatory bodies to request records older than 3 years. (See 24:14.)

23:18 Who is the recipient of the findings from the facility inspection and program review?

Reg. In accordance with PHS Policy (IV,B,3) and the AWAR (§2.31,c,3), the IACUC must submit a report of the inspection and program review to the IO.

Opin. It also is useful to submit a copy of the report to those that are responsible for correcting deficiencies such as the facility management, AV, and the physical facilities maintenance department. Members of the IACUC should receive a copy of the report as well.

23:19 Should deficiencies and possible corrective actions be discussed at the time of the inspection?

Reg. Neither the AWAR nor PHS Policy mandate when corrective actions should be discussed. The PHS Policy (IV,B,3) does require that IACUC semiannual program reviews that note program or facility deficiencies contain a reasonable and specific plan and schedule for correcting each deficiency. The AWAR (§2.31,c,3) contain similar language.

Opin. In general, yes, although opinions are not unanimous. If the deficiencies are major (significant), the investigator, facility manager, or veterinarian should be immediately informed of the problem so that they can correct it as soon as possible. Delaying notification about a significant deficiency until the entire committee is informed only prolongs the problem. The IACUC can always request additional action be taken to correct a deficiency at a later date. By discussing the deficiency with the appropriate persons at the time of the inspection, the inspecting subcommittee can make sure that their recommended course of action and date for correction is workable before bringing it to the entire IACUC for review. (See 23:15.)

There are times when it might be prudent to withhold discussion at the time of the inspection, pending a full IACUC hearing. If the reviewing subcommittee is unsure of the nature of the deficiency, it is advisable to withhold judgment. On the other hand, if a problem is serious but corrective action is not immediately required, then full committee review is warranted. An example is if the inspecting subcommittee finds that an investigator is conducting research that was not authorized, but the animals are now only being observed prior to final disposition. Although very serious, in terms of the animals' well-being no immediate action may be required by the reviewers. Nevertheless, the reviewers should immediately report the finding of unauthorized research so that the IACUC can assess the situation and, if warranted, take immediate action.

23:20 What criteria are used to categorize deficiencies as minor or major (significant)?

Reg. The AWAR and PHS Policy are in concordance: "A significant deficiency is one which ... in the judgment of the IACUC and the Institutional Official, is or may be a threat to the health or safety of the animals." (AWAR §2.31,c,3; PHS Policy IV,B,3.)

Opin. The IACUC has broad discretion in applying this standard. Deficiencies should be discussed at a regularly convened meeting of the

IACUC. At that time, a final determination as to the nature of the deficiency should be made. Formal minority opinions should be attached to the final report if desired by the dissenting committee member (see 23:16). Discrepancies found in the program or the facilities must be identified on the report as either minor or major (significant). The IACUC must devise a "reasonable and specific plan and schedule for correcting each deficiency." (AWAR §2.31,c,3; PHS IV,B,3.) (See 23:22.)

23:21 Who decides the appropriate resolution of specific deficiencies? How is this done?

Reg. Both the PHS Policy (IV,B,3) and AWAR (§2.31,c,3) require the IACUC to report on the deficiencies found during the facility and program review and to categorize them as either minor or significant (major). The IACUC is required to develop a reasonable and specific plan and schedule for correcting each deficiency (see 23:19). A deficiency is resolved when the IACUC determines that it has been corrected.

Opin. The IACUC has the ultimate responsibility for determining what actions are needed to correct a deficiency. However, it is best to involve the persons that ultimately will be affected by the committee's mandate. As appropriate, the IACUC should consult the facility's management, veterinarian, investigator, and maintenance department. Such involvement makes for a more collegial atmosphere and ultimately a better solution. (See 23:22, 23:23.)

23:22 What regulatory and operational differences must the IACUC address when evaluating minor and significant deficiencies found during the animal facility inspection?

Reg. In accordance with the PHS Policy (IV,B,3) and AWAR (§2.31,c,3), a reasonable and specific plan and schedule must be given for correction of each deficiency. It does not matter if the deficiency is designated as minor or significant. Furthermore, "any failure to adhere to the plan and schedule that results in a significant deficiency remaining uncorrected shall be reported in writing within 15 business days by the IACUC, through the IO, to APHIS/AC and any Federal agency funding that activity." (AWAR §2.31,c,3.)

Opin. IACUCs should devise plans that are attainable within the timetable it establishes. Significant deficiencies must be handled immediately and plans formulated to ensure the health and well-being of the animals. The criteria for determining appropriate deadlines for correction are set in concert with a reasonable timetable for the type of correction required. Thus, for significant deficiencies, there may be two timetables — one to immediately ensure the animal's health and well-being and a second to correct the deficiency. The type of deficiency discovered will determine the appropriate timetable for correction. For example, a facility issue (brick and mortar) may take longer to correct than a programmatic issue in which a correction can be put into motion just by changing a policy. An issue that can be corrected by the IACUC, animal care staff, or investigative staff may be addressed more rapidly than a problem that involves outside contractors and facility issues, which can involve a long procurement process or allocation of funding. Temporary "quick fix" solutions may be put in place while planning and implementing the final solution.

Technically, the IACUC can revise its plan for correction of deficiencies and thereby avoid the AWAR reporting requirement for uncorrected deficiencies. *The reader is cautioned, however, that APHIS/AC disagrees with this interpretation,[7] noting that the AWAR (§2.31,c,3) require the plan to be "specific" and contain "dates for correcting each deficiency." The section further states that "any failure to adhere to the plan and schedule that results in a significant deficiency remaining uncorrected" must be reported.* The main concern in managing deficiencies must be that actions taken by the IACUC are reasonable, defensible, and protect the health and well-being of the animals in question.

23:23 How should the IACUC address minority opinions relative to inspection and review?

Reg. Both PHS Policy (IV,F,4) and AWAR (§2.31,c3) mandate that minority reports be attached to the semiannual report.

Opin. Minority reports are very few. Most IACUCs are able to devise reasonable compromises to difficult problems. Dissenting members generally do so by voting "no" on an issue and requesting that a notation be made in the regular committee minutes. However, a member may write a full minority opinion if so desired. This opinion then becomes part of the official report.

23:24 What Internet resources are available to help with conducting evaluations?

Opin. A number of useful Websites are available. Those listed below are particularly useful because they provide links to a wide body of information related to the care and use of animals in biomedical research.

- *http://rants.nih.gov/grants/oprr/oprr.htm*: This NIH/OPRR Web page gives access to three useful sections:

- The NIH/OPRR Laboratory Animal Welfare page has information such as the PHS Policy, policy guidance, articles, OPRR Reports, *IACUC Guidebook,* and other related materials.
- The PHS Policy Tutorial. This training material is especially useful to new committee members. From the section, "Acronym Glossary and Additional Resources," it is possible to link with a large number of sites related to animal care.
- Sample Documents, including forms for conducting the semiannual review.

- *http://www.aphis.usda.gov/oa/new/aw.html*: The APHIS/AC Animal Welfare home page.
- *http://erebus.rutgers.edu/~labanim/*: Rutgers University animal facility home page. Other major research universities often have laboratory animal care Websites that may prove useful.

Acknowledgment

The assistance of Barbara McLouth in the preparation of this document is gratefully acknowledged.

References

1. NIH/OPRR Tutorial is available on the World Wide Web at: *http://www.nih.gov/grants/oprr/tutorial/iacuc.htm*
2. Committee to Revise the Guide for the Care and Use of Agricultural Animals in Agricultural Research and Teaching, *Guide for the Care and Use of Agricultural Animals in Agricultural Research and Teaching,* 1st Revised ed., Federation of Animal Science Societies, Savoy, IL, 1999.
3. National Research Council, *Occupational Health and Safety in the Care and Use of Research Animals,* National Academy Press, Washington, D.C., 1997.
4. U.S. Public Health Service, Centers for Disease Control and Prevention, Biosafety in Microbiological and Biomedical Laboratories, 4th ed., U.S. Government Printing Office, Washington, D.C., 1999.
5. Canadian Council on Animal Care, *Guide to the Care and Use of Experimental Animals,* Vol. I and II, Canadian Council on Animal Care, Ottawa, Ontario, Canada, 1980.
6. U.S. Department of Health and Human Services, Public Health Service, National Institutes of Health, Institutional Animal Care and Use Committee Guidebook, NIH Publication No. 92-3415, 1992.

7. DeHaven, W.R., personal communication, 1998.
8. Potkay, S., Garnett, N., Miller, J.G., Pond, C.L., and Doyle, D.J., Frequently asked questions about the Public Health Service policy on humane care and use of laboratory animals, *Contemp. Topics Lab. Anim. Sci.*, 36, 47, 1997.
9. Division of Animal Welfare, Office for Protection from Research Risks, National Institutes of Health, The Public Health Service responds to commonly asked questions, *ILAR News*, 33(4), 68, 1991.

24

Inspection of Animal Housing Areas

Patricia A. Ward

Introduction

Animal housing facilities must be inspected by the IACUC at least once every 6 months, according to the AWAR, PHS Policy, and the *Guide*. Nevertheless, aside from this simple directive, a few subsequent policy clarifications, and the detailed "standards" of animal housing and care provided in the AWAR and *Guide*, these regulatory authorities give IACUCs little direction in *how* to carry out this mission. An IACUC may approach implementation of the inspection requirement in a variety of ways. The information provided in this chapter is intended to assist IACUCs in developing an effective animal housing facility inspection program tailored to the needs of their individual institutions.

Information, ideas, and opinions expressed in this chapter were compiled from the results of two surveys of IACUCs at institutions located throughout the U.S. The first survey, conducted in 1997, gathered information from 93 IACUCs on how their inspections are managed and administered. A second survey of 22 IACUCs was conducted in 1998 and focused on how animal facilities are evaluated by IACUCs during inspections. Both surveys include responses from institutions of varying size, type, and culture.

While the information provided in this chapter may represent how many, or even most, IACUCs conduct animal housing and care inspections, the common practices described should in no way be interpreted as the best practices for every individual institution. The methods that an institution's IACUC employs depend on the size, type, and culture of the institution. The "best" inspection methods are likely to be very different for large vs. small facilities, centralized vs. decentralized facilities, corporate organizations vs. educational institutions, etc. IACUCs should be encouraged to exercise their professional judgment and creativity in developing an effective animal housing facility inspection program tailored to the particular needs of their individual institutions.

24:1 What is the definition of an animal housing facility?

Reg. (See 15:1.)

24:2 What is the purpose of conducting IACUC inspections of animal housing facilities?

Reg. The IACUC must fulfill its regulatory responsibility to inspect animal housing areas at least once every 6 months to evaluate compliance with applicable guidelines (AWAR §2.31,c,2; PHS Policy IV,B,2; HREA sec. 495,b,3,A). The recommendations of the *Guide* (page 9) also specify that the IACUC should inspect the animal facilities at least once every 6 months.

Opin. The IACUC must represent its institution both by advancing the institutional mission and by serving as the institutional conscience. By inspecting animal housing areas at least twice a year, the IACUC can ensure that the animal housing and care program is furthering the institutional mission by providing quality animals (through proper attention to animal welfare), maintaining a suitable environment for research activities, and safeguarding the health and safety of its personnel. In addition to ensuring the institution's compliance with applicable regulations, the IACUC must be confident that the animal housing and care program is accomplishing these goals in a scientific, humane, and ethical manner.

24:3 Should an institutional guideline, SOP, or policy be developed for IACUC inspection of animal housing facilities?

Opin. There is no regulatory requirement for IACUCs to establish guidelines, SOPs, or policies regarding inspection of animal housing facilities. However, many IACUCs have found such documents to be helpful in the administration and management of the inspection process. Clarifying expectations for IACUC members conducting inspections, veterinary and administrative staff supporting inspections, and managers of facilities undergoing inspection can promote consistent and thorough inspections and enhance the rapport between all parties. Issues addressed in the guidelines, SOPs, or policies can include:

- Who conducts the IACUC inspections.
- The orientation or training inspectors receive.

- How inspections are scheduled.
- Whether inspections are to be announced or unannounced.
- How inspectors should prepare for each inspection.
- How inspections will be facilitated.
- Sites to be inspected.
- General issues to be considered during inspections.
- How findings will be communicated, documented, and distributed.
- How the IACUC will follow up deficiencies.
- How long inspection documents will be retained.

24:4 Who conducts the IACUC inspections of animal housing areas?

Reg. While both the PHS Policy (IV,B,2), the *Guide* (page 9), and the AWAR (§2.31,c,3) specify animal facility inspection as an IACUC function, the AWAR offer the most explicit "minimum" answer to this question: "No Committee member wishing to participate ... may be excluded. The IACUC may use subcommittees composed of at least two Committee members and may invite ad hoc consultants to assist... ."

Opin. At many institutions, particularly those with smaller or centralized animal facilities, the entire IACUC participates in the animal housing facility inspection. However, in larger or decentralized facilities, the IACUC may divide into subcommittees or appoint a single subcommittee to tackle the job. Such subcommittees must include at least two IACUC members (AWAR §2.31,c,3). Members of the veterinary or IACUC staff also may participate in the inspection to provide expertise, continuity, and support.

24:5 Should IACUC members receive orientation or training in preparation for conducting facility inspections?

Reg. The *Guide* (page 9) states that "it is the institution's responsibility to provide suitable orientation, background materials, access to appropriate resources, and, if necessary, specific training to assist IACUC members in understanding and evaluating issues brought before the committee."

Opin. Most IACUCs arrange for some form of orientation or training for incoming members that includes information about inspection of animal housing facilities. This training is especially beneficial to the nonscientist and nonaffiliated members of the IACUC. Information presented in the inspection training session or packet can include:

- An overview of the size and scope of the institution's animal housing facilities, including maps or floor plans.
- Institutional guidelines, SOPs, or policies regarding IACUC inspections.
- Copies of regulatory, national, and institutional animal housing and care standards.
- Guidelines for evaluating animal well-being.
- The institutional inspection checklist, if used.
- Sources of additional information (e.g., Websites or information about contacting key laboratory animal organizations).

Most IACUCs find it worthwhile to spend a significant portion of the training effort reviewing the various established standards that apply to the areas to be inspected. These include the animal housing and care standards provided in the AWAR, the *Guide*, the *Guide for the Care and Use of Agricultural Animals in Agricultural Research and Teaching* (*Ag Guide*),[1] and the various animal management documents produced by the Institute of Laboratory Animal Resources (ILAR) and the National Research Council (NRC), as applicable.[2-6] A working understanding of these standards not only enables IACUC members to evaluate regulatory compliance issues in the facilities inspected, but also helps to develop their professional judgment with regard to the more important issues of animal welfare, research integrity, and personnel health and safety.

24:6 What time periods are appropriate for IACUC inspections of animal facilities?

Reg. Both the AWAR (§2.31,c,2) and PHS Policy (IV,B,2) specify that animal facilities be inspected at least once every 6 months. (See 23:5 to 23:7.)

Opin. For the most part, meeting this requirement is not difficult for institutions where the animal facilities can be inspected in a relatively short period of time. For larger or decentralized facilities, however, meeting this requirement can be more problematic. If the inspection will take several days to complete, the IACUC is faced with management of a burdensome time commitment. Under these circumstances, most IACUCs will establish regular "inspection windows" at 6-month intervals and break the inspection up over several days within each window. Alternatively, an IACUC can opt to distribute inspection trips throughout the entire 6-month period. In this case, or if the inspection window spans more than a few weeks, care must be taken to inspect the various sites in approximately the same order for each 6-month period. Otherwise, the inspection interval for some sites might exceed the 6-month maximum. This could occur if a particular site is inspected early in one 6-month period and late in the next. In practice, NIH/OPRR does not ordinarily question the interval between inspections of individual sites unless they exceed the 6-month mark by more than 30 days. (See 23:5 to 23:7.)

24:7 Should inspections be announced or unannounced?

Opin. (See 23:14.)

24:8 How should IACUC inspectors prepare for each inspection?

Opin. Before embarking on each inspection, most IACUC inspectors want to review the sites to be visited with particular respect to entry and exit restrictions, previous inspection findings or deficiency history, and experimental activities being conducted. Entry and exit restrictions may affect how inspectors dress for the inspection, the order in which various areas are visited, and the extent to which exposure of inspectors to certain other animals is prohibited for a specified period of time before or after the inspection. Reviewing the previous inspection findings or deficiency history of the sites to be inspected may help IACUC inspectors focus attention on problem areas or remind them to express appreciation for improvements made. Some IACUC inspectors may wish to review experimental protocols approved for use in the areas to be inspected. Doing so can alert them to animal use procedures of particular concern, such as those requiring special consideration during protocol review (*Guide*, page 10), and allow IACUC inspectors to make pertinent observations or inquiries during the inspection visit. Alternatively, many IACUC inspectors choose to wait until after the inspection to review the experimental protocols of animals noted during the inspection.

24:9 What strategies can be used to facilitate inspections?

Opin. Every IACUC looks for ways to facilitate the inspection process. Including administrative or veterinary staff on the inspection team is one popular strategy used to provide expertise, continuity, and support. Another is the use of the previous inspection report or an inspection checklist for reference, or to record findings during the inspection. NIH/OPRR has made a sample checklist available on its Website (*http://grants.nih.gov/grants/oprr/sampledoc/checklist.htm*), and institutions may download and modify it to suit their own programs. Some IACUC inspection teams will carry copies of pertinent regulations and standards, or equipment for recording findings (photo, audio, or video) or validating conditions (temperature or humidity reader, light meter, smoke bottle, etc.). (See 24:11.)

24:10 What sites should be included during the inspection of animal housing facilities?

Reg. Species regulated by the AWAR (§1.1, Animal) include all warm-blooded animals used for research, teaching, testing, experimentation, or exhibition, *except* birds, laboratory rats and mice (*Rattus* and *Mus*), and farm animals used in agricultural research. The PHS Policy (III,A) applies to any vertebrate animal used in research, research training, experimentation, or biological testing. Most IACUCs resolve this discrepancy simply by including the housing areas of all vertebrate animals, regardless of species or type of use, in their inspections. Some IACUCs elect to inspect the housing areas of invertebrate species as well.

What constitutes an animal housing area is also a matter of regulatory definition (and see 15:1). The AWAR (§2.31,c,2) specify that animal facilities, *excluding* wild habitats but *including* study areas, must be inspected. While a definition of *animal facility* is not provided in the AWAR, a *study area* is defined as any area where animals are housed for more than 12 hours (§1.1).

The PHS Policy (III,B) defines an animal facility as "any and all buildings, rooms, areas, enclosures, or vehicles, including satellite facilities, used for animal confinement, transport, maintenance, breeding, or experiments inclusive of surgical manipulation. A satellite facility is any containment outside of a core facility or centrally designated or managed area in which animals are housed for more than 24 hours." The PHS Policy definition is quite comprehensive, including support and experimental surgery areas along with housing areas. (See 25:1.)

Opin. Many IACUCs, particularly those wishing to obtain or maintain accreditation by AAALAC, want to comply with the animal housing inspection requirements of the AWAR, PHS Policy, and the *Guide*. To accomplish this, virtually all IACUCs visit all areas where animals are housed or kept (including study areas, satellite facilities, and laboratories) for more than 12 hours (AWAR-covered species) or 24 hours (all other vertebrate species) during their semiannual inspections. In the survey of facilities conducted by this author, this indicates that areas where rats and mice were kept were not inspected if those animals were held there for less than 24 hours. Temporarily unoccupied animal housing areas also are visited by most IACUCs. Corridors and anterooms contiguous with these areas are included in most inspections. Field research locations are not usually inspected by IACUCs unless animals are held in captivity for more than 12 hours, or the IACUC is particularly interested in observing the field research procedures or setting. Cage wash and other sanitation facilities are to be included in the semiannual inspections, as are diet preparation and food and bedding storage areas. Most IACUCs also inspect areas where caging, equipment, and supplies are stored. Loading docks and transport vehicles, particularly those dedicated to traffic of animals and animal-related supplies and equipment, are usually inspected, as are carcass storage and disposal areas. Animal procedure areas contiguous with the animal housing facilities, including surgical suites, procedure rooms, and euthanasia stations, also may be inspected in conjunction with the semiannual inspection of the animal housing and support areas. (See 23:9.)

24:11 What general issues should be considered when inspecting animal housing facilities?

Reg. The AWAR (§2.31,c,1) require the IACUC to use the AWAR as a guide for semiannual inspections. PHS Policy (IV,A,1) requires the IACUC to use the *Guide* as a basis for evaluation.

Opin. Paramount among inspection issues are animal well-being, research integrity, and personnel health and safety. Proper management of animal care and use promotes all three and IACUC inspectors seek to evaluate the success of the program in this context. To do this during an inspection, most IACUCs try to organize the effort and attention they give to the various aspects of the animal care and use program. Many IACUCs will utilize currently available tabulations, such as the Table of Contents from the *Guide* or the Facility Inspection Checklist[7] available from the NIH/OPRR, to provide such structure. Some IACUCs prefer to focus their attention on issues of particular concern in the type of animal facilities at their institutions. (See also 24:9.)

No matter how the inspection is structured, the many issues that should be considered by IACUC inspectors can be overwhelming once they actually get into an animal housing area to conduct an inspection. One popular approach is for IACUC inspectors to begin with the condition of the animals and expand their focus to the following general categories in any convenient order:

- Primary enclosures, environmental conditions, physical plant.
- Housekeeping, disinfection, and sanitation.
- Water, food, bedding, and other supplies.
- Animal euthanasia, carcass storage, and carcass disposal.
- Labels, signage, and records, and personnel.

24:12 How should inspection findings be communicated, documented, and distributed?

Reg. Both the AWAR (§2.31,c,3) and PHS Policy (IV,B,3) require that reports of inspection findings be forwarded to the IO at the close of every 6-month inspection period. This is usually done in conjunction with the IACUC's semiannual evaluation of its institutional program for humane use and care of animals. Inspection findings may be summarized in the semiannual report to the IO or copies of detailed site-by-site inspection records may simply be attached to the report. In either case, the inspection findings must distinguish significant deficiencies from minor deficiencies (significant representing a threat to the health or safety of animals, see 23:20), and include a plan and schedule for correction (AWAR §2.31,c,3; PHS Policy IV,B,3).

Opin. In addition to reporting inspection findings to the institutional official, IACUCs generally communicate their findings to animal facility managers in order to assist them in correcting deficiencies and obtain full compliance. This communication can occur through many means, depending on the culture of the institution. If inspections are arranged with animal facility managers in advance, most IACUC inspectors ask questions and discuss their concerns with those managers throughout the inspection visit. Alternatively, and particularly if facility managers are not present during the inspection, IACUC inspectors may arrange for a verbal interview with the managers either at the conclusion of the inspection or at a later time. In addition to verbal communication, most IACUCs will present facility managers with a written report of the inspection findings. This may be as simple as a handwritten list of deficiencies or completed checklist presented at the conclusion of the inspection, or a more formal inspection report sent at a later date. In either case, these written inspection reports generally contain the following information:

- Date and site location.
- Names of IACUC inspectors.
- Deficiencies observed and whether each deficiency was significant or minor.
- A directive and date for correction.

In addition, many IACUCs also will reference one or more of the following in their inspection reports:

- The administrative unit responsible for the facility inspected.
- Names of facility representatives present at the inspection.
- A list, including description and contents, of all rooms inspected.

- Whether deficiencies observed were first-time or previously cited.
- The specific standard or regulation violated by each deficiency.
- A general summary, comment, or commendation.

24:13 How should the IACUC follow up on deficiencies?

Reg. Both the AWAR (§2.31,c,3) and PHS Policy (IV,B,3) require that deficiencies be included in the semiannual report to the IO, that a plan and date for correction be indicated for each deficiency, and that minor deficiencies be distinguished from significant deficiencies (deficiencies which may threaten the health or safety of animals). Additionally, the AWAR require that failure to adhere to the correction schedule for significant deficiencies be reported within 15 days by the IACUC through the IO to APHIS/AC and any federal agency funding the cited activity (AWAR §2.31,c,3). PHS Policy (IV,F,3,a) requires prompt reporting to NIH/OPRR of serious or *continuing* noncompliance with the Policy.

Opin. When an IACUC encounters a situation that represents a threat to the health or safety of animals, prompt corrective action should be taken, and a thorough followup conducted in order to ensure that the situation does not recur. However, the vast majority of deficiencies cited by the IACUC on inspection are minor and the nature of followup action depends on the culture of the institution. Many IACUCs reinspect the cited facility at the correction deadline to verify resolution of the deficiency. Most IACUCs specifically assess the correction of previously cited deficiencies at subsequent inspections. Occasionally, an IACUC encounters previously cited minor deficiencies that have either gone uncorrected or have recurred. These deficiencies do not become "elevated" to major deficiencies unless the uncorrected minor deficiency has become a threat to the health or safety of animals. When there are recurrent or uncorrected minor deficiencies the IACUC can employ a variety of strategies to obtain compliance. Examples include:

- Increased frequency of IACUC inspection.
- IACUC interview of responsible persons.
- Notification of superiors in the chain of command.
- IACUC-issued verbal or written warning.
- Withdrawal of IACUC approval to house or use animals (which may require notification of sponsoring agencies).
- Discipline of personnel in accordance with institutional policy.

(See 23:18.)

24:14 How long should inspection documents be retained?

Reg. Both the AWAR (§2.35,f) and PHS Policy (IV,E,2) require that semiannual reports to IO, which include the IACUC's animal facility inspection findings, be retained by the institution and be available for inspection for at least 3 years.

Opin. If additional inspection records are created, the length of time they are retained is at the discretion of the IACUC. Many IACUCs retain these records for the same 3 years as the semiannual reports to the IO. Most, however, find both the semiannual reports to the IO and additional inspection records to be a valuable documentation of animal facility management history. These IACUCs are likely to retain such records for an extended period of time. IACUCs at institutions subject to federal or state Freedom of Information Acts (FOIA) should consult their institutional FOIA Officer for advice on their records retention program. (See 23:17.)

24:15 How should the IACUC assess the well-being of animals during inspection?

Reg. IACUC inspectors should endeavor to assess the physical and psychological well-being of animals. According to the *Guide* (page 22), "Animals should be housed with a goal of maximizing species-specific behaviors and minimizing stress-induced behaviors." To accomplish this, the structural environment, social environment, and activity pattern of the animals must be considered (*Guide*, pages 37 and 38). The standards of the AWAR (§3.8, 3.81) include specific provisions for the exercise of dogs and environmental enhancement of nonhuman primates.

Observation of animal activity that "is repetitive, is nongoal-oriented, and excludes other behavior" (*Guide*, page 38) can indicate to IACUC inspectors that the psychological well-being of the animals demonstrating the behavior, and perhaps the colony in general, has not been adequately addressed.

Opin. During animal housing facility inspections, the first thing to capture the attention of IACUC inspectors is likely to be the condition of the animals housed. By virtue of training or experience, most IACUC inspectors will readily distinguish normal healthy animals from those exhibiting signs of illness, injury, pain, or distress. Discovery of animals in the latter group should prompt further inquiry into the cause of the condition and how it is being managed. If the condition is known to be experimentally induced, IACUC inspectors may wish to verify that the condition and its management are consistent with information provided in the IACUC-approved protocol. If the condition is not related to the experimental procedure, inquiry into the colony health status and other recent illnesses or injuries may be in order, and management practices questioned if undesirable patterns are identified. In either instance, IACUC inspectors should be assured that all animals are checked daily, that veterinary care is available around the clock if needed, and that the veterinary staff is promptly notified of animal health problems and closely monitors the effectiveness of the implemented treatment and management plan (*Guide*, pages 59 and 60; AWAR §2.33,b).

If the animal housing and care program is adequately providing for the psychological well-being of animals, IACUC inspectors should expect to see primary enclosures with features that accommodate the animals' species-specific postures and motor activities, management practices that cater to species-specific needs for cognitive stimulation and social interaction, and compatible social groupings where appropriate. (See 27:1.)

24:16 How should the IACUC evaluate animal enclosures during inspection?

Reg. Both the *Guide* (pages 23 to 28) and the AWAR (§3.6, §3.28, §3.53, §3.80, §3.101, §3.104, §3.128) provide specific standards for primary enclosures. To evaluate compliance with these standards, IACUC inspectors generally ask themselves the following questions when examining an animal enclosure:

- Is the enclosure of appropriate design for the comfort, security, and observation of the animals housed?
- Is it constructed of durable and sanitizable materials?
- Is space adequate for the size and number of animals contained?
- Is the enclosure clean and in good repair?
- Are food and water readily accessible?
- Does the enclosure in any way jeopardize the well-being of the animals housed therein?

24:17 What environmental parameters should the IACUC evaluate during inspection of an animal facility?

Reg. Specific standards for proper animal housing environmental conditions are provided in the *Guide* (pages 28 to 36) and AWAR (Part 3). The PHS Policy (IV,A,1) uses the *Guide* as a basis for the development and implementation of an animal care and use program.

Opin. IACUC inspectors generally find some environmental parameters, such as extremes of temperature and humidity, inadequate ventilation, excessive odor, and chronically loud noise, to be readily assessable the moment they enter an animal housing room. Other parameters require a little more investigation. For example, to adequately ascertain the lighting cycle and magnitude and frequency of temperature fluctuations, it requires more than momentary sensory perception on the part of the IACUC inspectors. Additionally, depending on the type of primary enclosure used, the environmental conditions experienced by the animals within the enclosure may be very different from those perceived by the IACUC inspectors in the larger room environment. Most IACUC inspectors interview animal facility managers about the monitoring methods used for such situations, and review the monitoring documentation to satisfy themselves that the appropriate environmental conditions are being provided.

24:18 How should the IACUC evaluate the animal facility's physical plant during inspection?

Reg. As indicated in 24:17, IACUC inspectors should consult the *Guide* and AWAR for specific standards regarding the animal housing facility physical plant.

Opin. In addition to ensuring compliance with these standards, IACUC inspectors generally evaluate the facilities in terms of design, construction, building systems, fixtures and equipment, security, maintenance, and housekeeping. To satisfy most IACUC inspectors, the facility physical plant should meet the following criteria:

- The design of the structure should be conducive to achieving the goals of animal well-being, research integrity, and personnel health and safety.
- Construction should be structurally sound, surface finishes should be sanitizable, and physical barriers should be in place to prevent entry of vermin.
- Building systems (electrical, plumbing, heating, ventilation, air conditioning, etc.) should be reliable and sufficient to support facility demands.
- The facility should be outfitted with fixtures and equipment appropriate to the housing, care, and use of the species housed. The surfaces of these items will be sanitizable.
- Equipment requiring periodic service or certification should be properly maintained.
- Appropriate security should include provisions for preventing both escape of animals and entry of unauthorized personnel.
- The facility structure and building systems should be maintained in good repair.

- In general, the facilities should be clean and free of unnecessary clutter.

24:19 How should the IACUC assess the adequacy of housekeeping, disinfection, and sanitation practices during inspection?

Reg. The *Guide* (pages 42 to 44) recommends frequent bedding changes and cleaning and disinfection of primary enclosures and animal housing rooms. The AWAR (§3.11, §3.31, §3.56, §3.84, §3.106, §3.107, §3.131) provide specific minimal requirements for cleaning, sanitization, housekeeping, and pest control in animal housing facilities.

Opin. During inspections, most IACUC members evaluate these issues through direct observation and examination of records. Animals are examined to assure that they are clean and dry. Bedding should not be excessively soiled. Primary enclosures should not exhibit an accumulation of soil. Likewise, animal housing rooms should neither exhibit an accumulation of soil or waste, nor be excessively cluttered (most IACUC inspectors discourage storage of equipment and supplies that are not used in the routine care and use of the animals). All surfaces should be sanitizable and agents used for sanitation and disinfection should be appropriate for the purpose and be used according to the manufacturers' directions. IACUC inspectors should see evidence of an effective vermin control program. Waste material should be removed from the animal rooms in a timely fashion and handled in a manner that minimizes the potential to introduce contaminants or attract vermin to the animal facility. Many IACUC inspectors also evaluate the effectiveness of an animal housing facility's housekeeping, disinfection, and sanitation program in terms of colony health. They inquire about recent incidences and spread of disease, and identify questionable housekeeping, disinfection, or sanitation practices that may contribute to colony health problems.

Inspection of facilities for washing and sanitizing caging and equipment also are included in the IACUC's evaluation of sanitation practices. Most IACUC inspectors want to see that these facilities do not exhibit an accumulation of soil and that caging and equipment are processed in such a way as to minimize the recontamination of clean items. Chemical agents and automatic washing equipment used for sanitation and disinfection of caging and equipment should be appropriate for the purpose used and be used according to manufacturers' directions. Most IACUC inspectors want to verify the effectiveness of the sanitation process. This can be done by inspecting records of validation test results (e.g., cage washer temperature indicator strips, cultures of sanitized surfaces, etc.). If the process calls for sterilization of caging or equipment, the performance of sterilizers used is similarly evaluated by most IACUC inspectors.

24:20 How should the IACUC evaluate the adequacy of the water, food, bedding, medications, and other supplies provided to or used for animals?

Reg. Prior to inspection, IACUC inspectors are advised to consult the *Guide* (pages 38 to 41) and AWAR (Part 3) for specific standards regarding the quality, storage, and provision of water, food, bedding, medications, and other supplies.

Opin. IACUC inspectors should be assured that all supplies that come into contact with animals are appropriate, safe, and effective. Food, bedding, medications, etc., can usually be evaluated by direct observation during inspection. Acceptable animal food must be appropriate for the species, within its recommended shelf life, and stored under conditions that minimize the potential for premature deterioration, spoilage, contamination, or infestation. It should be available to animals *ad libitum* or provided in an appropriate ration unless veterinary or IACUC-approved research protocols dictate otherwise. Bedding should be absorbent, nonnutritive, nontoxic, and stored in a manner that minimizes the potential for spoilage, contamination, or infestation. Food and bedding should be free of foreign objects and normal in appearance, texture, and odor. Medications, treatments, and other veterinary or experimental supplies should be medical grade, within designated expiration dates, and stored according to package directions (IACUC inspectors should be aware that special storage and recordkeeping requirements are imposed by the regulations of the Drug Enforcement Agency for controlled substances).[8] Unless veterinary or IACUC-approved research protocols dictate otherwise, water should be available to animals *ad libitum*, potable, and free of contaminants. To assess the latter, most IACUC inspectors inquire about the source of the water provided (municipal, well, purified, etc.) and the integrity of the system that delivers the water to the animals.

24:21 What aspects of animal euthanasia, carcass storage, and carcass disposal should concern the IACUC during inspection?

Opin. Most IACUC inspectors visit areas for animal euthanasia, carcass storage, and carcass disposal in conjunction with their inspection of animal housing facilities. Euthanasia areas should be clean and the methods employed should be consistent with the recommendations of the Report of the American Veterinary Medical Association (AVMA) Panel on Euthanasia.[9] (See Chapter 17.)

Carcasses are generally stored in dedicated refrigerators or freezers, with special provisions for the labeling and storage of carcasses contaminated with infectious, radioactive, or chemical hazards, including ether. Whether carcass disposal is handled by the institution or contracted to a professional waste management company, IACUC inspectors should be apprised of the disposal methods and be assured that these methods comply with all applicable regulations, including local requirements and restrictions. IACUC inspectors also may wish to review any facility records associated with animal euthanasia and carcass disposal, particularly those required by the AWAR (§2.35,b) for documentation of the final disposition of dogs and cats. Standards on animal euthanasia, carcass storage, and carcass disposal also can be found in the *Guide* (pages 44 and 45, 65, 77).

24:22 What labels, signage, and records should be evaluated by the IACUC during inspection?

Opin. In an animal housing and care facility managed in accordance with the AWAR, PHS Policy, and the recommendations of the *Guide*, most IACUC inspectors can expect to see labels, signage, and records associated with nearly every aspect of the program. Animal enclosures should be labeled with pertinent information (see the *Guide*, page 46) about the animals and their intended use (e.g., PI and IACUC-approved protocol number). Animal food should be labeled with the species formulation and the milling or expiration date. Various cleaning, husbandry, veterinary, and experimental substances found in the animal facility should be labeled with the name and opening or expiration date of each item. Usage records should accompany controlled substances as required by the regulations of the Drug Enforcement Administration.[8] Hazardous substances and contaminated animals or materials should be clearly identified and labeled with precautions. Special instructions should be prominently posted to assist personnel to comply with nonroutine or critical procedures. Animal housing areas should be posted with information on personnel to be contacted in case of emergency. Husbandry activities should be recorded on log sheets. Environmental readings also are logged or will be recorded on automatic monitoring system printouts. Animal monitoring records (postprocedural, food and water consumption, enrichment, exercise, etc.) are usually present at the animal housing location and should be reviewed. Animal health, clinical, veterinary treatment records, and animal receiving and disposition records are to be maintained in a central location and inspected by the IACUC. Records of sanitation and disinfection, including cage washer and sterilizer performance, also should be available for review by IACUC inspectors.

While not all of these labels, signs, and records are specifically required by the AWAR, PHS Policy, and recommendations of the *Guide* — and certainly not every activity in the animal housing facility need be recorded — many of these records and documents will prove to be useful animal facility management and inspection tools. Some IACUCs choose to require that extensive records be kept of animal housing activities and parameters, particularly to document to outside inspectors that there is compliance with standards. Others rely on the state of their facilities and animal care programs to demonstrate compliance, and forego recordkeeping not specifically required by applicable regulations and standards.

24:23 What personnel issues should concern the IACUC during inspection?

Opin. One of the highlights of the IACUC animal housing facility inspection process is the opportunity for dialog between IACUC members and personnel working in the animal facility. Such personnel can include members of the research staff, as well as facility managers and animal care and support personnel. In their conversations, most IACUC inspectors try to evaluate the extent to which personnel are qualified, either through experience or training, to perform their duties, and whether or not experimental procedures and work practices are consistent with IACUC-approved animal use protocols, institutional guidelines, policies, and SOPs.

Some of the most important personnel issues IACUC members consider during their inspections of animal housing facilities concern the occupational health and safety of animal care, use, and support personnel. IACUC inspectors should look for evidence that personnel are informed about and protected from the various health and safety risks associated with exposure to animals and the hazardous experimental agents and procedures used with animals. Examples of such evidence include elements of facility design and management (such as space delineation and air pressure differentials) that separate personnel areas and activities from those of animals, especially where hazardous agents are present. Compliance of personnel with appropriate personal hygiene and protective equipment recommendations also can indicate an effective occupational health and safety program. (See Chapter 20.)

References

1. *Guide for the Care and Use of Agricultural Animals in Agricultural Research and Teaching*, 1st rev. ed., Federation of Animal Science Societies, Savoy, IL, 1999.
2. Committee on Birds, National Research Council, Institute of Laboratory Animal Resources, *Laboratory Animal Management; Wild Birds*, National Academy of Sciences, Washington, D.C., 1997.

3. Committee on Standards; Subcommittee on Avian Standards, *Standards and Guidelines for the Breeding, Care, and Management of Laboratory Animals: Chickens*, National Research Council, Institute of Laboratory Animal Resources, National Academy of Sciences, Washington, D.C., 1966.
4. Committee on Standards; Subcommittee on Fish Standards, *Guidelines for the Breeding, Care, and Management of Laboratory Animals: Fishes*, National Research Council, Institute of Laboratory Animal Resources, National Academy of Sciences, Washington, D.C., 1974.
5. Committee on Standards; Subcommittee on Standards for Large (Domestic) Laboratory Animals, *Guidelines for the Breeding, Care, and Management of Laboratory Animals: Ruminants (Cattle, Sheep, and Goats)*, National Research Council, Institute of Laboratory Animal Resources, National Academy of Sciences, Washington, D.C., 1974.
6. Committee on Standards; Subcommittee on Amphibian Standards, *Guidelines for the Breeding, Care, and Management of Laboratory Animals: Amphibians*, National Research Council, Institute of Laboratory Animal Resources, National Academy of Sciences, Washington, D.C., 1974.
7. Office for Protection from Research Risks, National Institutes of Health, Public Health Service, Sample Semiannual Facility Inspection Checklist Animal Housing and Support Areas. Available on the World Wide Web at: *http://www.nih.gov:80/grants/oprr/sampledoc/chek2a.htm*
8. Office of the Federal Register, Controlled Substances Act, amended May 1, 1987, Title 21, Food and Drugs, Chapter 13, Drug Abuse Prevention and Control, 21 CFR §1301.11–§1308.15, Schedule of Controlled Substances, Washington, D.C., revised April 1, 1988.
9. AVMA Panel on Euthanasia, Report of the AVMA Panel on Euthanasia, *J. Am. Vet. Med. Assoc.*, 202, 229, 1993.

25
Inspection of Individual Laboratories

Neil S. Lipman and Scott E. Perkins

Introduction

One of the IACUC's principal roles is to ensure that animals used in research, testing, and teaching are used humanely by trained staff. Although IACUC review of proposals describing planned animal activity is an important mechanism for meeting this role, observation of animal use provides the most direct assurance that these goals are attained. At some institutions, animals are never removed from the animal facility as all activities are performed within animal holding or procedure rooms. Nevertheless, it is common for animals to be transported from the animal facility to an investigator's laboratory for experimental use. While animal resource personnel can readily observe the appropriateness of activities conducted within the animal facility, this task becomes significantly more difficult, especially at large institutions, when animal research is conducted in an investigator's laboratory.

Although there is no specific regulatory directive requiring IACUCs to inspect investigators' laboratories and observe ongoing activities (unless animals are held in these areas for greater than 12 hours, see 25:1), the IACUC is charged with ensuring that animal use is conducted humanely by appropriately trained personnel and that these activities have been approved, in advance, by the IACUC. It is this chapter authors' opinion that laboratories should be visited by the IACUC periodically in order to meet the spirit of applicable regulations, policies, and the *Guide*.

25:1 Is it necessary to inspect investigators' laboratories where research with animals is conducted?

Reg.   The AWAR (§1.1, Study area; §2.31,c,2) requires inspections if animals are maintained in the laboratory for more than 12 hours,

Opin. whereas the PHS Policy (III,B; IV,B,2) requires inspections if animals are maintained in the laboratory for more than 24 hours. (See 25:2.) Strictly speaking, neither the AWAR nor PHS Policy mandate inspection of investigators' laboratories where research with animals is conducted unless animals are maintained in the laboratory for greater than a specified time period, as noted above. However, many institutional IACUCs assume this responsibility in order to meet their regulatory and institutional charge of ensuring that animals are used humanely by trained personnel. Laboratory visitation frequently provides the added benefit of enhancing dialog between the committee and investigative staff, frequently improving the IACUC's understanding of proposals. (See 23:9, 24:10.)

NIH/OPRR[1] has addressed the IACUC inspections of laboratories where investigators use animals as follows:

> When considering IACUC responsibilities for semiannual review, it is important to keep in mind that the institution, usually acting through the IACUC and/or the facility veterinarian, is responsible for all animal-related activities of the institution, regardless of where animals are maintained *or the duration of their stay* [emphasis added]. The PHS Policy allows institutions some discretion regarding specific methods to assure compliance, but the institution is clearly responsible for what happens to animals in investigators' laboratories. The degree, method and frequency of IACUC oversight depends a great deal on the nature of the activity. For example, satellite holding facilities or areas where surgical manipulation is conducted should always be included in the semiannual review.
>
> Other activities, such as routine dosing, weighing, or immunization of animals in laboratories, may be monitored using other methods, such as random evaluation. Inclusion of these laboratories in the semiannual IACUC review would be another way to satisfy the PHS Policy requirements. In any case, the IACUC must have access to all investigators' laboratories for the purpose of verifying that activities involving animals are conducted in accordance with the proposal approved by that committee.

APHIS/AC supports the position that although laboratories do not have to be inspected if animals are held there less than 12 hours, IACUC inspection is encouraged to the extent that it helps the IACUC and the institution fulfill its responsibilities under the AWA to oversee *all* animal care and use activities at the institution.[2]

Surv. Which laboratories are inspected by your IACUC? More than one response is possible.

- Inspect all laboratories   7/15

- Inspect laboratories if the laboratory is used for
  studies on USDA-covered species   2/15
- Random selection of laboratories   2/15
- Those where animals are used in studies classified
  as APHIS/AC Annual Report category D or E   2/15
- IACUC is unfamiliar with the laboratory where
  research is planned   2/15
- Laboratories where survival surgeries on rodents
  are conducted   3/15
- Laboratories used for postoperative recovery   3/15
- Laboratories conducting studies deemed by the
  IACUC to be of increased concern   3/15

The frequency of inspections of these sites was commonly, but not always, every 6 months.

25:2 Is it necessary to inspect and evaluate *every* laboratory where animals are used at the time of each IACUC inspection?

Reg. AWAR (§1.1; §2.31,c,2) require that study areas (defined as any building, room, area, enclosure, or other containment outside of a core facility or centrally designated or managed area in which animals are housed for more than 12 hours) be inspected at least once every 6 months by a subcommittee of at least two IACUC members.

The PHS Policy (III,B) defines an animal facility as "any or all buildings, rooms, areas, enclosures, or vehicles, including satellite facilities, used for animal confinement, transport, maintenance, breeding, or experiments inclusive of surgical manipulation. A satellite facility is any containment outside of a core facility or centrally designated or managed area in which animals are housed for more than 24 hours." All facilities, including satellite facilities, must be inspected at least once every 6 months by the IACUC (PHS Policy IV,B,2). (See 25:1.)

Opin. As addressed in 25:1, many IACUCs conduct routine laboratory inspections in order to meet their oversight responsibility of the institution's animal care and use program. It should be noted that the AWAR (§2.31,c,2) require that the IACUC inspect *all* animal facilities where animals are kept for more than 12 hours, including animal study areas, at least once every 6 months. Other animal facilities include surgical rooms, food storage, cage wash, etc.[2] Some of these areas may not house animals for more than 12 hours, or may not house animals at all. Nevertheless, as we noted in the introduction to this chapter and in 25:1, it is our opinion that such inspections are appropriate.

25:3 How frequently should each laboratory where animals are used be inspected?

Reg. Research laboratories in which APHIS/AC-covered species are maintained for more than 12 hours must be inspected by a subcommittee of at least two IACUC members every 6 months (AWAR §1.1, Animal; §1.1, Study area; §2.31,c,2). IACUCs serving institutions required to comply with the PHS Policy (III,B; IV,B,2) must inspect laboratories in which animals are maintained for more than 24 hours at least once every 6 months.

Opin. As there are no other regulatory requirements mandating laboratory inspection, it is at the individual IACUC's discretion whether or not they should inspect animal research laboratories at their institution that do not meet these time requirements. It is the authors' opinion that at least select animal research laboratory inspections should be conducted by the IACUC in order to ensure compliance with both the content and spirit of applicable regulations and guidelines. (See 25:1.)

Surv. How often do you inspect animal research laboratories?

- Every 6 months   12/15
- Periodically, no defined frequency   2/15
- Do not inspect laboratories   1/15

25:4 Should a laboratory be approved by the IACUC before permitting an investigator to use the laboratory to conduct procedures on animals?

Reg. The AWAR (§2.31,d,1) requires the IACUC, as part of the protocol review process, to ensure that all components of animal care and use are in compliance with the AWAR prior to approval of the protocol. PHS Policy (IV,C,1) requires that the project will be conducted in accordance with the AWA and the *Guide.* (See 25:1.)

Opin. Other than as noted above for the AWAR and the PHS Policy, there are no regulatory requirements mandating the approval of laboratories by the IACUC. Nevertheless, it is the IACUC's responsibility to ensure that the proposed activity is conducted at a suitable location. It is the authors' opinion that there are significant advantages of having the IACUC inspect laboratories before providing approval for work with animals. Inspection of the laboratory provides the IACUC the opportunity to determine if the facilities and equipment are suitable for the proposed activity. In addition, it provides an opportunity for the investigative staff to gain a better understanding of IACUC functions, and provides IACUC members a more thorough understanding of the proposed project. At institutions where IACUC approval of sites is not feasible because of time constraints, the IACUC can delegate this responsibility to a member of the institution's animal resources staff.

Surv. Does your institution require IACUC approval of investigators laboratories before conducting procedures on animals in them?

- Approval of laboratory required but not inspected prior to granting approval   7/15
- Must inspect before approving the laboratory   2/15
- Approve only if specific procedures are done (e.g., survival surgical procedures for nonrodent mammals or areas meeting the AWAR or PHS definition of a study area or satellite facility)   3/15
- Do not approve laboratories   3/15

25:5 How does the IACUC determine the location of laboratories used for research with animals?

Opin. The most direct method for determining which laboratories are used to conduct research on animals is to request that information in the animal use proposal reviewed by the IACUC. This information is frequently used to generate a database. Additionally, laboratories may be identified based on information provided by the animal resources staff or may be available from the institution's administration.

Surv. How does your IACUC determine the location of laboratories used for research with animals?

- From the IACUC application form   15/15

Most respondents indicated that they used that information to develop a database of laboratories to visit at inspections.

25:6 Should the IACUC inspection be announced or unannounced?

Opin. There are advantages and disadvantages of both methods. An announced site visit generally ensures that a knowledgeable staff member is available to provide laboratory access, discuss animal research activities, and answer questions. Staff availability permits dialog, advantageous to both the IACUC members and research staff who frequently gain a better understanding of the IACUC's functions and activities. However, when conducting announced site visits, IACUC members may be exposed to an artificial environment not reflective of the normal state of the laboratory, personnel conduct, or animal use. Further, it is not uncommon for laboratories to cancel animal research activities when expecting a site visit.

Unannounced inspections permit the IACUC members to view the laboratory and its activities under normal operating conditions.

However, research staff may not be present, potentially limiting laboratory access, or if they are present, ongoing research responsibilities may limit them from interfacing with IACUC members.

Surv. A Are your IACUC inspections of laboratories announced or unannounced?

- Announced 7/15
- Unannounced 8/15

Institutions conducting announced inspections also will conduct unannounced inspections if the IACUC has specific concerns that may not be uncovered if inspections were announced.

Surv. B If unannounced, and there is nobody present, do you schedule a return site visit in the immediate time frame or do you wait until the next inspection cycle?

- Return for site visit in immediate time frame 5/8
- Wait for next inspection cycle 3/8

25:7 Is it necessary for a PI or his designee to be present during the inspection?

Opin. Although not mandated by the AWAR or PHS Policy, there are clear advantages in having a knowledgeable research staff member present during the inspection (see 25:6). If the inspections are announced, the IACUC may stipulate that the PI or his designee be present during the inspection. With unannounced visits, the IACUC cannot ensure that access can be gained to the area.

25:8 Prior to the inspection, should the IACUC members who are visiting the laboratory be briefed on procedures conducted in the laboratory?

Opin. It is advantageous for IACUC members to be briefed on procedures occurring in the laboratory prior to the inspection, either by the AV, a knowledgeable IACUC member, or by consulting the protocol. Alternatively, the PI or his designee can provide this information at the time of the inspection.

Surv. Prior to an inspection, is your IACUC inspection team briefed on procedures conducted in a laboratory?

- Yes, and protocols are provided prior to inspection 1/15
- Yes, but protocols not provided prior to inspection 2/15
- No briefing 12/15

25:9 What areas and issues should the IACUC focus on during inspection of the laboratory?

Opin. IACUC members should:

- Ensure appropriate facilities, personnel, and equipment are present to provide adequate veterinary care.
- Observe the laboratory and equipment used for cleanliness and adequate sanitation
- Assess safety issues such as:
  - Food or drink consumption within the laboratory
  - Anesthetic gas scavenging
  - Appropriate use of hazardous substances
  - Use of appropriate animal handling procedures
- Observe the condition of animals, if they are present.
- Confirm compliance with approved IACUC protocols including ascertaining that procedures and staff have been described and included in the approved protocol.
- Evaluate the adequacy of training.
- If performing surgery, determine adherence to aseptic technique, suitability and understanding of anesthesia monitoring, and postoperative care including pain relief.
- Review methods of euthanasia.
- Verify that all pharmaceuticals are handled and stored appropriately and are not expired.
- Review records, especially when conducting survival surgical procedures. (See 25:10.)
- If the laboratory is approved for animal housing, determine that the animals' husbandry needs are met and appropriate records are maintained.

25:10 What records should the IACUC inspection team review when visiting the laboratory?

Opin. Record evaluation is influenced by the activities that are conducted in the laboratory. If the laboratory is approved for housing and caring of animals, the PI's staff should maintain and the IACUC members should periodically examine those records indicating the frequency of cage changing, environmental monitoring (e.g., temperature and humidity), laboratory sanitation, and daily animal observations (including weekends). Records should be similar to those maintained by animal resource personnel in the animal facility. The following records, if applicable to the use of the laboratory, should be maintained and made available to the IACUC members during an inspection:

- Numbers of animals used per unit time (unit time should be a period in which the IACUC can compare actual animal usage to the number of animals requested in the protocol).
- Procedures conducted.
- Treatment or drug administration.
- Anesthetic, surgical, and postoperative care.
- Controlled substances log.
- Dosimetry for radiation exposure if using radionuclides or exposed to ionizing radiation.
- Any institutionally required documentation when using hazardous substances.
- Copies of institutionally required policies, handbooks, or guidelines.

Surv. Which records does your IACUC review during a laboratory inspection?

- Do not routinely review records during an inspection unless observations indicated the need to do so   10/15
- Surgical and postoperative care records   5/15

25:11 What criteria can the inspection team use to make sure that acceptable standards are being met in the laboratory?

Opin. There are many criteria that can and should be utilized to determine that acceptable standards are used in the research laboratory. IACUC members should utilize the *Guide* and the AWAR, as well as IACUC and institutional policies, as the basis for determining the suitability of the laboratory for conducting research with animals. Depending on the nature of the research conducted and the institution, the Good Laboratory Practices Act,[3] the Occupational Safety and Health Act,[4] the Bloodborne Pathogen Standard,[5] Standards for Protection Against Radiation,[6] Lab Safety Standard,[7] Controlled Substances Act,[8] the NIH Recombinant DNA Guidelines,[9] and the Centers for Disease Control and Prevention's *Biosafety in Microbiologic and Biomedical Laboratories*[10] also may be used as criteria.

25:12 If an IACUC inspection team misses a laboratory during an inspection or a member of the investigative staff is not present in the laboratory at the time of the inspection, should they go back at a alternative time to complete the inspection?

Opin. This is dependent on federal and state regulatory requirements and the regulatory and institutional charge of the IACUC. The AWAR (§2.31,c,2) and the PHS Policy (III,B; IV,B,2) require that study areas be inspected at least once every 6 months by the IACUC to maintain compliance. (See 25:1, 25:2.)

Surv. (See 25:6.)

25:13 How should the IACUC proceed after identifying an item of noncompliance in a laboratory?

Reg. The AWAR (§2.31,c,3) require the IACUC to identify deficiency items as significant or minor and develop a timetable for correction. The AWAR (§2.31,c,3) define a significant deficiency as "one which, with reference to Subchapter A, and, in the judgment of the IACUC and the Institutional Official, is or may be a threat to the health and safety of the animals." If the established timetable is not adhered to resulting in the significant deficiency remaining uncorrected, the deficiency must be reported to APHIS/AC and any Federal agency funding the activity within 15 days (AWAR §2.31,c,3). Similarly, the PHS Policy (IV,B,3) requires the IACUC to draft reports, which are submitted to the IO, of its inspection findings. As with the AWAR, identified deficiencies must be distinguished as significant or minor based on whether the deficiency is or may be a threat to the health and safety of the animals. If program or facility deficiencies are noted, the report to the IO must contain a reasonable and specific plan and schedule for correction of each deficiency. (See 26:16.)

Opin. The authors find it most effective if deficiencies are discussed with the responsible individuals or their designee, if available, at the time of the site visit. The ensuing dialog may clarify misinterpretations. Deficiencies should be reviewed by the IACUC and categorized as described above. A timetable for correction should be established. A written notice should be sent from the IACUC to the responsible individual, describing the stated deficiency and timetable provided for correction. The authors advise that the IACUC develop a mechanism by which the responsible individual can contest the deficiency if there is disagreement. Further, the responsible individual should be requested to provide written documentation that the deficiency is corrected.

Alternatively, the authors are aware of institutions which effectively use a multicopy form which is completed by the site visitors at the time of the site visit. The form contains a check list for the inspectors to use to identify compliance, or lack thereof. If deficiencies are noted, a timetable for correction is provided at the time of the site visit, based on preapproved criteria established by the IACUC. The form is signed by the responsible individual or their designee at the time of the site visit, and a copy is retained by the laboratory. The form stipulates that a formal response is required from the responsible individual for all noncompliant items.

Surv. How does your IACUC proceed after identifying an item of noncompliance during a laboratory inspection?

- Contact the responsible individual in writing, providing them a description of noncompliant items and a date for correction   6/15
- Response varies dependent upon degree and significance of noncompliant item; actions taken include verbal discussion, written warning, up to and including imposition of sanctions   9/15

All communications should convey the date of expected resolution. All of the IACUCs surveyed reinspected the area to determine if the item was corrected by the specified date or during the next scheduled inspection. Depending on the severity and repetition of the item of noncompliance, the IACUC may impose sanctions on the investigator (see Chapter 29).

References

1. Division of Animal Welfare, Office for Protection from Research Risks, National Institutes of Health, The Public Health Service responds to commonly asked questions, *ILAR News*, 33(4), 68, 1991.
2. DeHaven, W.R., personal communication, 1998.
3. Office of the Federal Register, Code of Federal Regulations, Title 43, Good Laboratory Practice Regulations, Washington, D.C., 1978.
4. Office of the Federal Register, Code of Federal Regulations, Title 29, Part 1910, Occupational Safety and Health Standards, Subpart G, Occupational Health and Environmental Control, Washington, D.C., 1984.
5. Office of the Federal Register, Code of Federal Regulations, Title 29, Part 1910.1030, Bloodborne Pathogens, Occupational Health and Safety Act, Washington, D.C., 1991.
6. Office of the Federal Register, Code of Federal Regulations, Title 10, Chapter 1, Nuclear Regulatory Commission Regulations, Washington, D.C., 1992.

7. Office of the Federal Register, Code of Federal Regulations, Title 29, Part 1910, Department of Labor, Occupational Safety and Health Administration, Occupational exposures to hazardous chemicals in laboratories; final rule, *Fed. Reg.,* January 31, Part 11, Washington, D.C., 1990.
8. U.S. Department of Justice, Drug Enforcement Administration, Pharmacist's Manual: An Informational Outline of the Controlled Substances Act of 1970, Washington, D.C., 1986.
9. U.S. Department of Health and Human Services, National Institutes of Health, Guidelines for Research Involving Recombinant DNA Molecules, 1997. Available on the World Wide Web at: *http://www.nih.gov/od/orda/toc/htm*
10. U.S. Department of Health and Human Services, Public Health Service, Biosafety in Microbiological and Biomedical Laboratories, 4th ed., U.S. Government Printing Office, Washington, D.C., 1999.

26
Inspection of Surgery Areas

Scott E. Perkins and Neil S. Lipman

Introduction

The AWAR (§2.31,c,2) require IACUCs to inspect "all of the research facility's animal facilities, including animal study areas" for species covered by the AWAR. PHS Policy (IV,B,2) requires any and all animal facilities and areas to be inspected at least every 6 months including satellite facilities and surgical areas. The PHS Policy (IV,B,1) also requires adherence to the *Guide* (page 9), which recommends inspection of the animal facilities and activity areas at least once every 6 months. The *Guide* and AWAR provide the reader with requirements for surgical facilities and procedures for conducting survival surgery on rodents and USDA-covered species. The IACUC must approve all surgical procedures involving animals to ensure that the research is conducted in a humane manner by appropriately trained individuals. This chapter is intended to help the reader to comply with these regulations and recommendations.

The survey information presented emanates from a mailing to 25 institutions, primarily academic. Not all institutions responded to all questions.

26:1 Should all surgical sites be visited and evaluated at the time of each IACUC inspection?

Reg. There are federal requirements for IACUCs to inspect areas used to conduct surgical procedures on APHIS/AC covered species.[1] The AWAR (§2.31,c,2) require all animal facilities, including study areas, to be inspected at least once every 6 months.

PHS Policy (IV,B,2) also requires any and all animal facilities to be inspected at least every 6 months. Specifically, PHS Policy (III,B) includes "any and all buildings, rooms, areas, enclosures, or vehicles, including satellite facilities, used for animal containment, transport, maintenance, breeding, or experiments inclusive of surgical manipulation." PHS Policy (IV,B,1) also requires adherence to the provisions of the *Guide* (page 9), which states, "The committee should review the animal-care program and inspect the animal facilities and activity areas at least once every 6 months."

Opin. The AWAR (§2.31,c,2) seem to imply that if the surgical facility is contained within the animal facility, it is required to be inspected every 6 months. Surgical facilities outside of a core animal facility, or centrally designated or centrally managed area, are considered Study Areas and, as such, are only required to be inspected if animals are maintained at these sites for more than 12 hours (AWAR §1.1 Study Area; §2.31,c,2). Nevertheless, *the reader is cautioned that APHIS/AC disagrees with the authors' interpretations of the regulations.*[2] *The agency considers all surgical facilities as part of the research facility's animal facilities and, thus, must be inspected by the IACUC (AWAR §2.31,c,2).* The PHS Policy (IV,B,2) requires semiannual inspection of surgical areas.[1]

Although our literal interpretation of the AWAR is as written in the previous paragraph, it is the opinion of this chapter's authors that all surgical areas should be inspected semiannually by the IACUC to maintain compliance with and meet the spirit of the AWAR, PHS Policy, and *Guide*. Both the AWAR (§2.31d,1,ix; §2.31,d,1,x) and the *Guide* (pages 60 to 64, 78 and 79) detail facility and procedural requirements for conducting survival surgical procedures on rodents or APHIS/AC covered species.

Surv. Does your IACUC visit and evaluate all surgical sites at each inspection?

- Inspect all surgical sites used for survival surgery on APHIS/AC-covered species   15/15
- Inspect all rodent or nonsurvival surgery sites   3/15
- Do not inspect rodent or nonsurvival surgery sites   9/15

26:2 What is the minimal frequency for evaluating surgical sites?

Reg. (See 26:1.)

Opin. The chapter authors recommend all survival surgical areas be inspected at least semiannually. As in 26:1 (and see APHIS/AC comment therein), the chapter authors interpret the AWAR to indicate that animal surgical sites contained within the core animal facility, or centrally designated or centrally managed area, be inspected every 6 months as specified in these regulations (AWAR §2.31,c,2). PHS Policy (IV,B,2) requires any and all animal facilities, including satellite and surgical facilities, to be inspected at least every 6 months. This includes sites used for survival and nonsurvival rodent surgery. Therefore, the authors recommend that IACUCs establish the policy of inspecting at least once every 6 months all sites in which deletion surgery is conducted.

26:3 What facilities are required for conducting survival surgery on nonrodent and rodent species?

Reg. The IACUC must differentiate between major and nonmajor surgical procedures when evaluating facilities for performing survival surgical procedures, as facility requirements depend on the procedure conducted (see 26:5 for definitions).

Both the AWAR (§2.31,d,1,ix; §2.31,d,1,x) and the *Guide* (pages 78 and 79) specify facility requirements for conducting major survival surgery on nonrodent species. The AWAR (§2.31,d,1,ix) require that "major operative procedures on nonrodents will be conducted only in facilities intended for that purpose which shall be operated and maintained under aseptic conditions." The AWAR do not require dedicated facilities when conducting major operative procedures at field sites; however, aseptic technique must be followed (§2.31,d,1,ix).

The *Guide* (pages 78 and 79) provides significantly more detail pertaining to facility requirements, recommending that nonrodent surgical facilities include a variety of functional components, usually but not always separated by physical barriers. These include areas for surgical support, animal preparation, surgeon's scrub, operating room, and postoperative recovery. The need to provide all of these areas depends on the size and scope of the surgical program. With respect to ventilation, the *Guide* (page 76) states "areas of surgery should be kept under relative positive pressure with clean air."

Opin. It is the chapter authors' opinion that the minimum requirement for most operative suites are

- Area for animal preparation.
- Area for surgical scrub.
- Operating room.

The surgical suite should be designed and operated in a manner to control microbial contamination. This can be accomplished by:

- Minimizing traffic in the area.
- Adhering to a rigorous sanitation and disinfection program.
- Limiting equipment present within the operating room to only essential movable equipment (such as surgical and support tables, and anesthesia and monitoring equipment).

- Outfitting the surgical suite with a ventilation system providing air flows which maintain the operating room at positive pressure with respect to the animal preparation and surrounding areas.

Facility requirements for conducting nonmajor surgery on nonrodents or major survival surgery on rodents are less stringent. The AWAR (§2.31,d,1,ix) do not require dedicated facilities for conducting these procedures, but do specify the use of aseptic technique. Areas used for such procedures can be a room or portion of a room that is easily sanitized and is dedicated for that purpose at the time or surgery. The laboratory bench or surgical table should be disinfected prior to and after use, be free of ancillary equipment, be located in an area free of drafts, and have a suitable temperature.

26:4 What facilities are required for conducting nonsurvival surgery on animals?

Opin. There are no facility specific requirements when conducting nonsurvival surgical procedures. The area used for nonsurvival surgery should be a room or portion of a room which is easily sanitized and, at the time of surgery, dedicated for that purpose. The laboratory bench or surgical table should be disinfected prior to and after use, be free of ancillary equipment, and be located in an area free of drafts with a suitable temperature. In general, adherence to aseptic technique is unnecessary. The surgical site should be clipped, the instruments clean, and the surgeon should wear gloves. The IACUC should evaluate the duration of nonsurvival surgery as procedures can occasionally last for sufficient periods of time for sepsis to develop. In these cases, adherence to aseptic technique may be necessary. Additionally, the IACUC should evaluate the need for aseptic technique or "aseptic" facilities, if certain nonsurvival procedures are being performed (e.g., harvesting an organ for transplantation).

26:5 How does one differentiate a major from a minor surgical procedure?

Reg. The AWAR (§1.1) define a major operative procedure as "any surgical intervention that penetrates and exposes a body cavity or any procedure which produces permanent impairment of physical or physiological functions."

The PHS Policy itself provides no definition of major or minor surgical procedures, but it does require adherence to the provisions of the *Guide* (PHS Policy IV,B,1; *Guide,* pages 61 and 62), which defines major survival surgery as one which "penetrates and exposes a body cavity and produces substantial impairment of physical or physiological functions." It defines minor survival surgery as one in which the procedure "does not expose a body cavity and causes little or no physical impairment."

Opin. Although both the AWAR and the *Guide* provide a definition of a major operative procedure, the classification of some surgical procedures requires IACUC interpretation, based on the procedure's extent, effects, and the potential for complications to develop. Major operative procedures include laparotomy, thoracotomy, craniotomy, and limb amputation. Minor operative procedures include wound suturing, and peripheral vessel cannulations, and certain routine farm animal procedures.

26:6 Assume an animal is covered by the AWAR and the PHS Policy. Under what circumstances can a major survival surgical procedure be performed in an area other than a dedicated surgical suite approved by the IACUC?

Reg. The AWAR (§2.31,d,1,ix) require that "major operative procedures on nonrodents will be conducted only in facilities intended for that purpose which shall be operated and maintained under aseptic conditions." The AWAR (§2.31,d,1,ix) do not require dedicated facilities when conducting major operative procedures in field studies, although aseptic procedures are required.

PHS Policy (IV,B,1) requires adherence to the *Guide*. The *Guide* (page 62) states, "In general, unless an exception is specifically justified as an essential component of the research protocol and approved by the IACUC, nonrodent aseptic surgery should be conducted only in facilities intended for that purpose." Nevertheless, the *Guide* (pages 62 and 63) does provide exceptions for emergencies and field studies, stating, "Emergency situations sometimes require immediate surgical correction under less than ideal conditions" and "generally, farm animals maintained for biomedical research should undergo surgery with procedures and in facilities compatible with the guidelines set forth in this section. However, some minor and emergency procedures that are commonly performed in clinical veterinary practice and in commercial agricultural settings may be conducted under less-stringent conditions than experimental surgical procedures in a biomedical research setting."

Opin. It is the authors' opinion, in accordance with the *Guide*, that major operative procedures can be performed in a nondedicated surgical suite when scientifically justified (i.e., unique instrumentation in the laboratory) and approved by the IACUC. In the event of an emergency where the animal cannot be transported to a dedicated operating room, this should be considered a clinical medical requirement, determined by veterinary medical judgment, and does not require IACUC approval.

Surv. Under what circumstances can a major survival surgical procedure be performed in an area other than a dedicated, IACUC-approved surgical suite?

• Not addressed by the IACUC	5/12
• Not permitted for nonrodent species	3/12
• Allowable if provided with strong justification, e.g., the need for critical equipment that could not be brought to the surgical suite	3/12
• Field studies (only institution conducting field studies)	1/12

26:7 Can nonsurvival surgery on nonrodent species or survival or nonsurvival surgeries on rodents be performed in survival surgical facilities used for nonrodents? If yes, are there any special procedures that should be followed?

Opin. IACUCs may permit the conduct of survival rodent surgery or nonsurvival surgical procedures in survival surgical facilities if either the following conditions are met:

- Nonsurvival procedures are conducted utilizing the same standards that apply to survival procedures, or
- The operating room is thoroughly sanitized and disinfected prior to conduct of a survival surgical procedure in nonrodents

26:8 What areas and issues should the IACUC focus on during inspection and evaluation of surgical areas?

Reg. PHS Policy (IV,B,2) requires any and all animal facilities (as defined in PHS Policy III,B) to be inspected at least every 6 months including "any and all buildings, rooms, areas, enclosures, or vehicles, including satellite facilities, used for animal containment, transport, maintenance, breeding, or experiments inclusive of surgical manipulation." PHS Policy (IV,B,1) requires compliance with the *Guide* (pages 61 to 66) which includes IACUC evaluation of aseptic procedures, postoperative evaluation and care, anesthetic monitoring, etc. Other sections of the *Guide* provide further information on occupational health and safety (*Guide,* pages 14 to 18), design of facilities for aseptic surgery (*Guide,* pages 78 and 79), etc. The AWAR (§2.32), PHS Policy (IV,C,1,f), U.S. Government Principle VIII, and

Opin. the *Guide* (page 61) all address the need for proper qualifications and experience of personnel working with animals. (See 26:1.)

Although the AWAR (§2.31,d,1,ix; §2.31,d,x) and the *Guide* (pages 60 to 64, 78 and 79) describe facility and procedural requirements for conducting survival surgical procedures in nonrodents, neither specifically mandate inspection of *all* surgical areas. *See explanatory and opposing comments in 26:1.*

In the authors' opinion, minimal evaluation of standards for surgical areas should include:

- Assessment of personnel training.
- Adherence to aseptic technique.
- Scavenging of waste anesthetic gases.
- The correct use of unexpired pharmaceuticals.
- The proper use of equipment and procedures for perioperative monitoring and care.
- Maintenance of perioperative records.
- The availability of appropriate equipment and pharmaceuticals for emergencies.
- Proper disposal of sharps.

Surv. What areas and issues does your IACUC focus on during inspection and evaluation of surgical areas?

- Inspect every 6 months surgical areas in which major survival surgery on APHIS/AC-covered species are performed   15/15

Although there was not a consensus among the respondents, the IACUCs surveyed indicated that they focus on some or all of the following issues during an inspection:

- Physical plant   7/15
- Facilities for postoperative recovery   3/15
- Personnel training and surgeons' competence   5/15
- Scavenging of anesthetic waste gases   3/15
- Traffic patterns within the surgical suite   3/15
- Availability of appropriate equipment and pharmaceuticals for emergencies   4/15
- Expired pharmaceuticals   2/15
- Drug storage and inventory control   4/15

- Cleanliness and ease of surgical room disinfection   9/15
- Methods and records of disinfection   1/15
- Procedures for sterilizing instruments and surgical supplies   5/15
- Noting dates when instruments were sterilized   1/15
- Adherence to aseptic technique including use of appropriate surgical garb   8/15
- Posting of emergency phone numbers   1/15
- Equipment and procedures for perioperative care   6/15
- Perioperative monitoring and records   6/15

26:9 What reference sources are available to help evaluate the issues in Question 26:8?

Opin. The IACUC should utilize the AWAR, the *Guide*, applicable state regulations, and institutional policies to develop acceptable standards for surgical facilities and their operation. The IACUC also can consult references, as well as expert consultants, to assist in the development of standards for conducting surgery. Additional references include: veterinary and human surgery and anesthesia texts, texts and references on experimental surgery, the Good Laboratory Practices Act,[3] the Occupational Safety and Health Act,[4] the Bloodborne Pathogen Standard,[5] Standards for Protection Against Radiation,[6] Laboratory Safety Standard,[7] Controlled Substances Act,[8] Recombinant DNA Guidelines,[9] *Biosafety in Microbiologic and Biomedical Laboratories*,[10] and references on disinfection and sanitation control.

26:10 What records should the inspection team review when visiting nonrodent and rodent surgical areas?

Opin. The IACUC should confirm that procedures and staff training, experience, and qualifications have been accurately described and included in the approved animal care and use proposal. The IACUC should review pertinent animal health and experimental records including:

- Perioperative records describing the surgical procedure performed.
- Experimental agents administered.
- Dose and route of anesthetics and analgesics administered.

- Frequency and type of intraoperative monitoring procedures performed.
- Postoperative monitoring and observations.
- Quality assurance records for sterilization of materials also may be periodically reviewed.

All records should be made available for IACUC review whether they are maintained with the animal, in the surgical facility, or by the investigator.

Surv. Does your IACUC review records when inspecting surgical areas?

- Records routinely reviewed during inspections   8/15
- Records not routinely reviewed during inspections   7/15

Records for nonrodent, APHIS/AC-covered species, and for currently active cases were reviewed most frequently.

26:11 How detailed should surgical and postoperative records be? What are the essential elements of an adequate record?

Reg. The *Guide* (page 46) states, "Clinical records for individual animals also can be valuable, especially for dogs, cats, nonhuman primates, and farm animals. They should include pertinent clinical and diagnostic information, history of surgical procedures and postoperative care, and information on experimental use." The AWAR (§2.33,b,5) require "adequate preprocedural and postprocedural care in accordance with current established veterinary medical and nursing procedures." The HREA (495,a,2,B) requires "appropriate presurgical and postsurgical veterinary medical and nursing care for animals" used in biomedical and behavioral research.

Opin. It is the opinion of the authors that the AWAR and HREA regulations could be interpreted as requiring surgical records for documenting adequate veterinary care. Records should be detailed and contain all relevant information including, but not limited to:

- Animal identification.
- Project identification.
- Detailed description of the surgical procedure.
- Experimental agents administered.
- Dose and route of anesthetics and analgesics administered.
- Frequency (e.g., time interval) and type (e.g., blood pressure, heart rate) of intraoperative monitoring procedures performed.

- Postoperative monitoring and observations recorded.
- Postoperative treatments administered.

Surv. How detailed does your IACUC require surgical and postoperative records to be?

- Require detailed records   11/15
- Does not require detailed records   4/15

Three of the 11 IACUCs requiring detailed records utilized specific forms for documenting the necessary information.

26:12 Should surgical records be kept for all species or only certain ones?

Opin. The *Guide* (page 46) states that records are especially valuable for dogs, cats, nonhuman primates, and farm animals. Although there are no specific recordkeeping guidelines in the AWAR or PHS Policy, APHIS/AC inspectors expect records, in accordance with established veterinary medical practice and as interpreted from the AWAR (§2.33,b,5), to be maintained. PHS Policy also requires adherence to the requirements of the AWAR (PHS Policy II) and the *Guide* (PHS Policy IV,B,1).

It is the opinion of the authors' that surgical records should be maintained for all species to provide documentation that the research was conducted in a humane manner by appropriately trained individuals.

Surv. For which species does your institution require surgical records?

- For APHIS/AC-covered species and nonrodents   10/12
- All species, including rodents   2/12

26:13 Should surgical records be kept for each individual animal or can they be maintained by groups?

Opin. There are no regulatory requirements for having individual surgical records for each individual animal. The *Guide* (pages 46 and 47) notes that clinical records for individual animals can be useful and states that they should include history of surgical procedures and postoperative care information. It is the opinion of the authors that records can be maintained by groups for rodents, but nonrodents should have surgery records maintained for the individual animals. (See 26:10, 27:10.)

26:14 Do records for survival surgery differ from those maintained for animals undergoing nonsurvival surgical procedures?

Opin. Neither the AWAR nor the *Guide* stipulate that recordkeeping should be less stringent or different for animals undergoing nonsurvival procedures. Records should be maintained in accordance with established veterinary medical standards as noted in 26:10 and 26:12.

Surv. Are record requirements different for survival vs. nonsurvival surgery?

• No policy	2/15
• No records kept for nonsurvival survery	8/15
• Keep records of anesthetics, dosages, and monitoring criteria for survival surgery	13/15

26:15 What are the recordkeeping requirements when performing surgery on rodents?

Reg. The *Guide* (page 46) states that clinical records for animals are valuable and describes the kind of information to be maintained. Rodents are not excluded from these statements.

Opin. At the authors' institutions, records are required for rodents undergoing surgical procedures, although the frequency of recording monitoring procedures and details for intraoperative monitoring procedures are not required. Additionally, the records may be maintained for groups. Rodent records should include:

- A description of the surgical procedure conducted.
- Experimental agents administered.
- Dose and route of anesthetics and analgesics administered.
- Postoperative monitoring and observations.
- Treatments administered during the perioperative period.

26:16 How should the inspection team proceed after identifying an item of noncompliance in a surgical area?

Reg. The IACUC should identify the item as significant or minor and develop a timetable for correction as specified in the AWAR (§2.31,c,3) and the PHS Policy (IV,B,3). (See 25:13.)

Opin. The policies for contacting the appropriate individual varies dependent upon the magnitude of the item. IACUC options for notification include:

- Briefing the responsible individual from the laboratory by an IACUC member at the time of inspection, or later by telephone.
- Sending a copy of the inspection report to the responsible individual.
- Sending a formal letter documenting the noncompliant item.

All communications should convey the date of expected resolution. Depending on the severity and repetition of the item of noncompliance, the IACUC may impose sanctions on the investigator such as a letter of reprimand from the IO, cessation of research for a specified time or permanently, mandatory training, or increased inspection frequencies. The IACUC may revisit the area prior to the next site visit to confirm correction. (See 25:13, Chapter 29.)

Surv. How does your inspection team proceed after identifying an item of noncompliance in a surgical area?

- Contact the PI, AV, or facility manager and give them a description of the problem and correction timetable — 6/15
- Response varies dependent upon degree and significance of noncompliant item; actions taken include verbal discussion, written warning, up to and including imposition of sanctions — 9/15
- Reinspect the area to determine if the item has been corrected by the specified date, or sometimes during the next scheduled inspection — 15/15

26:17 Can unrelated major survival surgeries be conducted simultaneously in a single operating room on multiple nonrodent animals of the same or different species? Does the response differ for rodent species?

Opin. Neither the AWAR nor the *Guide* have restrictions on the performance of more than one surgery, on the same or different species, in the same operating room. There are unique situations (e.g., inter- and intraspecies transplantation) when conducting surgery on more than one animal in the same operating room is justifiable.

Because of the potential for infectious agent transmission and behavioral incompatibility, it is not advisable to conduct survival surgical procedures on more than one species simultaneously unless the species are of similar microbiological status, e.g., specific pathogen-free rodents. Conduct of simultaneous survival surgeries on the same species is acceptable as long as the nature of each surgical procedure conducted does not place the other animals undergoing surgery at increased risk from situations such as different microbiological status of animals or an infectious disease study.

26:18 Can a major survival and a nonsurvival operative procedure be performed simultaneously on nonrodent species in the same operating room?

Opin. When justifiable, as described above in 26:17, survival and nonsurvival surgical procedures can be conducted in the same operating room. However, both surgeries should be conducted using aseptic technique as if they were survival procedures because strict aseptic technique should be followed to provide the highest level of sanitation for the animal undergoing the survival procedure.

26:19 What level of training should the IACUC require for individuals performing survival surgical procedures? Should the IACUC verify if individuals have adequate training and skills to perform major operative procedures?

Reg. The IACUC has the responsibility for ensuring that individuals conducting surgery, irrespective of species used, have the necessary skills to successfully execute the surgical procedure. Both the AWAR (§2.32) and PHS Policy (IV,c,1,f) require that all personnel involved in animal care, treatment, and use are qualified and trained to perform their duties. The U.S. Government Principle VIII stipulates that "investigators and other personnel shall be appropriately qualified and experienced for conducting procedures on living animals." In addition to the conduct of the surgery, individuals associated with the activity must be skilled at administering and monitoring anesthesia and have the capability of providing suitable postoperative care.

Opin. Although the IACUC's responsibility is clear, neither the PHS Policy nor the AWAR describe implementation methods. There are a variety of mechanisms which the committee can use to meet this responsibility. IACUCs frequently garner information pertaining to an investigators training and skills via required descriptions of the same in the initial proposal submitted to the IACUC. Although not all individuals on a specific project may have suitable skills to carry out the project, the IACUC must determine if there are sufficient skilled staff available to meet the project's goals. Clearly, the variety and scope of surgical procedures conducted in a research or training setting varies considerably, requiring the IACUC to make this determination on a case-by-case basis. The IACUC may request additional information or invite the investigative staff to meet with them to review their experience and training.

At many institutions, the animal resource program personnel are responsible for the oversight of a centralized surgical facility. The committee frequently relies on the animal resource program's veterinary or paraprofessional staff for information pertaining to an individual's skill or training. Alternatively, the committee may request that one or several of its members observe procedures on a select number of animals before giving the project full approval. Adequacy of training should be ascertained prior to the start of the project; however, it is important for continuing assessment to be conducted. Although used infrequently, the committee may request the aid of a consultant to make an assessment for projects of a sensitive nature or requiring highly specialized staff.

26:20 What constitutes aseptic technique for major operative procedures in rodents?

Reg. The AWAR (§2.31,d,1,ix) and the *Guide* (pages 60 to 64) require that major survival surgical procedures on all animals be conducted utilizing aseptic technique.

Opin. Aseptic technique is a body of practices employed to prevent microbial contamination of living tissues or sterile materials by excluding, removing, or killing microbial organisms. Aseptic technique requires strict adherence to practices in several areas. The laboratory bench or surgery table should be clean, disinfected, and free of ancillary equipment. Devices or equipment (e.g., animal restraining devices, monitoring equipment, stereotaxic devices) that will be required in the surgical field also should be disinfected. These practices reduce or eliminate potentially infectious organisms and the substrates on which they grow.[11,12]

All surgical instruments, implantable devices, and materials that will contact the surgical site or are implanted in the animal must be sterilized. The sterilization method selected is dependent on time considerations, specialized equipment available, and the composition of the material to be sterilized. Sterilization monitoring devices should be routinely utilized to validate sterilization techniques. Surgical instruments and supplies should be prepared prior to sterilization so that they can be opened without touching them.[13]

The regulatory requirements for preparation of the surgeon for rodent surgery are less rigorous than what is required for nonrodents. Either a clean or disposable laboratory coat or a surgical scrub shirt should be worn by the surgeon and assistants. A surgical mask (which serves to prevent contamination of the wound by droplets of saliva from the surgical team) and a surgical cap, for individuals with long hair that potentially can fall into the surgical field, should be worn. Surgeons should wash their hands with a disinfectant and don sterile surgical gloves using a method that maintains sterility.[14]

Preparation of the surgical site involves removal of fur by clipping or using a depilatory, and disinfection of the skin by scrubbing with suitable agents. The surgical site should be covered with a sterile drape. The goal of aseptic technique is to prevent the surgeon, the surgical site, and all instruments and equipment utilized from causing infection and subsequent disease.[15] (See 18:8 to 18:14.)

26:21 If performing major operative procedures on multiple rodents, can a single set of surgical instruments be used on multiple animals?

Opin. Groups of rodents are frequently surgically manipulated during a single session. Care must be taken to avoid contaminating one animal from another. Sterilized instruments should always be used when initiating a surgical session. New sterile gloves should be donned and, preferably, separate sterile surgical instruments provided for each animal subject. When a limited number of surgical instruments are available and separate sterile surgical packs are not feasible, the following options may be considered:

- Use two sets of instruments. One set is used for incising and manipulating the skin, which is considered a potentially contaminated site because of microbial flora. Once the initial incision and skin manipulation is complete, these instruments are set aside and not used again without prior sterilization. A second set of instruments is used to manipulate deeper, sterile tissues. The second set of instruments (which is shared) should be rinsed with sterile saline between animals. Care must be exercised to maintain the sterility of the surgical gloves by avoiding contacting tissues with fingers when utilizing this option.

- Utilize a glass bead sterilizer to sterilize instrument tips between animals. Recognize that only the instrument tips are sterilized. Care must be taken to avoid touching tissues with instrument handles or other nonsterile instrument parts. Because nonsterile instrument handles are held with gloved fingers, contact of tissues with the fingers is avoided.

- Have two or more sets of sterile instruments available so that the contaminated set can soak in cold sterilant solution for the minimum time required by the solution manufacturer, typically 15 minutes, to kill vegetative bacteria prior to reuse. Instruments must be rinsed thoroughly with sterile saline or water prior to reuse.

(See 18:10.)

References

1. Division of Animal Welfare, Office for Protection from Research Risks, National Institutes of Health, The Public Health Service responds to commonly asked questions, *ILAR News*, 33(4), 68, 1991.
2. DeHaven, W.R., personal communication, 1999.
3. Office of the Federal Register, Code of Federal Regulations, Title 43, Good Laboratory Practice Regulations, Washington, D.C., 1978.
4. Office of the Federal Register, Code of Federal Regulations, Title 29; Part 1910, Occupational Safety and Health Standards, Subpart G, Occupational Health and Environmental Control, Washington, D.C., 1984.
5. Office of the Federal Register, Code of Federal Regulations, Title 29, Part 1910.1030, Bloodborne Pathogens, Occupational Health and Safety Act, Washington, D.C., 1991.
6. Office of the Federal Register, Code of Federal Regulations, Title 10, Chapter 1, Nuclear Regulatory Commission Regulations, Washington, D.C., 1992.
7. U.S. Department of Labor, Occupational Safety and Health Administration, 29 Code of Federal Regulations, Part 1910, Occupational exposures to hazardous chemicals in laboratories; final rule. *Fed. Reg.*, January 31, Part 11, Washington, D.C., 1990.
8. U.S. Department of Justice, Drug Enforcement Agency, Pharmacist's Manual: An Informational Outline of the Controlled Substances Act of 1970, 1986.
9. U.S. Department of Health and Human Services, National Institutes of Health, Guidelines for Research Involving Recombinant DNA Molecules, *Fed. Reg.*, 59 (127), 34496, 1997.
10. U.S. Department of Health and Human Services, Biosafety in Microbiological and Biomedical Laboratories, 4th ed., U.S. Government Printing Office, Washington, D.C., 1999.
11. McCurnin, D.M. and Jones, R.L., Principles of surgical asepsis, in *Textbook of Small Animal Surgery*, 2nd ed., Slatter, D., Ed., W.B. Saunders, Philadelphia, 1993, Chap. 10.
12. Hobson, H.P., Surgical facilities and equipment, in *Textbook of Small Animal Surgery*, 2nd ed., Slatter, D., Ed., W.B. Saunders, Philadelphia, 1993, Chap. 14.
13. Berg, J., Sterilization, in *Textbook of Small Animal Surgery*, 2nd ed., Slatter, D., Ed., W.B. Saunders, Philadelphia, 1993, Chap. 11.
14. Wagner, S.D., Preparation of the surgical team, in *Textbook of Small Animal Surgery*, 2nd ed., Slatter, D., Ed., W.B. Saunders, Philadelphia, 1993, Chap. 12.
15. Fries, C.L., Assessment and preparation of the surgical patient, in *Textbook of Small Animal Surgery*, 2nd ed., Slatter, D., Ed., W.B. Saunders, Philadelphia, 1993, Chap. 13.

27
Assessment of Veterinary Care

Mark A. Suckow and Bernard J. Doerning

Introduction

Animals used in research or educational programs may experience spontaneous illness or, on occasion, problems related to specific procedures they have undergone. For these reasons, the availability of proper veterinary care is essential to assure the humane treatment of such animals. It must be emphasized that specific treatment of ill animals should always be guided by the professional judgment of the veterinarian.

The overall goal of the program of veterinary care is to assure that the animals remain healthy and are used in a humane and judicious manner. In this regard, the AV has responsibilities to the animals to assure their well-being, to the institution to promote regulatory compliance, and to the investigator to facilitate research. Of these, the well-being of the animals is paramount.

A challenge for the IACUC is to identify means to evaluate the scope and adequacy of the program of veterinary care. The IACUC should recognize that a sound program of veterinary care relies on the professional judgment of the AV. Several useful parameters for the IACUC include the accuracy and scope of records that allow for retrospective evaluation of the program, the number and training of individuals available to assist the AV and investigators in delivery of acceptable care, and the mechanism for reporting problems and the response of the AV and his or her staff to such information. Periodic evaluations of the program of veterinary care by peer groups (such as AAALAC) also can be useful in providing an objective programmatic review.

27:1 What is meant by "adequate veterinary care?"

Reg. Adequate veterinary care is an essential part of every animal care program. The AWAR (§2.33,a) states that "each (registered) research facility shall have an attending veterinarian who should provide adequate veterinary care to its animals." The PHS Policy (IV,C,1,e) requires the IACUC to assure that "medical care for animals will be available and provided as necessary by a qualified veterinarian." Specific descriptions of proper veterinary care are not contained within the PHS Policy; however, the *Guide* (page 56), which is used as a basis for evaluation by the PHS Policy, lists the following as components of effective veterinary care:

- Preventive medicine.
- Surveillance, diagnosis, treatment, and control of disease, including zoonotic disease.
- Management of protocol-associated disease, disability, or other sequelae.
- Anesthesia and analgesia.
- Surgery and postsurgical care.
- Assessment of animal well-being.
- Euthanasia.

The *Guide* also indicates that the program of veterinary care is the responsibility of the AV.

Opin. Adequate veterinary care implies a minimal acceptable standard; however, institutions should strive to exceed that level. General aspects of adequate veterinary care can be considered under three headings: animal husbandry, animal health, and study coordination (see 24:15, 27:6).

Animal husbandry is much more than cleaning cages and feeding the animals. Prevention of disease and injury is a primary objective of the animal husbandry program. Areas of concern as they relate to veterinary care are listed below.

Animal procurement (see 27:15). Newly acquired animals can introduce disease into established colonies if appropriate effort is not invested into the selection of vendors. A health surveillance program to screen or monitor incoming animals or potential vendors can be used to select appropriate sources of animals. All animals should be observed upon arrival at the facility for signs of illness.

Acclimation and quarantine (see 27:15). In general, animals should be allowed a period of acclimation and physiologic stabilization after arrival at the facility. This allows animals to recover from shipping stress and permits them to adapt to their new surroundings. Animals known or suspected to be from contaminated sources should be quarantined in such a way that risk to other animals is minimized. Quarantine often involves special precautions, including restriction of personnel access, maintenance of air pressure differences between facility areas, and strict disinfection procedures.

Separation of species and source. Different species and animals from different sources are best maintained in separate housing units. Separating animals by species is useful in reducing anxiety due to interspecies conflict and in minimizing the possibility of interspecies disease transmission. Similarly, animals from different sources should be separated to control possible transmission of infectious disease between groups. Separation can be achieved by housing animals in separate rooms, in cubicles, or various forms of isolators.

Daily animal care. All animals should be housed in cages of appropriate size and style for the species and cleaned and maintained in accordance with the standards of the AWAR and the *Guide*. In addition, it is imperative that all animals are observed every day to assure that their health and living conditions remain acceptable.

An SOP describing methods and equipment to be used in animal care helps to specifically define the standards being followed. It is further useful to utilize a check-off sheet to document when animals or room conditions have been evaluated and when procedures associated with animal care have occurred. This provides a permanent record that can be used to evaluate the consistency of animal care over time.

Animal health is intricately related to veterinary care. Ill or injured animals must receive prompt medical attention when needed (see 27:17). Important issues to address are listed below.

Observation. All animals must be observed daily and any unexpected deaths and deviations from normal behavior or appearance must be reported to the veterinarian. Sick or injured animals should receive prompt attention.

Reporting animal health problems (see 27:15). A mechanism to report and record progress of animals in need of veterinary attention must be defined. Notations in notebooks and "animal health forms" can be used to accomplish this. Documentation by the veterinarian in writing of his or her findings, recommendations, and the progress of the animal is an effective way to assist the IACUC in evaluation of the adequacy of veterinary care.

Anesthesia and analgesia. Animals undergoing procedures that might cause more than momentary pain or distress must receive appropriate anesthetics or analgesics according to the AWAR (§2.31,d,iv,A) which states, "Procedures that may cause more than momentary or slight pain or distress to the animals will be performed with appropriate sedatives, analgesics, or anesthetics, unless withholding such agents is justified for scientific reasons. ..." The PHS Policy (IV,C,1,b) states essentially the same as the AWAR. Similarly, the *Guide* (page 64) states, "The proper use of anesthetics and analgesics in research animals is an ethical and scientific imperative." It is the duty of the veterinarian to advise the investigator on proper use of these drugs and to assure that animals are being adequately medicated. Use of such agents should be carefully documented in writing. The IACUC may find evaluation of such records to be useful. (See Chapter 16.)

Survival surgery. A major responsibility of the veterinarian is to assure that survival surgical procedures are conducted in proper facilities by trained individuals using acceptable methods. Again, records of procedures performed and documentation of qualifications of personnel can be used by the IACUC to evaluate this aspect of veterinary care. (See Chapter 18.)

Postoperative and postprocedural care. Additional care and attention is often needed for animals that have undergone surgical or other invasive procedures which involve anesthesia. For example, care immediately following surgery might include administration of supportive fluids, analgesics, and other drugs as required; monitoring of vital signs such as temperature, pulse, and respiration; provision of supplemental heat to prevent the animal from becoming hypothermic; and remaining with the animal until it is awake and ambulatory.

Longer-term care might include administration of antibiotics or other drugs, evaluation of the surgical site for signs of dehiscence or infection, and overall continued attention to the general medical needs of the animal. For all postoperative or postprocedural care, it is important that accurate records be maintained for retrospective evaluation if problems are found. This also allows the IACUC to evaluate this aspect of veterinary care.

Study Coordination: *Prestudy veterinary consultation.* As a component of adequate veterinary care, it is useful for investigators to discuss research proposals with the veterinarian and it is required when the research may result in more than momentary pain or distress to the animal (§2.31,d,1,IV,B). The AWAR (§2.33,b,4) states that the program of veterinary care must include "guidance to principal investigators and other personnel involved in the care and use of animals regarding handling, immobilization, anesthesia, analgesia, tranquilization, and euthanasia."

Training and assessment of skills. Adequate veterinary care is greatly augmented by training and evaluation of skills for anyone planning to conduct animal research. As part of the program of veterinary care, the AV should assure that a program is in place to accomplish this.

27:2 Must the AV be a laboratory animal specialist?

Reg. The AWAR (§2.31,b,3,i) and the PHS Policy (IV,3,b,1) indicate that the IACUC must include at least one member who is a Doctor of Veterinary Medicine with training or experience in laboratory animal science or medicine who has direct or delegated responsibility for activities involving animals at the research facility. By definition, this individual is usually the AV, but the AWAR indicate that another veterinarian with delegated program responsibility for activities involving animals can serve as the veterinary representative to the IACUC (§2.33,a,3). The AWAR (§1.1, Attending Veterinarian) also mandate that the AV be a person who has graduated from a veterinary school accredited by the American Veterinary Medical Association's Council on Education, or has a certificate issued by the American Veterinary Medical Association's Education Commission for Foreign Veterinary Graduates, or has received equivalent formal education as determined by the USDA. The PHS Policy (IV,C,1,e) states that the IACUC must determine that "medical care for animals will be available and provided as necessary by a qualified veterinarian," although the Policy does not specifically define "qualified."

The *Guide* (page 56) indicates that the AV must be an individual who has the Doctor of Veterinary Medicine degree and has training or experience in laboratory animal science and medicine. The *Guide* (page 9) does not specifically require the AV to serve as the veterinary member of the IACUC, but does state that the veterinary member of the IACUC also must have training or experience in laboratory animal science and medicine or in the use of the species in question.

Opin. The answer to this question is "yes," if a "specialist" is defined as someone who has specialized training or experience. Individuals may have several areas of specialization, including laboratory animal science and medicine. Thus, the use of consultants as the AV is acceptable even if they devote major effort to other areas of veterinary medicine. The American Veterinary Medical Association recognizes laboratory animal medicine as a specialty practice area of veterinary medicine.

Completion of an internship at a laboratory animal facility as part of the veterinary professional curriculum is one means by which individuals might gain minimal experience in laboratory animal medicine. Sometimes, individuals become specialists as a result of on the job experience under the supervision of an experienced AV. Alternatively, some choose to complete residencies in laboratory animal medicine at academic, military, or industrial institutions. Such residencies allow for concentrated study and experience in the specialty.

An individual who has completed an approved residency or has approved practical experience has an approved publication related to laboratory animal medicine or animal research, and has successfully passed the appropriate written and practical examinations is eligible for certification as a Diplomate by the American College of Laboratory Animal Medicine (ACLAM). Diplomate status within

ACLAM is widely viewed as evidence of significant training and experience in laboratory animal medicine.

Membership and involvement in relevant professional organizations can be interpreted as evidence of specialized interest in laboratory animal science and medicine. For example, activity in the American Association for Laboratory Animal Science (AALAS) or the American Society of Laboratory Animal Practitioners (ASLAP) would likely improve the professional knowledge of individuals active within the profession.

27:3 Is it necessary for all organizations housing or using laboratory animals to have an AV?

Reg. Any institution that is under the purview of AAALAC, the AWAR, or the PHS Policy is required to have an AV (see 27:1). Other organizations that do not fall under the purview of those standards may still choose to employ an AV on a full time or consulting basis.

Opin. The AV typically has several functions within the institution[1] (AWAR §2.33a; §2.33b). For example, it is the duty of the AV to assure that the husbandry of the animals is appropriate for the species. The AV is responsible for developing and implementing programs for preventive medicine to minimize the likelihood of animals developing clinical illness. Furthermore, the AV directs the diagnosis and treatment of clinical disease affecting the animals. In addition, the AV typically advises investigators and the IACUC on methods to relieve or alleviate pain and distress, perioperative care and surgical procedures. In short, the role of the AV is to assure that animals remain healthy and are used in a judicious and humane manner (see 27:6).

Several types of facilities are not required under the existing regulations to have an AV. For example, institutions which use only rats of the genus *Rattus* and mice of the genus *Mus* bred specifically for use in research or teaching are not required to have an AV if they lack AAALAC accreditation and do not receive PHS or other federal funds for research or teaching using animals. Such facilities should still develop a written plan for animal health in the event of unexpected illness.[2] In addition, if animals are to be used in surgical or other invasive procedures, it is advisable to identify an individual who can serve as the AV.

27:4 Under what circumstances are full-time, part-time, or consultant veterinarians used?

Reg. The AWAR (§2.33,a,1) require that institutions using only a consultant veterinarian have a written program of veterinary care and that the AV make regularly scheduled visits to the facility. The PHS Policy does not specify a requirement for a full-time vs. part-time veterinarian. However, the Policy (IV,C,1,e) requires that the IACUC assure that medical care be "available and provided as necessary." PHS Policy (IV,A,3,b,1) also requires that there is a veterinarian with direct or delegated program authority and responsibility who will serve on the IACUC.

Opin. PHS Policy (IV,C,1,e) suggests that the veterinary staff must be sufficient to provide appropriate medical care whenever the need arises. In this regard, the IACUC is charged with determining the level of veterinary service that must be procured.

The need for the services of the AV or other ancillary staff vary extensively with the size and scope of animal research-related operations of the institution. In this sense, it is not possible to outline specific standards for the need of a full-time or other veterinarian.

The American College of Laboratory Animal Medicine (ACLAM) recommends that a formal arrangement for veterinary care exist in all instances, while recognizing that the specific arrangement may vary with the number and species of animals involved and the nature of the experimentation at the institution.[1]

Institutions where research involving survival surgery or other invasive procedures are performed on a routine basis should seek greater veterinary involvement than institutions where this does not occur. Similarly, other conditions which merit increased veterinary involvement are shown in Table 27.1.

The IACUC should consider the factors listed in Table 27.1 and determine if the available veterinary staff can provide the necessary service and oversight. Each facility has its own unique characteristics and needs. Some small facilities may need a full-time veterinarian, while some larger ones may require only a part-time or consulting veterinarian. Experience of the authors has shown that institutions characterized by three or more of the above factors can generally justify the need for a full-time veterinarian.[2] Facilities which are characterized by few or none of the above listed factors may find a part-time (veterinarian who is a regular employee of the institution) or even a consultant veterinarian (one who provides veterinary care on a fee-for-service basis to the institution) to be sufficient.[2]

Veterinary support in addition to the AV may be needed as the number of animals in the facility and the number of relevant factors increase.[2] Such support can be in the form of additional veterinarians or, if appropriate, ancillary veterinary staff such as veterinary technicians or trained animal care technicians (see 27:7). Veterinary activities performed by technicians should be conducted under the direction of a qualified veterinarian. Communication between the veterinarian and technicians should occur regularly to ensure that

TABLE 27.1

Factors Which Increase Need for Veterinary Services

Animal-Related Factors	Research-Related Factors
Large number of animals in the colony	Use of biohazards
Use of phylogenetically higher animals (i.e., those of high sentience, such as mammals)	Use of chemical hazards
	Use of radiologic hazards
Use of immunodeficient animals	Procedures requiring the use of anesthetics or analgesics, particularly survival surgery
Use of animal models with special medical or biologic needs	Survival surgical procedures involving extensive perioperative care

the veterinarian is aware of any animal health or care issues and can advise the technical staff accordingly (AWAR §2.33,b,3; *Guide*, page 56). Under any arrangement, veterinary services must be available at all times, including evenings, weekends, and holidays (AWAR §2.33,b,2; and see 27:5). It is important that the information for contacting the veterinarian (telephone or pager numbers) is communicated to all research and animal care personnel.

27:5 If a consultant is used, how frequent should visits be and how can emergency care be provided if the consultant is not available?

Reg. The frequency of visits to a facility by a consultant can vary significantly with the size and scope of animal use activities (*Guide*, page 12). The AWAR (§2.33,a,1) state that "regularly scheduled" visits should be made to the facility.

Opin. Visits should be frequent enough to provide the same degree of oversight to each study as would be provided to studies in a larger facility employing full time veterinary staff.[2] The experience of the authors is that on-site visits, which occur less frequently than once a month, diminish the effectiveness of the AV if animals are housed continuously at the facility.[2] As the scope and complexity of the facility increases, so should the frequency of on-site visits by the consultant. Visits by the consulting veterinarian should be regularly scheduled and supplemented with additional visits as needed.

The availability of consultants to respond to emergency situations may be limited by time or distance constraints. Nonetheless, emergency veterinary care must be provided (AWAR §2.33,b,2; *Guide*, page 46). Trained veterinary or animal care technicians can provide such care under the direction of the veterinarian, through instructions communicated by such means as telephone, fax, e-mail, or prior written instructions and guidelines.[2] In contrast, some facilities make prior arrangements with a local, practicing veterinarian to provide occasional emergency care on-site. Any veterinarian employed in this capacity should be provided with initial training regarding the

27:6 What are the AV's responsibilities?

Reg. The principal responsibility of the AV is to develop and implement an effective program of veterinary care (see 27:1, 27:15). Such a program has multiple components including preventive medicine, proper use of anesthetics, and analgesics and tranquilizers, perioperative care, euthanasia, diagnostic procedures, provision of veterinary care, emergency care, and aspects of animal husbandry[1] (AWAR §2.33,b,1–§2.33,b,5; *Guide*, page 56). These responsibilities and associated aspects are summarized in Table 27.2.

TABLE 27.2
Responsibilities of the Attending Veterinarian

Type of Responsibility	Specific Aspects of Responsibility
1. Diagnostic procedures	1. Direct or perform procedures to identify disease etiologies
2. Provision of medical care	2. Prescribe treatments to treat ill animals
3. Preventive medicine	3. Develop programs for quarantine and isolation, monitoring of vendors, monitoring of colony health, and routine vaccinations and parasite control
4. Use of anesthetics, analgesics, sedatives, and tranquilizers	4. Advise investigators on and assure proper use of methods to relieve or reduce pain and distress
5. Perioperative care	5. Oversee presurgical preparation, surgical procedures, and postsurgical care of animals
6. Euthanasia	6. Assure that animals are euthanized when appropriate and in a humane manner
7. Animal husbandry	7. Assure that programs for disinfection, housing, nutrition, breeding, and environmental enrichment are appropriate
8. Use of hazards	8. Work with hazards oversight professionals to assure safe use of biologic, radiologic, and chemical hazards
9. Use of animals	9. Advise investigators and the IACUC on the appropriateness of specific techniques and methods in animals, and the availability of alternative animal and nonanimal models
10. Occupational health	10. Advise IACUC and health professionals on aspects of the occupational health program

Opin. Although the AV has a principal responsibility to assure the well-being of the animals being used at the institution, the AV also has responsibilities to the institution and to the investigators.[2] Ultimately, the AV should facilitate the ability of the investigator to perform animal research in a humane manner and consistent with regulatory principles. Typically, the IACUC and the AV have a close working relationship. The IACUC relies on the professional judgment of the AV in animal health issues, and the AV turns to the IACUC for support if resistance to proper animal care and use is encountered.[2]

27:7 What staff, other than veterinarians, are useful in providing adequate veterinary care? What level of support staff is considered adequate?

Opin. Ancillary personnel such as veterinary technicians, research technicians, and trained animal care technicians can provide valuable support to the program of veterinary care (see 27:4). Such individuals often interact closely with the animals on a daily basis and can serve as a primary point for observation of animal health.

The size of the staff needed to provide adequate veterinary care is completely dependent upon the size and complexity of the animal facility. Any animal housed in a research facility must be checked at a minimum of once per day, every day (AWAR §2.33,b,3; *Guide*, page 46). This need can increase based on the invasiveness or potential for untoward outcomes of the study. Unreported or untreated illness, insufficient feed or water, or signs of insufficient husbandry can all indicate inadequate observation (or response) of the animals and may signal a need for additional staff.

27:8 To what extent and under what circumstances can the AV delegate clinical responsibilities to other veterinarians or nonveterinarians?

Opin. The AV may delegate the responsibility for observation of animals and "hands-on" care to other veterinarians or technicians provided that it can be documented that such individuals are adequately trained in the care of the species and in the procedures involved. All animals should be observed every day by someone trained to evaluate the health and well-being of that species (AWAR §2.33,b,3; *Guide*, page 46; see 27:1). The ultimate responsibility for assuring proper clinical care remains with the AV and he should establish a mechanism for receiving animal health information that will ensure fulfillment of this responsibility (AWAR §2.33,b,3; *Guide*, page 56). (See 5:24, 6:7.)

27:9 Should the IACUC attempt to evaluate animal health? If so, how?

Opin. Yes, the IACUC should periodically evaluate the health of the animals, since the IACUC is responsible for evaluating the program of veterinary care (AWAR §2.31,c,1; §2.31,d,1,vii; PHS Policy IV,B,1; IV,C,1,e; *Guide*, pages 9, 12). This can be troublesome since the IACUC typically relies on the AV for advice with respect to animal health concerns, thereby creating a conceivable conflict of interest if the AV is asked to evaluate his or her own program of veterinary care.

The most basic way for the IACUC to evaluate animal health is to observe the animals during the semiannual inspections. Table 27.3 lists some clinical observations which might be evidence of an animal health problem. The number of animals with indications of illness is a valuable measure of animal health for the IACUC to use. Of course, animals found moribund or dead also can signal an animal health problem.

The IACUC should also examine animal health and surgical records for possible evidence of health-related problems[2] (see 27:10). In this regard, it is useful to spot-check those records detailing animal health over the last 6 months. This should include records of animal health monitoring such as routine fecal examinations for endoparasites or routine serologic monitoring of rodent colonies. It is worthwhile to query personnel involved with animal care to determine if they are aware of any animal health problems.

If the preventive care aspect of the program of veterinary care (including careful vendor selection) is effective, the IACUC should find few problems. However, it is not unusual for the IACUC to identify occasional problems, since spontaneous disease and consequences of experimental procedures can sometimes lead to clinical illness even in well-managed animal facilities. The key for the IACUC is to evaluate the *adequacy of the mechanism for reporting prob-*

TABLE 27.3

Some Indications of Illness in Animals

Body System	Abnormal Clinical Signs
1. Respiratory system	1. Nasal discharge, labored breathing, blue tint to mucous membranes
2. Gastrointestinal system	2. Vomiting, diarrhea, thin appearance
3. Integumentary system	3. Bleeding or seeping wounds, alopecia, observed persistent scratching, closed wounds
4. Nervous system	4. Inability to move normally, overly aggressive behavior, lethargy
5. Circulatory system	5. Edema in extremities, ascites, labored breathing

lems and the response to any problems[2] (see 27:11). For example, efforts to diagnose a problem, medications administered, supportive care provided, advice to investigators on technique refinement, and efforts to isolate sick animals to prevent disease spread could all be considered evidence of an appropriate response to a clinical problem. The precise response will vary with the specific problem.

27:10 Which veterinary care records should be kept? How detailed should these records be? (Also see 27:11, 27:12.)

Opin. The purpose of any recordkeeping system is to document the occurrence of a procedure, result, or event. All records must be detailed enough to adequately explain what occurred to a person knowledgeable in the subject. Records associated with adequate veterinary care serve two main purposes. They assist in facility management and document an adequate veterinary care program to regulatory officials. All records should be maintained as long as they may be useful to the facility.

Some regulatory documents have sections that pertain to the retention of records. The majority of such requirements relate to the product submission and approval process. In many cases, particularly for records related to studies conducted under Good Laboratory Practice (GLP) standards, permanent archiving of records is done (21 CFR part 58.195). In the case of the Federal Insecticide, Fungicide, and Rodenticide Act (FIFRA), records need to be retained for the life of the product (40 CFR Part 160.195). It stands to reason that these retained records should include those related to animal health. By law, certain records such as acquisition records of dogs and cats, are required to be retained for a minimum of 3 years after completion of the study (AWAR §2.35,f). As a very general rule of thumb, it is useful to maintain records for a minimum of 3 years unless regulatory or institutional requirements mandate a longer holding period.

Records having importance with respect to veterinary care include:

- *Standard procedures and animal health.* Activities such as observation of animals, provision of feed and water, and evaluation of environmental parameters are usually checked daily and should be documented. Many facilities have found it useful to develop an SOP detailing what is to be included in routine animal husbandry. A daily check-off form then can be used to document that a procedure has occurred. Many facilities employ a monthly room care sheet to document activities occurring less frequently such as cage changing or disinfection.
- *Veterinary care records* (see 27:1, 27:13). Any animal exhibiting abnormal behavior or signs suggestive of illness should be

reported to the veterinary staff. Written documentation should describe the veterinarian's findings, diagnostic procedures, treatments, followup care, and progress of the animal. In addition, preventive medicine efforts should be documented.

- *Colony animals* (see 27:15, 27:16). Maintenance of long-term animal colonies requires an additional level of recordkeeping. For example, production records in a breeding colony may offer clues on disease transmission or genetic abnormalities. It is typical to maintain individual animal records on dogs, cats, nonhuman primates, and, occasionally, farm animals. These individual records serve as a reference for growth, development, health-related observations, and preventive medical care.
- *Surgery and periprocedural care.* Records should detail preoperative preparation of the animal, use of anesthetics, analgesics and other medications, and monitoring of the animal during surgery and during recovery. Records should be sufficiently detailed and organized for the IACUC to readily determine that procedures are being conducted in a manner that minimizes the risk of pain or distress to the animals. In addition, it should be clear that the veterinary staff has responded expeditiously to any significant problems.

27:11 How can the IACUC assure that veterinary care records are adequate?

Opin. Records for veterinary care should be evaluated for completeness and for evidence of an adequate response to reported problems[2] (see 27:10, 27:12). For example, records should clearly identify the specific animal involved, the clinical history, the source of the animal, the investigator, age and gender of the animal, date of animal receipt, and the types and dates of experimental or therapeutic procedures which have been performed on the animals (*Guide*, page 46).

Appropriate diagnostic efforts and prescribed treatments should be documented. In addition, the record should indicate each time the animal is medicated or otherwise treated. The clinical progress of the animal should be noted until the clinical problem has resolved or the animal has been euthanized (see 27:14). Evidence that an animal languished until it died or was finally euthanized without reasonable attempts to diagnose and treat the problem can be considered as indicative of inadequate veterinary care.[2] (See 27:9.) In addition, proper records should reflect the substance of any discussions with investigators or other staff on ways to minimize problems in the future. Individuals examining animals or providing treatment or other actions should sign or initial their notations.

Records can be considered to be inadequate if the IACUC is unable to easily understand the steps taken to identify, report, diagnose, discuss, treat, and continue to evaluate an animal until the humane resolution of the case.

27:12 Should every animal in the facility have a health record or should every colony have one? (Also see 27:10, 27:11.)

Opin. All observed illness or abnormalities should be entered into a clinical record. However, records are frequently more detailed for some species than for others. For example, individual health records for dogs, cats, nonhuman primates, and farm animals are useful (*Guide*, page 46).

A useful rule of thumb is that individual records should be maintained for those animals which receive routine periodic individual health evaluations. For example, individual records should be maintained for dogs, detailing vaccinations, parasite evaluations, any treatments administered, and periodic physical examinations. Separate records are not needed for animals which may undergo periodic evaluation by examination of several representative individuals from the colony. Rather, the health information of such animals can be recorded as a colony health record. An alternative rule of thumb is that individual records should be maintained for long-lived animals (*Guide*, page 47). In either case, the interpretation is that individual records are useful for species such as dogs, cats, nonhuman primates, rabbits, and farm animals, but would be of less value for rodents or common nonmammalian species. Alternatively, individual surgical records can be very useful for animals that have undergone invasive procedures, regardless of species.

27:13 How sick or abnormal should an animal be to merit veterinary attention?

Opin. Every animal that is found to be ill, behaving abnormally, or is suspected of being ill should be reported to the AV or to the veterinary staff as directed by the AV. In some cases, the veterinary staff may determine that no specific treatment is merited, while in other cases treatment may be needed. It is the veterinary staff, under the direction of the AV, who should make a judgment regarding the severity of illness and subsequent course of action.

27:14 How often should notations be made in the animal healthcare records for an animal reported to be ill?

Opin. As stated in 27:11, clinically ill or abnormal animals should be followed until resolution, either through return to clinical normalcy or euthanasia or determination that the relevant condition is clinically minor with little likelihood of increased severity.[2] In this regard, records should document that followup care and evaluation have occurred.

The exact frequency at which followup evaluations should be made varies considerably with the clinical condition and depends upon the sound professional judgment of the veterinarian. Severe conditions might necessitate evaluation several times daily, while other conditions might be aptly evaluated once every several weeks. In general, follow-up evaluations should occur and be documented in the records at a frequency such that significant increases in the severity of the condition could reasonably be expected to be detected. For severe conditions, this usually involves at least daily evaluation with more frequent evaluations if necessary. For less threatening conditions, weekly or biweekly evaluations are often sufficient.[2]

Repeated instances of ill animals progressing to severe disease states in the absence of corresponding notations in the record may indicate that animals are not being examined frequently enough. Conversely, if notations exist for such animals, it could suggest that the decline in clinical condition is being duly noted but ignored by the veterinary staff. In either circumstance, the possibility that records are being maintained or utilized inappropriately exists, and the IACUC must further investigate the situation.

27:15 How can the IACUC evaluate programs for preventive medicine and health monitoring? What type of programs should be expected?

Opin. Preventive medicine and health monitoring programs are designed to assure that healthy animals are acquired initially and that these animals remain healthy. To accomplish this, regularly scheduled evaluations of animals and vendors should be performed. The IACUC should inspect the veterinary care records (see 27:11) for accordance with a written description of the preventive medicine program, as well as visually examine animals for general appearance as an indicator of health status. Further, the IACUC should ascertain that animal health evaluations indicate the animals remain healthy and a timely and appropriate response is taken when evaluations indicate a problem (see 27:1).

The program of preventive medicine should begin with evaluation of the vendor or source of the animals. Regular reports of the health status of the vendor's animals should be reviewed by the AV or his designate to determine that healthy animals are being acquired by the institution. If this review determines that health problems exist at the vendor, then the AV should coordinate arrangements to identify alternate vendors or other means which ensure that healthy animals are being used.[2]

TABLE 27.4

Typical Components of a Preventive Medicine Program

Aspect of Preventive Medicine	Typical Activity
1. Animal vendor surveillance	1. Review of vendor health status reports, on-site visits to vendor when possible
2. Animal transportation	2. Evaluate method of transportation for minimization of animal stress and risk of disease
3. Processing of newly received animals	3. Inspection and examination of animals, initial vaccinations, medications, parasite evaluations
4. Periodic re-evaluation (usually every 6 to 12 months)	4. Physical examination, vaccination, tuberculosis testing (nonhuman primates), dental prophylaxis
5. Rodent health surveillance	5. Serologic and histopathologic evaluation for microbial contaminants, pelt examination for ectoparasites, examination of gastrointestinal tract for endoparasites

At the time of arrival at the facility, animals should be inspected by the veterinary staff and then separated from other more established groups of animals (see 27:1) until, in the professional judgment of the veterinary staff, the newly arrived animals do not pose a threat to the established animals. For example, animals such as dogs, cats, and nonhuman primates often receive a physical examination upon arrival as part of the preventive medicine program, and are then separated from established animals during a quarantine period.

The program of preventive medicine also should include procedures such as appropriate vaccinations, parasite evaluations, and testing for tuberculosis for nonhuman primates. Livestock, such as sheep and cattle, should undergo routine deworming. Other procedures are often included depending on the facility and the species involved. These procedures should be repeated periodically, as per veterinary medical standards. Periodic, often annual, physical examinations should be performed for animals such as cats, dogs, and nonhuman primates. Routine physical examinations on other animals such as rabbits, guinea pigs, and other small animals can be useful for animals maintained for extended periods of time. Often, such examinations are brief and may be conducted by trained personnel other than the AV.

Rodent colonies are often evaluated for microbial contamination (*Guide*, page 60). In general, a small number of animals (often sentinel animals placed in the room specifically for the purpose of health evaluation) are sacrificed, blood is collected for serologic detection of microbial contaminants, tissues are evaluated histologically, and the gastrointestinal tract and pelt are evaluated for parasites. Recommendations for the exact procedures and number of animals evaluated in this way are presented in detail elsewhere.[3]

Examples of activities that can be associated with a preventive medicine program are shown in Table 27.4. The exact procedures and scope of activities vary significantly with the species and type of animals and the type of research. Activities listed in Table 27.4 are examples only.

27:16 How can the IACUC determine that the veterinary response to health monitoring findings is reasonable and adequate?

Opin. Having determined that the program for preventive medicine and health monitoring is properly designed and implemented, the IACUC must further ensure that results of these efforts are interpreted correctly and that appropriate action is taken when potential problems are identified. To accomplish this, the IACUC must evaluate the response of the veterinary staff, in particular the AV.

In general, the IACUC should expect evidence of a plan to expeditiously respond to identified perturbations in health monitoring and surveillance. The scientific objectives of the study, the potential consequences of infection with a particular pathogen, and the possible adverse effects of the infectious agent on other studies within the facility should all be considered when determining the appropriate response (*Guide*, page 60). The precise response can vary and can include continued monitoring of animals, isolation of suspect animals, specific treatment of animals, or euthanasia of all affected or exposed animals. Again, the appropriateness of the response depends on the type and species of animals, the nature of the research, and the professional judgment of the AV and the veterinary staff.

Additionally, the IACUC should further examine health records to determine if the response of the veterinary staff has been successful. If not, the IACUC should determine that the AV is developing other strategies to resolve the problem. The IACUC must be assured that problems are not being ignored and that the veterinary staff does not abandon the care of an animal.

27:17 Are PIs responsible for any aspect of adequate veterinary care?

Opin. Investigators play an indirect but significant role in veterinary care. It is critical to recognize that veterinary care is not exclusively the treatment of sick animals. It includes the proper care and use of animals in such a way that animals remain healthy and recover normally from experimental manipulations. Beginning with protocol initiation, an appropriate animal model must be chosen to ensure a project is successful. Potential adverse reactions or outcomes should be identified by the investigator and plans made to address them. Investigators and their technicians may frequently observe animals during the procedure and are, therefore, in an excellent position to report potential problems. For example, weight loss in a cage of group-housed animals may not be observable to a caretaker whereas the investigator's study notes with individual body weights of animals could detect the problem.

Depending on the staffing of the facility, investigators or their technicians may be responsible for administering treatments prescribed by the veterinarian. As part of this, they should be instructed on procedures to contact the veterinary staff when animals are found in poor clinical condition.

For these reasons, it is critical to view the role of the investigator and his or her staff, along with the AV, as providers or facilitators of veterinary care. Although the AV is ultimately responsible for provision of proper veterinary care, the IACUC should recognize that there are others who have an ethical, if not a regulatory, responsibility to positively impact the health and welfare of research animals.

References

1. American College of Laboratory Animal Medicine, *Report of the American College of Laboratory Animal Medicine on Adequate Veterinary Care in Research, Testing and Teaching*, American College of Laboratory Animal Medicine, Cary, NC, 1996.
2. Opinion of authors based on informal survey of 10 academic and commercial institutions.
3. Committee on Infectious Diseases of Mice and Rats, Institute of Laboratory Animal Resources, National Research Council, Health surveillance programs, in *Infectious Disease of Mice and Rats*, National Academy Press, Washington, D.C., 1991, Chap. 4.

28

Laboratory Animal Enrichment

Kathryn A. L. Bayne

Introduction

In 1985 Congress passed amendments to the AWA (PL 99-198) which included a provision mandating an environment "adequate to promote the psychological well-being of nonhuman primates." Six years later, the USDA published implementing regulations pertaining to this amendment.[1] In December 1996, APHIS/AC published the results of an internal survey entitled "USDA Employee Opinions on the Effectiveness of Performance-Based Standards for Animal Care Facilities"[2] which included questions pertaining to environmental enrichment programs to improve the psychological well-being of nonhuman primates. Approximately 45% of respondents to the survey indicated that the criteria for primate enrichment are not sufficiently clear, and about 50% said the criteria were not useful. APHIS/AC inspectors responding to the survey indicated that there should be a clearer definition of the requirements and enhanced documentation to ensure that enrichment plans were being followed. Some suggested stricter guidelines or policies pertaining to group housing, cage space, and enrichment. The USDA team assigned to review the survey results made several recommendations. Those relevant to environmental enrichment programs for nonhuman primates included the following: "For primates, the enrichment options should be grouped into elements and a set of examples given for each category. Facilities should be required to provide something in each category. They should document consideration of different species' needs, efforts on behalf of individuals with behavior problems, periodic review of the results of plan implementation, and improvements made after review."[2]

The 7th edition of the *Guide* includes a section on "Behavioral Management." It describes potential enhancements to the animal's cage environment, social environment, and activity level. The National Research Council's report on "The Psychological Well-Being of Nonhuman Primates"[3] and pending additional APHIS/AC policy implementation on environmental enrichment highlight the increasing need for IACUC involvement in, and oversight of, laboratory animal enrichment programs.

The field of environmental enrichment is still evolving. Empirical measures of assessing animal well-being are not completely defined and the roles of genetic, developmental, and environmental influences on well-being require further study. The *Guide* recommends expanding the enrichment program to all laboratory animals rather than restricting enrichment provisions to nonhuman primates as required by law. The institution must avoid a minimalist approach to implementing the enrichment program(s) and the IACUC must take a proactive role in its oversight.

28:1 What is meant by laboratory animal enrichment?

Opin. Environmental enrichment is an environment in which complex stimuli are provided to alleviate the occurrence of abnormal behaviors.[4] It is a component of animal husbandry that is designed to enhance the quality of captive animal care by identifying and providing the environmental stimuli necessary for optimal psychological and physiological well-being. In practice, this covers a multitude of innovative and imaginative techniques, devices, and practices aimed at keeping captive animals occupied, increasing the range and diversity of behavioral opportunities, and providing more stimulating and responsive environments.[5]

The physical environment in the primary enclosure must be enriched by providing means of expressing noninjurious species-typical activities. Species differences should be considered when determining the type or methods of enrichment. Examples of environmental enrichments include perches, swings, mirrors, and other cage complexities; providing objects to manipulate; varied food items; using foraging or task-oriented feeding methods; and providing interaction with the care giver or other familiar and knowledgeable person consistent with personnel safety precautions.[1]

Perhaps enrichment should be considered as anything that the animal finds pleasing at that moment in time. Enrichment might include provision of human interaction, a cagemate, a nesting box, opportunity to forage for food in deep bedding or having to work for food through manipulation of "food puzzles," or control over some aspect of the environment, such as visual access to other animals, music, or video. Of course, provision of the presumed enrichment must be well thought out and with a rather detailed knowledge of the species' normal behavior, constraints imposed by the research, safety considerations, and the person's level of experience.[6]

Surv. What is meant by laboratory animal enrichment?

Responses to this question included no definition of the term. Rather they referenced language in the *Guide* (pages 36 to 38), and also included alterations to the environment that promote physical and psychological well-being, and promoting species-typical behavior.

- No formal definition 3/8
- Promotes species-typical behavior 2/8
- AWAR definitions/promote psychological well-being 2/8
- Cage complexities 1/8

28:2 Besides the federal mandate for canine exercise (AWAR §3.8) and improving the psychological well-being of nonhuman primates (AWAR §3.81), what other requirements for laboratory animal enrichment must be met?

Reg. The *Guide* (page 36) states, "Depending on the animal species and use, the structural environment should include resting boards, shelves or perches, toys, foraging devices, nesting materials, tunnels, swings, or other objects that increase opportunities for the expression of species-typical postures and activities and enhance the animals' well-being." It further states, "It is desirable that social animals be housed in groups; however, when they must be housed alone, other forms of enrichment should be provided to compensate for the absence of other animals, such as safe and positive interaction with the care staff and enrichment of the structural environment. ... Animals should have opportunities to exhibit species-typical activity patterns."

Opin. It also should be noted that institutions which are accredited by AAALAC or receive PHS funds are required to conform with the *Guide* (PHS Policy IV,A,1). Although the provisioning of environmental enrichment to laboratory animals other than nonhuman primates is a relatively new endeavor, there are a number of good references to use when developing such a program.[7-9]

Surv. Besides the federal mandate for canine exercise and improving the psychological well-being of nonhuman primates, what other requirements for laboratory animal enrichment must be met at your institution?

Respondents to the survey were divided between having no additional mandates for enrichment and having an internal institutional policy to do so. Two respondents indicated that providing appropriate animal care, by implication, included providing enrichment.

- No other requirements 3/8
- Internal policy 3/8
- Moral/ethical requirements 2/8

28:3 What should be the goals of a laboratory animal enrichment program?

Opin. Mench[10] has suggested that enrichment programs should provide opportunities for exploration for species that are generalists or "are adapted to environments that are highly variable in terms of resource availability." Enrichment also is needed for species that exhibit "complex antipredator behaviors" and species that have a complex social order. Mench encourages viewing enrichment from the perspective of the information-gathering needs of the animal. She points to successful enrichments that allow animals to carry out appetite-related components of behavior (e.g., foraging boards).

Goals of an enrichment program might include:

- Promoting the well-being of the research animal and providing a more refined animal model for the research.
- Increasing knowledge of the environment's impact on behavior and stress reduction for personnel.
- The enrichments should be meaningful to the animals and should not compromise personnel safety, animal health and safety, or the research aims.

Goals should not include provision of enrichments that are aesthetically pleasing but have no species relevance (e.g., a brightly colored ball may not prove to be behaviorally stimulating to some species of animals).

In *The Psychological Well-Being of Nonhuman Primates*[3] a sample of goals and aims of the enrichment program (for nonhuman primates) is offered. In addition to those objectives described above, this report suggests providing cognitive stimulation and opportunities for animals to alter their environment as well as decreasing self-injurious behavior. The report emphasizes training of personnel in the natural history, behavior, and husbandry of the species as an essential means of achieving enrichment program goals.

Surv. What should be the goals of a laboratory animal enrichment program?

Goals described by respondents included meeting regulatory requirements, promoting species typical behavior, allowing some control over the environment, decreasing abnormal behaviors, and buffering stressful situations. In a few instances, the IACUC did not approve formal goals; goals were established in those cases by the veterinary support program or an enrichment committee.

- No formal goals 4/8
- To promote species-typical behavior 2/8
- Meet regulatory requirements 1/8

• Provide improved environment	1/8

28:4 What species should be included in a laboratory animal enrichment plan?

Reg. The regulatory requirement for enrichment is currently restricted to promoting nonhuman primates (AWAR §3.81) and providing canines the opportunity to exercise (AWAR §3.8). The *Guide* (pages 36 to 38) strongly recommends addressing the structural environment, social environment, and activity of laboratory animals as they pertain to behavioral management. PHS Policy (II; IV,A,1) requires institutions to comply with the AWAR and to follow the *Guide*.

Opin. Reference documents[7,11,12] are available that contain numerous suggestions and literature citations pertaining to environmental enrichment of nonhuman primates, birds, cats, dogs, farm animals, ferrets, rabbits, and rodents. The large number of references contained in this resource, as well as more recent publications,[13-20] suggest that whether or not laboratory animals other than primates are covered by an enrichment plan, enrichment strategies are being developed for them. (See 28:5.)

Surv. What species should be included in a laboratory animal enrichment plan?

A variety of species are provided environmental enrichment, though not all may be covered by a formal enrichment plan. Species reported to be included in an enrichment program were nonhuman primates, dogs, cats, rabbits, guinea pigs, and farm animals. There may be more than one response per respondent.

• Nonhuman primates	7/8
• Dogs	5/8
• Cats	3/8
• Rabbits	2/8
• Guinea pigs and other rodents	2/8
• Farm animals	1/8
• Most species	1/8

28:5 Where can help be found to develop a laboratory animal enrichment plan?

Opin. Assistance in developing an enrichment plan may be found in Appendix A of the ILAR report on nonhuman primate psychological well-being.[3] The report includes two sample plans that provide general information on creating an enrichment plan for a large, complex, nonhuman primate program (e.g., many animals, diverse species, diverse research uses) and a smaller or less diverse primate program (e.g., few animals or one species). A checklist of items the plan should address is also presented in the report (pages 21 to 24). (See 28:4.)

The AWAR (§3.81) has identified several key points which must be considered when developing an enrichment plan for nonhuman primates.[1] At a minimum, the plan must contain "specific provisions" to address the social needs of nonhuman primates in accordance with currently accepted professional standards. The plan also must address the physical environment of the primates in terms of environmental enrichment. Consideration must be given to the special needs of some categories of primates such as infants and young juveniles, those that show signs of being in distress, those which experience protocol-related restricted activity, those housed in a manner which precludes seeing or hearing primates of their own species, and the needs of great apes weighing over 50 kg. The use of restraint devices for medical treatment or IACUC-approved research activities and a contingency for at least one continuous hour of unrestrained activity when primates are held in restraint devices for more than 12 hours also should be indicated in the plan. Finally, identification and monitoring of any animals that are exempted from the plan because of health or well-being reasons or because of research constraints. Also available is the National Institutes of Health "Nonhuman Primate Intramural Management Plan"[21] which may be obtained from the National Agriculture Library (Federal Register, Vol. 56, No. 32, Friday, February 15, 1991).

28:6 What are the types of enrichment animals might receive?

Opin. Regardless of the species for which the enrichment plan is being considered, strategies can be categorized as social or nonsocial. Social enrichment includes both contact and noncontact elements and can be applied to interactions between animals and personnel, animals of the same species, and animals of different species. Nonsocial enrichment techniques include cage furniture, bedding, feeding techniques, toys, and other environmental conditions (e.g., lighting, sound). The enrichment technique(s) selected should be safe for both the animal and personnel, be species-relevant, and should not conflict with the goals of the research study.

Surv. What are the types of enrichment animals might receive?

Species mentioned by the eight respondents were nonhuman primates, dogs, cats, rabbits, guinea pigs, rats, mice, hamsters, chinchillas, chickens, pigeons, frogs, turtles, sheep, and pigs. Both social and nonsocial enrichment strategies were described. Enrichment provided, other than conspecific contact, is itemized in Table 28.1. Excellent references for enrichment techniques are available.[3,7-9,11,12]

TABLE 28.1
Enrichment Provided to Laboratory Animals

Animal	Enrichment
Nonhuman primates	Perches, wooden gnawing blocks, varied food items, chew toys, mirrors, foraging boards, puzzle feeders, video, positive human interaction, nesting box (if appropriate for species), music, ladders/swings, ice cubes, perfume samples, sun lamps (marmosets)
Dogs	Toys, exercise outside of cage, positive human interaction, occasional food treat, resting shelf, wood shavings or shredded paper bedding
Cats	Multilevel resting surfaces, climbing posts, toys, food treats, positive human interaction, music, scratching posts, "cat condos," group exercise
Rabbits	Toys (cat bell-ball, syringe cases, plastic soft drink bottle), food treats, positive human interaction, exercise areas, music, chewing/gnawing toys (Booda Yapple®, flavored wood blocks, flavored rawhide bones), chain in cage
Guinea pigs	Toys (see rabbit examples), food treats, positive human interaction
Rats, mice, hamsters, chinchillas	PVC tubes, Nylabones®/carrot bones for chewing, Nestlets®, chew blocks (chinchillas), paper towels if housed on wire bottom cages
Chickens, pigeons	Perches, pecking objects, complex diet for seed searching, hanging toys
Frogs and turtles	Large cage, dry and wet surfaces, blocks/rocks for climbing
Pigs	Toys (e.g., balls, chains), back-scratch device, positive human interaction, chew toys

28:7 What criteria are used to identify animal exemptions from the enrichment plan? Who authorizes the exemption(s)?

Reg. The AWAR (§3.81,e,1) state that "the attending veterinarian may exempt an individual nonhuman primate from participation in the environment enhancement plan because of its health or condition, or in consideration of its well-being. The basis of the exemption must be recorded by the attending veterinarian for each exempted nonhuman primate. Unless the basis for the exemption is a permanent condition, the exemption must be reviewed at least every 30 days by the attending veterinarian." The IACUC also may exempt an individual nonhuman primate for scientific reasons (AWAR §3.81,e,2).

PHS Policy (II; IV,A,1) requires institutions to comply with the AWAR and the *Guide*. The *Guide* (page 37) also addresses circumstances where exceptions might be necessary.

Opin. Exemptions from an enrichment program should not be considered as an "all or nothing" participation. An animal may be exempted from only a portion of the program, while still enjoying the benefits of other elements of the program. The ILAR report[3] affirms that an animal's role in research involving infectious diseases, atypical rearing conditions, physical restraint, surgery, pain, substance abuse, or aggression should not be excluded *a priori* from the enrichment program. Indeed, the authors of the report note that protocols which may pose restrictive environments for animals should be re-evaluated periodically to determine if new technologies may be available which could reduce the restrictions placed on the animal.

The report recommends four criteria for assessing animal well-being in addition to general physical health.

- The animal's ability to cope effectively with environmental changes.
- The animal's ability to engage in beneficial species-typical behavior.
- The absence of maladaptive or pathological behavior.
- The presence of a balanced temperament.

When evaluating an animal for an exemption based on health or behavior, it is important to understand fully what is "normal" for an animal in that condition (e.g., an aged animal frequently has reduced perceptual and locomotor capabilities[22] and, thus, the enrichment program may need to be modified to accommodate these limitations).

Surv. What criteria are used to identify animal exemptions from the enrichment plan? Who authorizes the exemption(s)?

Three reasons were provided for exemption from the enrichment program: (1) a scientific justification based on research goals, (2) an animal health reason, or (3) a behavioral justification (specifically for social housing exemption). Exemptions are authorized by either the IACUC or the AV. Exemption from the enrichment program based on a scientific justification is typically handled on the IACUC protocol form. More than one answer per respondent is possible.

- Experimental exemption   7/8
- Health exemption   7/8
- Behavioral exemption   1/8

28:8 Who is responsible for keeping the enrichment plan current, making adjustments to the plan based on experience, and adjusting the plan?

Opin. For the enrichment program to be implemented in a coordinated and efficient manner, there must be a responsible party. If an *individual* is made responsible for the program, that person should seek input from the scientific, veterinary, and animal care staff. A *committee* comprised of these same categories of personnel also will achieve the diversity of input requisite for the program to be successful. A committee may gain the added benefit of greater involvement and vesting in the enrichment program by committee members if they assist not only in developing the enrichment program but participate in keeping it current and vital.

Surv. Does your institution have an enrichment committee? If so, who serves on it? If not, who is responsible for keeping the plan current, making adjustments to the plan based on experience, and adjusting the plan?

Most respondents' institutions did not have an enrichment committee. In general, they relied on the veterinary staff to design and implement the enrichment program. Enrichment committees had scientific, veterinary, and animal care staff representation. Some committees also had a representative from the IACUC serving as a member.

- No enrichment committee 6/8
- Have an enrichment committee 2/8

28:9 Is it necessary to maintain enrichment records? If so, what do the records reflect and how long do you keep them?

Reg. The AWAR (§3.81) require that dealers, exhibitors, and research facilities develop, document, and follow an environment enhancement plan adequate to promote the psychological well-being of nonhuman primates. They further require that records of exemptions of animals from the plan be maintained (§3.81,e,3) and be made available to USDA officials or representatives of federal funding agencies upon request.

PHS Policy (II; IV,A,1) requires institutions to comply with the AWAR and to follow the *Guide*. Institutions must keep records of departures from the *Guide* and IACUC approval of any departures (PHS Policy IV,B,3).

Opin. Documentation serves not only as a verifiable record of enrichment activities, but also validates the effectiveness of ongoing enrichments or provides evidence of the need to modify the enrichment program. It has been suggested[3] that nonhuman primate enrichment programs should include a mechanism whereby there are "protocols for diagnosing the cause of physical impairments and abnormal behavior, determining when remediation is necessary, developing remediation plans, assessing the effectiveness of remediation, and maintaining appropriate records."

Surv. Is it necessary to maintain enrichment records? If so, what do the records reflect and how long do you keep them?

All respondents indicated they retained records for the enrichment program. Different records are maintained at different institutions. Records reflecting rotation of toys through the animal colony, food treat usage, records of animal exemptions from the program, and records of observed effects of the enrichments were described by respondents to the survey. Records were commonly kept for 3 years, but retention reportedly ranged from 1 year to permanent archiving.

Records were frequently maintained in conjunction with individual animal health records.

28:10 How are resources allocated to support the environmental enrichment program?

Opin. Since the provision of an enrichment program is mandated by federal law (see 28:4) for some species, it is critical that adequate resources be identified for this institutional responsibility. An ongoing budgetary commitment to the enrichment program will allow for the program to evolve as new information becomes available and to stay dynamic. Inadequate funding can result in a program that is ineffective and does not meet regulatory requirements or recommendations of the *Guide*. The Institutional Administrator's "Manual for Laboratory Animal Care and Use"[23] states that "sustained and visible support from institutional officials is absolutely essential to establishing and maintaining a high quality animal care and use program. ... They can assure sufficient monetary and personnel resources are allocated to the institution's program." The enrichment program should not be viewed as separate or an "extra" program, but rather must be considered as an integral component of the entire animal care and use program.

Surv. How are resources allocated to support the environmental enrichment program?

Fiscal support for enrichment programs frequently is derived from per diem charges or from the laboratory animal resource department's budget. Personnel resource requirements for implementing the enrichment program are often overlaid on existing job requirements of animal care staff. Less frequently, a dedicated animal enrichment technician is employed. In academic environments, students often play an active role in implementing the program.

28:11 How are investigators notified of the necessity to consider enrichment for their animal subjects?

Opin. Keeping investigators informed about the enrichment program through the protocol form, newsletters, Web pages, training programs, and dialog fosters their involvement in the program and will likely result in improvements to the program due to their diverse scientific expertise. As the successful enrichment program depends on a strong team approach among various institutional sectors (administrators, investigators, and animal care personnel), the better informed these parties are about the program, the more effective the enrichments will be for the animals.

Surv. How are investigators notified of the necessity to consider enrichment for their animal subjects?

Most respondents to this question indicated that investigators were informed of the enrichment program and offered the opportunity to request an exemption from it via the IACUC protocol form. In one case, the survey respondent indicated that information pertaining to the enrichment program was discussed with the investigator during the veterinary consultation that occurs before submission of the protocol form. Another format for disseminating information regarding the institution's enrichment program was through formal training sessions investigators are required to attend. Enrichment program information also was provided via institutional intranets or electronic bulletin boards. In a few cases, no structured means of providing notification was described. In these cases the animal resource staff provided enrichment unless instructed otherwise.

- E-mail/electronic bulletin boards   2/8
- Veterinarian   2/8
- Training program   2/8
- Protocol form   1/8
- Not routinely informed   1/8

28:12 How are staff trained to implement the enrichment program and to identify animals requiring special consideration?

Reg. The AWAR (§2.32,b; §2.32,c) require training and instruction of personnel to include "humane methods of animal maintenance." The PHS Policy (IV,A,1,g) stipulates inclusion of a "synopsis of training or instruction in the humane practice of animal care and use ..." in the NIH/OPRR Animal Welfare Assurance document. Personnel qualifications and training also are emphasized in the *Guide* (page 13) that states that "personnel caring for animals should be appropriately trained ..., and the institution should provide for formal or on-the-job training to facilitate effective implementation of the program and humane care and use of animals." The *Guide* lists behavioral management as one area in which personnel with expertise may be required.

Opin. Regarding nonhuman primate enrichment, the ILAR report[3] notes that the success of an enrichment program is dependent on personnel having knowledge of and experience with nonhuman primate behavior. The report also states that "periodic training of staff to acquaint them with advances in the field is essential." Safety training also may be considered an essential topic in a behavioral management training program. Seminars at local, regional, or national meetings; scientific publications, autotutorial slide sets, and videos; consultants; and on-the-job training are good venues for providing staff training.

Surv. How are staff trained to implement the enrichment program and to identify animals requiring special consideration?

Given the regulatory requirements for training and strong emphasis training received in the *Guide* and other reports, it is no surprise that the eight respondents to the survey uniformly indicated that their staff received training in enrichment. In some cases, formal training sessions were held at the institution or employees were sent to local, regional, or national meetings, while in other cases staff were trained by their supervisors (e.g., research staff were trained by the PI and animal care staff were trained by the veterinarian). In one case, the environmental enrichment committee members were designated as trainers when new enrichment activities were implemented.

28:13 How is the efficacy of the enrichment program assessed? Does the IACUC make judgments on the efficacy of the program?

Reg. The AWAR (§2.31,c,1) and PHS Policy (IV,B,1) require that the IACUC review at least once every 6 months the program for humane care and use of animals. The *Guide* (page 9) states that the IACUC has oversight and evaluation responsibility for the animal care and use program and the components of the program described in the *Guide*. The *Guide* (page 9) echoes the regulatory requirement for IACUC review of the program and facilities at least once every 6 months. (See Chapter 23, Chapter 27.)

Opin. Morgan et al.[24] reviewed various methods of assessing enrichment programs. Methods include reliance on others to inform us, common sense, and empiricism. They note that a significant disadvantage to reliance on others is that acceptance is often based on the "authority" of the other person (the tendency to accept that a point of information is true simply because it comes from an authority figure). They suggest that common sense is a good starting point for evaluating enrichment, although its main disadvantage is that common sense is grounded in our own interpretations and biases. However, they strongly argue that empirical testing of enrichment strategies is necessary to validate their use both from the standpoint of the value to the animal and for judicious expenditure of limited resources.

An assessment strategy specific to the psychological well-being of nonhuman primates is available.[3] Four factors are proposed to consider when assessing psychological well-being:

1. The animal's ability to cope with day-to-day changes in its environment.
2. The animal's ability to exhibit beneficial species-typical activities.

3. The absence of maladaptive or pathological behaviors that could result in self-injury or other "undesirable consequences."
4. The presence of a balanced temperament.

These assessment variables may be sufficiently broad to be applicable to other laboratory animal species.

The IACUC has a responsibility to review the animal care and use program semiannually and to ensure its completeness, appropriateness, and currency.

Surv. How is the efficacy of the enrichment program assessed? Does the IACUC make judgments on the efficacy of the program?

Respondents most commonly reported that no mechanism, or informal mechanisms, were in place to assess the success of the enrichment program. Animal care staff were noted as providing "input" into the assessment and subjective evaluations by animal care staff were described. The IACUC was not reported as making systematic evaluations of the efficacy of the enrichment program during their semiannual program reviews. The IACUC was reported to inquire farther into the enrichment being provided specific animals if a behavior problem was noted during the semiannual facility inspections. One respondent indicated that behavioral data are collected and evaluated with reports made to the IACUC regarding the enrichment program.

- Veterinarian/animal care staff   3/8
- Not assessed   3/8
- IACUC   2/8

28:14 How is the IACUC trained so that the enrichment program may be appropriately evaluated?

Reg. The *Guide* (page 9) states that the IACUC "is responsible for oversight and evaluation of the animal care and use program and its components described in this *Guide*." The *Guide* (page 9) further states that "it is the institution's responsibility to provide suitable orientation, background materials, access to appropriate resources, and, if necessary, specific training to assist IACUC members in understanding and evaluating issues brought before the committee."

Opin. The *Guide* clearly places responsibility for all elements of the animal care and use program with the IACUC (as designated by the IO). The laboratory animal enrichment program is one component of the entire program and, thus, should be included in the semiannual program review performed by the IACUC. To accomplish this task, the

IACUC should be provided with the tools to evaluate the behavior of the animals they see during the facility inspection, to understand the rationale for the enrichment program's design, and to assess the program's efficacy.

Surv. How is the IACUC trained so that the enrichment program may be appropriately evaluated?

Answers to this question ranged from no training is provided to the IACUC to presentations specific to the subject of enrichment. In one case, the IACUC had a liaison member on the environmental enrichment committee. It was reported that the IACUC frequently relies on advice from the veterinary staff or other related expertise (e.g., a research psychologist, primatologist).

- Not trained   7/8
- Presentations   1/8

28:15 What resources are used to develop and update the enrichment plan?

Surv. Information used for developing and revising enrichment plans was derived from a blend of anecdotal and scientific resources. Electronic discussion groups such as CompMed, Primate Science (formerly Primate Talk), and Primate Enrichment Forum were cited. Published literature from *Contemporary Topics, Lab Animal, Laboratory Primate Newsletter,* as well as from textbooks were also mentioned. The Animal Welfare Information Center bibliographies were stated to be a valuable resource, as was informal conversation among colleagues. Visiting other facilities and viewing their enrichment techniques was indicated to be especially helpful. Consultants also were used. Most of the eight respondents stated that presentations at local, regional, or national meetings were a consistent source of very current information. More than one response per respondent is possible.

- Printed material   5/8
- Web information   4/8
- Presentations/consultations   3/8
- Colleagues from outside the institution   2/8
- Veterinarians   1/8

28:16 Is the enrichment plan reviewed by the IACUC semiannually as part of the program review? Are other enrichment records reviewed for this purpose also? If so, what records?

Reg. (See 28:13.)

Opin. To fully conform with AWAR and *Guide* recommendations, the IACUC should include an assessment of the enrichment plan and program in its semiannual review. The review should include an evaluation of the efficacy of the enrichment program and whether it is current with the literature.

Surv. Is the enrichment plan reviewed by the IACUC semiannually as part of the program review? Are other enrichment records reviewed for this purpose also? If so, what records?

Most respondents indicated that the enrichment plan was reviewed by the IACUC. The frequency at which this review was reported to occur varied and included once (the initial establishment of the plan), only when modified, once a year, and semiannually. Typically, no other records were reviewed; however, in one case, SOPs pertaining to the enrichment program were reviewed and, in a second case, animal health records were examined.

- Reviewed semiannually   5/8
- Not reviewed semiannually   3/8

References

1. Office of the Federal Register, 9 Code of Federal Regulations, Part 3, Animal Welfare; Standards; Final Rule, *Fed. Reg.*, 56 (32), February 15, 1991, 6369.
2. U.S. Department of Agriculture, USDA Employee Opinions on the Effectiveness of Performance-Based Standards for Animal Care Facilities, U.S. Department of Agriculture, Animal and Plant Health Inspection Service, Animal Care, Riverdale, MD, December 1996.
3. *The Psychological Well-Being of Nonhuman Primates*, National Research Council, Washington, D.C., 1998.
4. Bayne, K., Dexter, S., Mainzer, H., McCully, C., Campbell, G., and Yamada, F., The use of artificial turf as a foraging substrate for individually housed rhesus monkeys (*Macaca mulatta*), *Anim. Welf.*, 1(1), 39, 1992.
5. Shepherdson, D.J., Tracing the path of environmental enrichment in zoos, in *Second Nature: Environmental Enrichment for Captive Animals*, Shepherdson, D.J., Mellen, J.D., and Hutchins, M., Eds., Smithsonian Institution Press, Washington, D.C., 1998, 1.
6. Wolfle, T.L., Psychological well-being: The billion dollar solution, in *Through the Looking Glass: Issues of Psychological Well-Being in Captive Nonhuman Primates*, Novak, M.A. and Petto, A.J., Eds., American Psychological Association, Washington, D.C., 1991, 119.
7. Environmental Enrichment Information Resources for Laboratory Animals: 1965–1995, AWIC resource series, No. 2, Animal Welfare Information Center, National Agriculture Library, USDA, Washington, D.C., 1995.

8. *Guide to the Care and Use of Experimental Animals*, Vol. 1, 2nd ed., Olfert, E.D., Cross, B.M., and McWilliam, A.A., Eds., Canadian Council on Animal Care, Ottawa, 1993.
9. BVAAWF/FRAME/RSPCA/UFAW Joint Working Group on Refinement, Refinements in rabbit husbandry, second report of the BVAAWF/FRAME/RSPCA/UFAW Joint Working Group on Refinement, *Lab. Anim.*, 27, 301, 1993.
10. Mench, J.A., Environmental enrichment and the importance of exploratory behavior, in *Second Nature: Environmental Enrichment for Captive Animals*, Shepherdson, D.J., Mellen, J.D., and Hutchins, M., Eds., Smithsonian Institution Press, Washington, D.C., 1998, 30.
11. Environmental Enrichment Information Resources for Nonhuman Primates: 1987–1992, Animal Welfare Information Center, National Agriculture Library, USDA, Washington, D.C., 1992.
12. Reinhardt, V., Reinhardt, A., and Selig, D., *Environmental Enrichment for Nonhuman Primates: An Annotated Bibliography for Animal Care Personnel*, Animal Welfare Institute, Washington, D.C., 1998.
13. Armstrong, K.R., Clark, T.R., and Peterson, M.R., Use of corn-husk nesting material to reduce aggression in caged mice, *Contemp. Topics Lab. Anim. Sci.*, 37 (4), 64, 1998.
14. Arnold, C. and Westbrook, R.D., Enrichment in group-housed laboratory golden hamsters, *Anim. Welf. Inform. Cent. Newsl.*, 8 (3–4), 22, Winter 1997/1998.
15. Task Force Report No. 130, *The Well-Being of Agricultural Animals*, Council for Agricultural Science and Technology, Ames, IA, 1997.
16. Coviello-McLaughlin, G.M. and Starr, S.J., Rodent enrichment devices — evaluation of preference and efficacy, *Contemp. Topics Lab. Anim. Sci.*, 36 (6), 66, 1997.
17. Lidfors, L., Behavioral effects of environmental enrichment for individually caged rabbits, *Appl. Anim. Behav. Sci.*, 52, 157, 1996.
18. Turner, R.J., Held, S.D.E., Hirst, J.E., Billinghurst, G., and Wootton, R.J., An immunological assessment of group-housed rabbits, *Lab. Anim.*, 31, 362, 1997.
19. van de Weerd, H.A., Baumans, V., Koohaas, J.M., and van Zutphen, L.F.M., Nesting material as enrichment in two mouse strains, *Scand. J. Lab. Anim. Sci.*, 23 (1), 119, 1996.
20. van Loo, P.L.P., van de Weerd, H.A., and Baumans, V., Short- and long-term influence of an easy applicable enrichment device on the behavior of the laboratory mouse, *Scand. J. Lab. Anim. Sci.*, 23 (1), 113, 1996.
21. Office of Animal Care and Use, National Institutes of Health, Nonhuman Primate Intramural Management Plan, Bethesda, MD, 1991.
22. Bayne, K., Qualitative observations of idiosyncratic behavior in old monkeys, in *Behavior and Pathology of Aging in Rhesus Monkeys*, Davis, R.T. and Leathers, C.W., Eds., Alan R. Liss, New York, 1985.
23. Institutional Administrator's Manual for Laboratory Animal Care and Use, Office for Protection from Research Risks, National Institutes of Health, Bethesda, MD, NIH Publication No. 88-2959, 1988.
24. Morgan, K.N., Line, S.W., and Markowitz, H., Zoos, enrichment, and the skeptical observer: the practical value of assessment, in *Second Nature: Environmental Enrichment for Captive Animals*, Shepherdson, D.J., Mellen, J.D., and Hutchins, M., Eds., Smithsonian Institution Press, Washington, D.C., 1998, 153.

29

Animal Mistreatment and Protocol Noncompliance

Jerald Silverman

Introduction

One of the most contentious roles of the IACUC is its mandate to review allegations of animal mistreatment and protocol noncompliance. This must be done carefully and with the utmost concern for due process and confidentiality as careers may be severely hurt by unfounded accusations and unsubstantiated rumors. Nevertheless, the welfare of animals cannot be compromised. This leads to a thin line on which the IACUC walks. The intent of this chapter is to provide the reader with guidance to many of the problems that IACUCs face when confronted with allegations of animal mistreatment and protocol noncompliance. It is important to remember that allegations remain nothing more than allegations until proven otherwise.

Because of the seriousness and sensitivity of this IACUC function, it is very important that the procedures it will use to review and investigate complaints be formalized ahead of time. The time to hunt for the most appropriate operating procedure is not when a complaint reaches the IACUC. Know what you will do ahead of time, know what authority you are given under the AWAR, PHS Policy, and through your own institution. The AWAR and PHS Policy do not provide all the answers to the myriad problems the IACUC faces when complaints are brought to it. Although IACUC policies can never account for all contingencies, it is wise to have a reasonable number available.

No IACUC works in a vacuum. Rules, regulations, policies, and the like require a conscientious IACUC and strong institutional support. Without the latter, the most conscientious IACUC is severely handicapped when reviewing and investigating allegations of protocol noncompliance and animal mistreatment.

The surveys presented in this chapter emanate primarily from academic institutions from different parts of the country. A small number of private research institutions, private laboratories, and pharmaceutical companies also are represented. Possible answers were provided, as was space for possible additional answers from respondents.

29:1 What is meant by animal mistreatment and protocol noncompliance?

Opin. Animal mistreatment is physical or psychological wrongful or abusive treatment of an animal.[1] Examples include hitting animals, taunting animals, or not providing food for punitive reasons.

Protocol noncompliance indicates that procedures or policies approved by the IACUC are not being followed.[1] Examples include performing unauthorized surgery, unauthorized persons participating in a research project, or injecting drugs that the IACUC has not approved. It is not unusual to have some overlap between mistreatment and protocol noncompliance. For example, the unauthorized restraint of a nonhuman primate for an unusually long time can potentially entail both animal mistreatment and protocol noncompliance.

When faced with protocol noncompliance the IACUC's first step, if possible, should be to find a way to bring the protocol into compliance. On the other hand, when faced with animal mistreatment, the IACUC must take immediate action to stop the mistreatment.

29:2 What role does an institution's administration have in the overall concept of animal mistreatment or protocol noncompliance?

Reg. Under the PHS Policy, the IO, by signing an Animal Welfare Assurance with NIH/OPRR, commits the institution to compliance with the PHS Policy. Noncompliance (such as animal mistreatment or protocol noncompliance) is ultimately the responsibility of the IO who represents the institution's administration (PHS Policy III,G; IV,A).

Opin. The IACUC is a regulatory committee that is an agent of the institution (PHS Policy IV,B; AWAR §2.31,c). The institution must make it known that the IACUC not only helps to assure animal welfare, but it is an indispensable link in helping protect the researcher against unwarranted accusations and assuring the integrity of the institution.

Because regulatory committees are often disparaged by investigators, it must be understood that an IACUC cannot properly function without general institutional ethical and administrative support. The institution, by actions and words, must stand behind the IACUC. The IO can be a powerful ally and should have a clear understanding of his or her responsibilities and authority under the AWAR and PHS Policy, and communicate this support to the IACUC and investigators.

29:3 What is the initial responsibility of the IACUC relative to allegations of animal mistreatment or protocol noncompliance as per the AWAR and the PHS Policy?

Reg. Under AWAR (§2.31,c,4), the IACUC must "review and, if warranted, investigate concerns involving the care and use of animals at the research facility. ..." This includes complaints from the public and from laboratory or research facility personnel or employees. The PHS Policy (IV,B,4) is slightly broader and states that the IACUC must "review concerns involving the care and use of animals at the institution."

Opin. A key phrase in the AWAR is "if warranted," as not all complaints need to be fully investigated. The IACUC chairperson, the entire IACUC, or appropriate designees should make such a decision based on the nature of the complaint. For example, a complaint of "the dogs prefer Brand A dog food to Brand B" requires far less action than "Dr. X is performing unapproved splenectomies on dogs." Nevertheless, all concerns and complaints must be at least reviewed. In both the AWAR and PHS Policy, it is prudent for the IACUC to interpret "institution" and "research facility" as being any site where the IACUC has jurisdiction.

29:4 Who can bring allegations of animal mistreatment or protocol noncompliance to the IACUC?

Reg. The AWAR (§2.31,c,4) state that the general public and institutional employees may make complaints. The PHS Policy is silent on this issue; however, it does state that "all institutions are required to comply, as applicable, with the Animal Welfare Act and other Federal statutes and regulations relating to animals" (PHS Policy, II). Further, there is nothing in the PHS Policy that precludes the IACUC acting upon concerns raised from any source. In addition, it is not unusual for APHIS/AC veterinary medical officers to identify such problems to the IACUC Chair or the AV.

Surv. At your institution, who can bring allegations of animal mistreatment or protocol noncompliance to the IACUC (check all appropriate answers)?

- No policy in place   4/54
- Any employee   50/54
- The public in general   32/54
- Students   4/54

29:5 What communication pathways should be followed in order to bring allegations of animal mistreatment or protocol noncompliance to the attention of the IACUC?

Reg. The AWAR (§2.32,c,4) state that research facility personnel must receive training on how to report alleged deficiencies in animal care and treatment, but there are no specific statements in the AWAR as to how to comply with this requirement. PHS Policy is silent on this question.

Opin. Institutions have developed various procedures of their own and most are focused on providing open and easy communication. Any allegation must eventually reach the Chair of the IACUC. If the allegation is against the Chair, a written policy should state who will help adjudicate the complaint. As an example, this might be the Vice Chair of the IACUC, the AV, or the IO.

Typical communication methods include direct verbal conversations or written letters to the AV, IACUC Chairperson, IO, IACUC members, college deans, etc. This information must be efficiently passed on to the committee Chair. It is helpful to have a written IACUC-initiated policy on how to do this.

Surv. In your institution, how can people bring allegations of animal mistreatment or protocol noncompliance to the attention of the IACUC? Check all appropriate answers.

- No mechanism in place   2/54
- Through the Attending Veterinarian   45/54
- Through any veterinarian/animal facility director   38/54
- Through the IACUC Chairperson   48/54
- Through an IACUC member   46/54
- Through the Institutional Official   29/54
- Through an institutional administrator (e.g., dean)   24/54
- Through a telephone hotline   4/54
- Other   9/54

29:6 How are employees, students, researchers, and others trained and informed that they have the right to bring complaints to the IACUC about animal care and use?

Reg. The AWAR (§2.32,c,4) state that research facility personnel (i.e., all persons, not just those in the animal housing areas) must receive training on how to report alleged deficiencies in animal care and treatment. The means of providing this training is left up to the institution. PHS Policy is silent on this question.

Opin. It is suggested that training be provided for all institutional personnel even if the research facility only comes under the auspices of PHS Policy.

Surv. At your institution, how are people notified of the means to bring allegations of animal mistreatment or protocol noncompliance to the attention of the IACUC? Check all appropriate answers.

• No mechanism in place	6/54
• During a formal training session	37/54
• From handouts given to investigators	24/54
• Through a newsletter	10/54
• Through word of mouth	19/54
• Through posted signs in the animal facility	23/54
• Written Standard Operating Procedures (SOPs)	3/54
• Computer via a Local Area Network (LAN)	2/54
• Other	4/54

29:7 Must a complainant be identified to the IACUC or the IACUC Chair?

Opin. In general, no. There are no federal animal welfare laws, regulations, or policies requiring the identification of a complainant. Required identification might potentially deter certain people from bringing forth complaints. Some academic or other institutions have policies that require the identification of a complainant before action can be taken. Although this can lead to a conflict between the IACUC and the institution, in practice it need not be so. The IACUC itself can act as the complainant in order to maintain the confidentiality of the true complainant (whether known or not). It is prudent for the IACUC and the institution to develop policies for handling such procedural conflicts well before the problem arises. It is suggested that confidentiality be maintained when requested by the complainant.

Surv. At your institution, must a complainant identify himself/herself before the IACUC is willing to consider a complaint?

• No policy in place	13/54
• Must openly identify himself/herself	3/54
• Need only be identified to the IACUC chairperson	8/54
• Need only be identified to the AV or IACUC chairperson	1/54
• Need not be identified	29/54

29:8 Is there protection against repercussions if a person makes a complaint to the IACUC; that is, are "whistle-blowers" protected?

Reg. The AWAR (§2.32,c,4) provide specific protection for employees, IACUC members, or laboratory personnel against discrimination or other reprisals for reporting violations of the AWAR or the AWA itself. Although the PHS Policy makes no definitive statements about reprisals, it does state that all institutions are required to comply, as applicable, with the AWA and other federal statutes and regulations (PHS Policy II; U.S. Government Principle I).

Opin. It is advisable for institutions to develop similar guidelines against repercussions.

29:9 How common are investigations of allegations of animal mistreatment or protocol noncompliance?

Opin. In this author's experience, allegations are not common. Although it is appropriate to handle problems as expeditiously as possible, there are times when problems should be brought to the attention of the IACUC (e.g., unauthorized surgery). To do otherwise makes the IACUC somewhat of a paper tiger and defeats one of the intents of the AWAR and PHS Policy. Thus, the fact that some institutions in the survey below resolve problems without IACUC input should not be construed as being fully appropriate.

Surv. At your institution, has the IACUC ever been involved with investigating allegations of animal mistreatment or protocol noncompliance?

• No, the issue has never arisen	10/53
• No, the issue has arisen but the attending veterinarian or others have resolved the problem without its reaching the IACUC	8/53
• Yes, but very rarely	19/53
• Yes, occasionally	14/53
• Yes, fairly often	2/53

29:10 To what extent should allegations of animal mistreatment or protocol noncompliance be documented?

Reg. The PHS Policy (IV,E,1,b) requires the institution to maintain minutes of IACUC meetings, activities of the IACUC, and IACUC deliberations. To the extent that allegations are considered by the IACUC at meetings or acted upon by the IACUC, records of such actions must be maintained. The AWAR (§2.35,a) have a similar requirement.

Opin. It is unlikely that all complaints brought to the IACUC will be fully documented in writing and signed. Thus, the person who receives the complaint (typically the AV or IACUC Chair) must get as much information as possible. If the complaint is verbal, questions such as: did you see this yourself, when did this happen, where did this happen, and so on are appropriate. If possible, obtain the name of a contact person (such as the complainant) and a means of contacting that person.[2]

Surv. At your institution, how much documentation from the complainant does the IACUC require before starting an investigation?

- Not applicable, has never happened   7/52
- No policy in place   7/52
- We will act on any hearsay or other skimpy documentation   17/52
- We will act on hearsay or other skimpy documentation only if it suggests a significant problem   15/52
- We require sufficient documentation to support an allegation   6/52
- We require heavy documentation   0/52

29:11 When should the IACUC take investigative action on allegations of animal mistreatment or protocol noncompliance?

Reg. The IACUC is responsible for determining whether an activity is in accordance with the AWA, the *Guide*, the institution's Assurance, the PHS Policy (PHS Policy IV,C,6), or the AWAR (AWAR §2.31,d,6). Thus, the IACUC must consider all allegations of noncompliance and determine if there is sufficient reason to investigate further.

Opin. Although the IACUC must review all allegations, it need not investigate all of them (see 29:3). There are at least four circumstances when IACUCs should consider becoming actively involved.[2]

- When an allegation should be but is not satisfactorily resolved at the local level.
- When any reasonable person would consider that gross mistreatment or noncompliance has occurred. This is based on the allegation itself, although it may subsequently be found to be untrue.
- When there are repeated minor instances of noncompliance or mistreatment from the same person or research group. These complaints often come from the AV.
- When the IACUC cannot clearly decide if an allegation is worth investigating, it is probably better to investigate.

The NIH/OPRR provides examples of the above.[3]

29:12 What types of documentation might an IACUC need to fully investigate a complaint?

Opin. This varies with the specific situation. Assuming the IACUC as a whole or the designated person (e.g., the Chair) determines that further investigation is warranted, the investigating persons can determine what information they must collect to properly investigate an allegation. It may be necessary for the IACUC to interview people, examine animals, and obtain surgical records, housing records, and even an investigator's research records (in the last example, confidentiality must be assured). Any records obtained should be directly related to the allegation made. Because the IACUC does not have any authority to subpoena people or records, generalized institutional support is crucial to the investigative process. (See 29:10.)

29:13 Should the IACUC let all concerned persons know the basis of a complaint and the procedures to be followed? Who might be "concerned persons?"

Opin. Assuming the IACUC has determined that a complaint should be fully investigated, it is prudent to inform all involved persons about the nature of the complaint and procedures to be followed. This will help assure due process (the protection of the legal and ethical rights of all concerned individuals).[2] Concerned persons include the alleged violator and that person's immediate superior if the alleged violator is not the PI. The IO is often informed of the problem at or before this time. Clearly, individual circumstances will dictate final decisions on whom to notify.

Surv. At your institution, when do you notify an investigator of an allegation of animal mistreatment or protocol noncompliance against him or his staff?

• No policy in place	8/52
• Notify only when an initial inquiry suggests reason to proceed; we do not provide details at this time	13/52
• Notify only when an initial inquiry suggests reason to proceed; full details provided at this time	14/52
• Notify immediately, irrespective of the quality of the allegation; no details provided at this time	2/52
• Notify immediately, irrespective of the quality of the allegation; full details provided at this time	14/52
• Depends on the situation	1/52

29:14 Once a complaint has reached the IACUC, what is the responsibility of the IACUC Chair?

Opin. The Chair oversees the efforts described in 29:3.

29:15 Is it necessary to fully investigate all complaints brought to the IACUC, even those that seem frivolous?

Reg. (See 29:11.)
Opin. No. (See 29:3.)

29:16 Should the IACUC approach each complaint on an individual basis or is a standard operating procedure for handling complaints more appropriate?

Reg. Although each complaint is likely to have its own unique needs, the AWAR and PHS Policy require the IACUC to review all complaints (see 29:3, 29:11).

Opin. It is suggested that the IACUC have an SOP in place to handle such complaints to help avoid careless errors in due process or confidentiality. It is important that the persons formulating the policy for handling complaints and investigations be representative of the institution's research community as a whole in order to achieve broad support for its recommendations.

29:17 Can the IACUC Chairperson appoint an investigative subcommittee?

Opin. Yes. There is nothing in the AWAR or PHS Policy which details how an IACUC should proceed with an investigation if the IACUC determines one is needed. The choice of procedures is entirely at the discretion of the IACUC. It is strongly recommended that an IACUC policy on this topic be established well before a potential problem arises.

Surv. Once a complaint has reached the IACUC, what procedures does the IACUC Chairperson follow? Check all appropriate answers.

• Not applicable, has never happened	12/53
• No policy in place	5/53
• Chairperson does his/her own preliminary investigation	18/53
• Chairperson establishes a subcommittee to investigate	21/53
• Chairperson informs Institutional Official (IO) who then establishes a subcommittee	4/53

- Chairperson assigns Attending Veterinarian (AV) to investigate 13/53
- Varies with the problem 3/53
- Entire IACUC investigates 1/53
- IACUC administrative staff investigates 1/53
- Other 1/53

29:18 What might be the composition of an appropriate investigative subcommittee?

Opin. There are no federal regulations or policies defining the composition on an investigative subcommittee of an IACUC. Investigative subcommittees can be composed of one or more persons, including non-IACUC members. Some IACUCs assign one person (often a veterinarian) to do all (or at least preliminary) investigations, while others establish a larger subcommittee (see 29:17). If the IACUC is small, the entire IACUC may act to investigate an allegation. Some IACUCs attempt to assure that a scientist be on an investigative subcommittee if the allegation is against a scientist. Most investigative subcommittees include a veterinarian. It is suggested that with a large IACUC there be broad enough representation on a subcommittee to help assure due process and acceptance of its recommendations.

Surv. If an investigative subcommittee (one or more persons) is established, what is a typical composition of this subcommittee? Check all appropriate answers?

- Not applicable at our institution 13/53
- We have a single investigative person who initially handles all initial allegation inquiries 6/53
- Our subcommittee always includes a veterinarian and other persons 26/53
- Our subcommittee always includes a representative from accused's department 4/53
- Our subcommittee tries to have very broad representation 10/53
- Our subcommittee always has a scientist 12/53
- Our subcommittee always has a nonscientist 7/53
- Our subcommittee can have persons not on the IACUC 11/53
- Rarely includes a veterinarian 1/53
- Always includes nonaffiliated member 1/53
- Three people only 1/53

29:19 Can the IACUC use non-IACUC members as part of an investigative subcommittee?

Opin. Yes (see 29:18). There is no federal regulation or policy prohibiting this. However, since the IACUC is charged with reviewing all claims of animal mistreatment or protocol noncompliance (see 29:11), it is suggested that only the IACUC appoint such member(s) and that they report back to the IACUC directly or through the IACUC subcommittee of which they are a part. The IACUC retains the responsibility for investigative subcommittees under its auspices and any subsequent IACUC actions or determinations.

29:20 Should the IACUC keep a complainant informed of the progress of an IACUC investigation?

Opin. Assuming the complainant is known, this becomes an IACUC and institutional choice. It is this author's opinion that limited feedback should be provided to a complainant during the course of an investigation. That feedback should be little more than updates (if requested) on where the IACUC is in the investigation and a possible date for the conclusion of the investigation. Full feedback, in this author's opinion, can potentially lead to unwarranted pressure on the investigative committee, and intentional or unintentional leaks of confidential information.

Surv. At your institution, is a complainant kept informed of the progress of an ongoing IACUC investigation?

• Not applicable because it has never happened	10/55
• No policy in place	10/55
• No feedback is provided	1/55
• We provide limited feedback	16/55
• We provide full feedback	16/55
• Depends on the situation	1/55
• Anonymous complainant can get limited feedback	1/55

29:21 Should the IACUC keep a complainant informed of the results of an IACUC investigation?

Opin. Assuming the complainant is known, this becomes an IACUC and institutional choice. In this author's opinion, providing limited results to a complainant once an investigation is completed is appropriate. It reinforces the credibility of the IACUC to both the institution and the public. There is a down side to full disclosure; the complainant may not be satisfied with the IACUC's determination,

Surv. At your institution, is a complainant informed of the results of an IACUC investigation?

• Not applicable, has never happened	9/53
• No policy in place	11/53
• We provide limited results to complainant	13/53
• We provide full results to complainant	9/53
• Depends on the situation	1/53
• No results are given to complainant	10/53

and he or she may initiate actions against the institution or an individual. Likewise, the alleged violator may do the same and also claim defamation of character. Institutional policy must be established well in advance.

29:22 How might the AV report activities not included in an approved IACUC protocol?

Opin. The nature of the activity often dictates actions. An extra toy placed in a cage by a researcher's technician might be welcomed rather than vilified and likely will lead to no report. If a protocol does not include walking a dog up and down a hallway in the animal facility, but it is being done for the benefit of the animal, the AV might suggest to the PI that a protocol addendum (amendment) be made. For significant activities not included in an IACUC protocol (e.g., a major surgical procedure) the most direct route is personal contact with the IACUC Chair followed by a written description of the alleged violation. The AV also may contact the IO or request another IACUC member to contact the Chair. If the AV is the IACUC Chair, he or she should call (if necessary) an emergency meeting of the committee. It is not necessary for the AV to be identified as the source of the complaint. (See 29:5.)

In order to rapidly correct the problem, immediate communication between the AV and PI is strongly advised, if at all possible.

29:23 What can the IACUC do to ensure confidentiality of research or other records during the course of an investigation?

Opin. Various procedures are in use. Some IACUCs keep no records of investigative proceedings, some keep limited records, and almost all instruct people involved with the proceedings to maintain confidentiality. It must be remembered that allegations remain allegations until proven otherwise, and unintentional leaks of information can adversely affect people's lives. It is suggested that records be kept as

Surv. they may be crucial for further reference, particularly in APHIS/AC or NIH/OPRR needs to conduct their own investigation. (See 29:10.) How does your IACUC ensure the confidentiality of research or other records during the course of an investigation? Check all appropriate answers.

- Not applicable at our institution   6/53
- No policy in place   15/53
- No recordings or other records are kept by the IACUC   3/53
- All records and notes are collected after a meeting   7/53
- Only a limited number of people have access to records   25/53
- Instructions are given not to discuss the proceedings   23/53

29:24 Is legal or other representation on behalf of the accused person or the institution allowed during the course of an IACUC investigation?

Opin. PHS Policy and the AWAR are silent on this question. Most IACUCs have not addressed this issue and there is no AWAR or PHS Policy guidance on the same. One suggestion is to allow legal or other representation as advisory only to the accused or the IACUC, not as an active participant in an investigation or hearing.

Surv. Is legal or other representation on behalf of the accused person or on behalf of your institution allowed during the course of an IACUC investigation?

- No policy in place   44/52
- We allow legal representation for the accused only   0/52
- We allow legal representation for the institution only   0/52
- We allow legal representation for all parties   7/52
- We do not allow any legal representation   1/52

29:25 What is meant by sanctions?

Reg. There is no definition of the word "sanction" in the AWAR. The word "sanction" does not appear in the PHS Policy. The PHS Policy (IV,C,6; IV,C,7) addresses "suspension of an activity" by the IACUC. NIH/OPRR has defined a suspension as any IACUC intervention that results in the temporary or permanent interruption of an animal activity.[3] If the IACUC places an ongoing project "on hold" or requires a "temporary cessation," these actions are synonymous with

Opin. suspension. Furthermore, PHS Policy, AWAR, and the *Guide* presume that all ongoing animal activities have received prospective review and approval. Accordingly, the IACUC's authority to suspend unauthorized activities is always implied, if not explicit.[4]

Opin. For the purposes of the questions that follow, sanctions refer to any penalty or coercive action taken by the IACUC to help ensure compliance with the AWAR or PHS Policy. Sanctions are only one of the many tools the IACUC can use to help assure the welfare of laboratory animals and they rarely should have to be the first used. The IACUC's suspension of a previously approved activity can be considered a type of sanction. In the discussions below, we often segregate out suspensions from other forms of sanctions.

The IACUC always has the authority to make recommendations to the IO regarding any aspect of the institution's animal care and use program (PHS Policy IV,B,5; AWAR §2.31,d,5). If the IACUC believes that the institution should impose institutional sanctions, e.g., a reprimand for repeated violations or revocation of a PI's privilege to conduct research with animals, the IACUC should make those recommendations to the IO.

29:26 What sanctions may the IACUC impose if allegations of animal mistreatment or protocol noncompliance are verified?

Reg. AWAR (§2.31,c,8) and PHS Policy (IV,B,8) only allow for suspension of an activity previously approved by the IACUC. Thus, any other sanctions imposed by the IACUC, the IO, or the institution are imposed through institutional policy. (See 29:25.)

Surv. Please provide examples of sanctions you have used, irrespective of the verified allegation:

- Not applicable because we have never used sanctions   19/54
- Examples:
    - Suspension of all of the investigator's protocols until problem(s) is corrected   12/35
    - Suspension of specific protocol during retraining 8/35
    - Denied access to animal facility   4/35
    - Permanent suspension of animal use privileges   4/35
    - Technical ability had to be approved by veterinarian 4/35
    - Written reprimand   4/35
    - Written warning   4/35
    - Other research staff must take over work   2/35
    - Must attend IACUC meetings   2/35

– Charge investigator for extra work done by veterinary or animal care staff	2/35
– Significant salary reduction for 6 months	1/35
– Large fine	1/35
– Dismissal of personnel	1/35
– Loss of research funds	1/35
– Must add research staff	1/35
– Other	3/35

29:27 If an allegation of animal mistreatment or protocol noncompliance is verified, must the IACUC apply sanctions?

Reg. PHS Policy (IV,C,6) states that the IACUC may suspend an activity if it determines that the activity is not being conducted in accordance with appliable provisions of the AWA, the *Guide*, the institution's Assurance, or IV,C,1,a–IV,C,1,g of the PHS Policy. The AWAR (§2.31,d,7) and the PHS Policy (IV,B,8) state that the IO, in consultation with the IACUC, must take appropriate corrective action in the event of a suspension of a previously approved activity.

Opin. If, in the opinion of the IACUC, sanctions are not appropriate, they need not be applied. A clearly minor and unintentional misinterpretation of an IACUC policy which has created no problem for an animal is an example of where a verified allegation of protocol noncompliance might lead to an explanation, not a sanction. For sanctions other than a suspension of a previously approved activity, institutional and IACUC policy should be established and followed. It is strongly suggested that any authority the IACUC, IO, or any other institutional representative has to impose sanctions (other than suspensions) be unambiguous.

Surv. If an allegation of animal mistreatment or protocol noncompliance is verified, does your IACUC apply sanctions?

• Not applicable because problem has never occurred	16/53
• We have never applied sanctions	5/53
• We have occasionally applied sanctions of some sort	17/53
• We always apply sanctions of some sort	13/53
• Other	2/53

29:28 Does the IACUC, the IO, or both decide on the sanctions to be taken?

Reg. The AWAR (§2.31,d,7) and the PHS Policy (IV,C,7) state that the IO, in consultation with the IACUC, must take appropriate corrective action after a suspension of a previously approved activity. (See 29:25.)

Opin. The wording of the AWAR and PHS Policy suggests that the IO has somewhat more say than the IACUC in the development of corrective actions after suspensions. In practice, most IACUCs decide what sanctions are to be taken and only the IACUC can suspend a previously approved activity (see 29:26). A smaller number of IACUCs recommend a sanction to the IO, who then imposes the same.

Surv. At your institution, who decides what sanctions are to be taken?

- Not applicable because problem has never arisen 10/54
- No policy in place 6/54
- The investigative subcommittee of the IACUC only 0/54
- The investigative subcommittee with full IACUC approval 4/54
- The full IACUC 23/54
- The IACUC Chairperson with IACUC approval 0/54
- The Institutional Official (IO) only 0/54
- The Institutional Official (IO) and the IACUC 9/54
- Other 2/54

29:29 If sanctions are imposed by the IACUC, do they occur via a formal full committee vote or can the IACUC Chair take action on behalf of the committee?

Reg. A formal full committee vote is required for the suspension of a previously approved activity (AWAR §2.31,d,6; PHS Policy IV,C,6).

Opin. Neither the AWAR nor PHS Policy address the question of sanctions (as defined in 29:25) other than suspension of a previously approved activity. The method for approval of the imposition of sanctions (other than a suspension) to help correct a problem should become a policy of the IACUC in consultation with the IO. It is suggested that any sanctions taken require an affirmative vote by a majority of the quorum present.

29:30 Can an appeal be made to the IACUC or other institutional authority if sanctions are imposed?

Reg. The AWAR and PHS Policy are silent on this issue. Nevertheless, officials of an institution cannot approve an activity that has not been approved by the IACUC (AWAR §2.31,d,8; PHS Policy IV,C,8).

Opin. An institution, as part of its own policy, may allow appeals to be made to the IACUC, the IO, or other persons. This may be appropriate as the IACUC is not infallible or (for example) may not have followed the correct procedures for suspending a protocol. Another pair of eyes or the presentation of a different perspective on a problem might suggest alternative solutions to the IACUC. If appeals are permitted through institutional policy, then that same policy can permit the IO or other persons to override a sanction (other than a suspension of a previously approved activity). Nevertheless, this is a dangerous policy, and not one that can be recommended.

For the IACUC to run smoothly and accomplish its task, it is strongly recommended that all policies concerning sanctions be clearly understood by the IACUC, IO, and all animal users.

Surv. Does your IACUC allow appeals of any sanctions imposed? (Some respondents checked more than one category below.)

- Not applicable because the problem has never arisen   21/54
- No policy in place   14/54
- Appeals allowed to IACUC only   8/54
- Appeals allowed to IO only   7/54
- Appeals allowed to IACUC and IO   4/54
- Appeals allowed to IO and higher administrators   1/54
- No appeals allowed   3/54

29:31 Can an appeal of sanctions imposed by the IACUC be made to an APHIS/AC Administrator or to the NIH/OPRR?

Opin. No. There are no provisions for such appeals. The PHS Policy (V,A,5) allows the NIH/OPRR to waive provisions of its policy, but that refers to general programmatic decisions, not decisions on a given sanction.

29:32 Can an institution or IO override the IACUC and impose lesser sanctions on a person?

Reg. An IO may not override an IACUC suspension (IACUC withdrawal of approval). In the event of an IACUC suspension, the IO is required to take appropriate corrective action and report that action with a full explanation to NIH/OPRR (PHS Policy IV,C,7) or APHIS/AC (AWAR §2.31,d,7).

Opin. Although the IO may not override an IACUC suspension, other sanctions do potentially open the door to overriding. As noted in 20:30, this is not a recommended procedure for the IO. (See 9:54.)

29:33 Can an institution or IO impose sanctions which are more severe than those imposed by the IACUC?

Reg. The PHS Policy and the AWAR do not preclude any institutional or IO sanctions. Therefore, these may be even more severe than those imposed by the IACUC.

Opin. Yes. There is nothing to prohibit the institution or its authorized representative from imposing a sanction more severe than that recommended by the IACUC. The AWAR (§2.31,d,7; §2.31,d,8) imply that the research institution may do this as does PHS Policy (IV,C,7; IV,C,8).

29:34 Can an IACUC impose sanctions if the vertebrate animals used are not currently covered by the AWAR (e.g., common laboratory rats) and funding for the study comes from nonfederal sources?

Opin. The AWAR (§1.1, Animal) and PHS Policy (III,A) clearly state which nonhuman animals are covered by their respective agencies. Most IACUCs, as an institutional policy, oversee the care and use of all vertebrate animals in biomedical research, teaching, and product testing, whether or not those animals fall under the auspices of pertinent federal regulations or policies. Indeed, many institutional Animal Welfare Assurances to the NIH/OPRR are worded in a manner that includes all vertebrate animals, not just those studies funded by the PHS. If, as part of an institutional policy, your IACUC is empowered to apply sanctions (in addition to the suspension of a previously approved activity), then in many instances those sanctions can be applied to all vertebrate animals.

29:35 Can the IACUC Chair apply sanctions (other than suspension of a previously approved protocol) without full IACUC approval if the IO is in agreement with the sanctions?

Opin. The only sanction the IACUC is empowered to employ under the AWAR and PHS Policy is the suspension of a previously approved activity. Thus, the application of other sanctions is a matter of institutional policy. (See 29:30.) It is this author's opinion that the IACUC Chair should not unilaterally act on actions (other than suspension, which requires approval by a quorum of the IACUC) without the approval of the IACUC.

29:36 Can the AV suspend an ongoing, IACUC-approved research project or related activity?

Reg. There is nothing in the AWAR or PHS Policy that specifically authorizes the AV to suspend a research project. However, the AWAR (§1.1

Attending Veterinarian; §2.33,a,2) state the AV must have direct or delegated authority for activities of animal care and use, and must be able to provide adequate veterinary care. If this means stopping an activity to provide adequate veterinary care, then the AV has that authority.

The PHS Policy has a similar statement (IV,A,3,b,1). The IACUC clearly has authority to suspend a research protocol (AWAR §2.31,d,6; PHS Policy IV,B,8).

Opin. These regulations have been interpreted by some IACUCs as giving the AV authority to suspend an ongoing project, subject to IACUC review. Such an in-house policy can become part of an Adequate Veterinary Care statement for APHIS/AC (for consulting AVs) or the PHS Assurance statement. It should be disseminated to investigators. The IACUC should be immediately notified if a suspension occurs. In this author's opinion, the AV should have the authority to suspend an ongoing study, subject to rapid review by the IACUC.

Surv. In your institution, does the Attending Veterinarian (AV) have the authority to suspend a research or teaching project?

- No policy   2/52
- No   12/52
- Yes, and IACUC will rubber stamp the veterinarian   6/52
- Yes, but the IACUC will quickly meet and review   31/52
- Yes, and review at next IACUC meeting   1/52

29:37 Can a veterinarian other than the AV suspend a previously approved research or teaching project?

Opin. There is nothing in the AWAR or PHS Policy that specifically authorizes the AV or any other veterinarian to suspend a previously approved activity. Any such authorization is a matter of institutional policy. (See 29:36.) In this author's opinion, the veterinarian with primary responsibility for laboratory animal science issues should have the authority to suspend an ongoing project, subject to subsequent rapid review by the IACUC.

Surv. In your institution, does any other veterinarian have the authority to suspend a research or teaching project? (Some respondents provided more than one answer.)

- No policy   2/52
- No   29/52
- Yes, and IACUC will rubber stamp the veterinarian   6/52
- Yes, but the IACUC will quickly meet and review   31/52
- Yes, and IACUC will review at next monthly meeting   1/52

29:38 Can the IACUC Chair suspend an ongoing IACUC-approved research project prior to a meeting of the full IACUC? (Also see 29:30, 29:35.)

Opin. There is nothing in the AWAR or PHS Policy that gives such authority to the IACUC Chair. Nevertheless, an institution may delegate such authority to the IACUC Chair pending review by the full committee. Such an in-house policy should become part of the PHS Assurance statement and should be disseminated to investigators. It also should become part of an Adequate Veterinary Care statement for the APHIS/AC (if a consulting veterinarian is used). In the opinion of this author, the IACUC Chair should have the authority to suspend an ongoing study, subject to subsequent rapid review by the IACUC.

Surv. In your institution, does the IACUC Chairperson have the authority to suspend a research or teaching project?

- No 14/50
- Yes, and IACUC will rubber stamp the chairperson 3/50
- Yes, but the IACUC will quickly meet and review 33/50

29:39 Can the IACUC suspend a previously approved research project?

Reg. Yes. The AWAR (§2.31,d,6) and PHS Policy (IV,B,8; IV,C,6) specifically allow the IACUC to suspend a previously approved activity after review of the matter at a convened quorum of the IACUC, and with a suspension vote of a majority of the quorum present.

29:40 Must the IACUC suspend an entire previously approved activity or can it suspend parts of a previously approved activity?

Opin. The AWAR and PHS Policy are silent as to whether the IACUC has the authority to suspend only portions of a previously approved activity. It, therefore, may be interpreted to allow the IACUC the authority to suspend either a full activity or a portion of such an activity. As an example, an IACUC can determine that only a surgical component of an ongoing research protocol need be suspended.

29:41 How long should a suspension of a previously approved activity last?

Opin. That is at the discretion of the IACUC. It is unusual for a protocol to be permanently suspended. Most protocols are suspended for the length of time which the IACUC feels will prevent recurrence of the problem and which allows any remedial actions to occur. Any suspension must include the enactment of corrective actions to help prevent future problems (AWAR §2.31,d,7; PHS Policy IV,C,8).

29:42 If a protocol is suspended, how is it reactivated?

Opin. The IACUC should lift a suspension only after it has determined that the activitiy is, or will be, in full complaince with the PHS Policy, the NIH/OPRR Animal Welfare Assurance, AWAR, and the *Guide*. The IACUC should establish a policy to provide for reactivation. If the suspension is for a finite period of time, the committee might vote to have it automatically lifted when that time period ends. If the suspension is open ended until a particular action occurs, the IACUC can vote (at the time of the suspension) for the IACUC Chair to be empowered to reactivate the protocol at his discretion, or it can vote that the committee itself must approve reactivation.

29:43 If sanctions are imposed, which persons or agencies must be informed and who does the informing?

Reg. PHS Policy (IV,C,7; IV,F,3) and the AWAR (§2.31,d,7) refer only to the suspension of previously approved activities. Under that circumstance, the IO and the IACUC must promptly report the suspension along with a full explanation to NIH/OPRR. PHS Policy (IV,F,3) also requires the IACUC, through the IO, to report to NIH/OPRR any serious or continuing noncompliance with the PHS Policy or serious deviations from the provisions of the *Guide*. If the species used comes under the auspices of the AWAR, the suspension and a full explanation must be sent to APHIS/AC and any federal agency funding the suspended activity.

Opin. Sanctions taken by the IACUC (other than a suspension of a previously approved activity) need not be reported to federal agencies. However, some institutions routinely report any sanctions, even minor ones, to the IO and in certain cases other officials of the institution.

29:44 What types of problems should be reported to the NIH/OPRR and APHIS/AC?

Reg. PHS Policy (IV,F,3) requires the IACUC, through the IO, to report to NIH/OPRR any serious or continuing noncompliance with the PHS Policy or serious deviations from the provisions of the *Guide*. The requirement to report suspensions of previously approved activities (PHS Policy IV,C,7; IV,F,3; AWAR §2.31,d,7) has been previously noted (see 29:43). NIH/OPRR offers examples of actions that require reporting. They include serious or continuing noncompliance with the PHS Policy and serious deviations from the provisions of the *Guide,* such as shortcomings in programs of veterinary care, occupational health, or training.[3]

The AWAR (§2.31,c,3) require reporting of any uncorrected significant deficiency (found on semiannual review) to be reported to APHIS/AC and any federal funding agency.

29:45 If only parts of a previously approved activity are suspended by the IACUC (e.g., the surgical portion of a large project), must the IACUC notify the NIH/OPRR or APHIS/AC?

Reg. PHS Policy (IV,F,3) requires the IACUC, through the IO, to promptly report any serious or continuing noncompliance with PHS Policy or a serious deviation from the provisions of the *Guide* or any suspension of an activity by the IACUC, along with a full explanation of the circumstances and actions taken by the IACUC.

The AWAR (§2.31,d,6) allow the IACUC to suspend an activity if it is not being conducted in accordance with the description that was provided by the principal investigator and approved by the IACUC. The AWAR (§2.31,d,7) require a full explanation as part of the reporting.

Opin. Many suspensions of a part of an ongoing approved activity would meet one or more of the above concerns and, therefore, would be reportable even if the entire protocol was not suspended. In some instances (e.g., excessive mortality or morbidity), the IACUC might suspend part of an ongoing activity until more information can be gathered. In this author's opinion, and under this circumstance, reporting the suspension would not be required. *Nevertheless, the reader is cautioned that this opinion is inconsistent with the PHS Policy.* (See 29:51.)

29:46 Must the IACUC inform a federal funding agency for any sanction that is imposed or only if certain classes of sanctions are imposed?

Reg. (See 29:43, 29:45.)

29:47 Must a federal funding agency be informed if allegations are verified, but sanctions are not imposed?

Reg. No. A federal funding agency must be informed only if a suspension of a previously approved activity occurs and the species involved fall under the auspices of the AWAR (§2.31,d,7). The NIH/OPRR must be informed of certain circumstances that may not necessarily evoke sanctions. (See 29:43 to 29:45.)

Opin. This requirement should not be confused with the need to notify appropriate federal funding agencies and the NIH/OPRR if major deficiencies were noted on an inspection of animal care and use areas which were not corrected in the appropriate time period (AWAR §2.31,c,3; PHS Policy IV,F,3). (See 29:44.)

29:48 Must nonfederal funding agencies be informed of full or partial suspensions of a previously approved activity?

Opin. Nonfederal funding agencies may establish whatever criteria are appropriate to their needs. These criteria may be binding on your institution. The IACUC may be requested by the nonfederal agency to provide information to it about suspensions or other sanctions.

29:49 If an IACUC suspension of a previously approved activity occurs, what information should be provided to federal funding agencies?

Reg. (See 29:43, 29:45.)
Opin. It is suggested that a clear statement of the problem be provided along with the planned (or completed) corrective action. Details of the suspension (e.g., length of time) should be included.

29:50 My institution has an IACUC but it does not house or use animals that fall under the auspices of the AWAR or PHS Policy. Must we inform APHIS/AC and NIH/OPRR of any suspensions of a previously approved activity?

Reg. No. If no activity involving PHS-supported animals is conducted at the institution, the institution does not have reporting obligations to NIH/OPRR. Likewise, if animal use does not include species covered by the AWAR, there is no reporting obligation to APHIS/AC.
Opin. Under the circumstances described above, the applicable federal regulations and policies do not pertain to your institution. The words "Institutional Animal Care and Use Committee" do not guarantee that a facility has indirect federal oversight of its animal care and use activities. Nevertheless, your NIH/OPRR Assurance may include vertebrate animals in studies that are not PHS funded. Under those circumstances, NIH/OPRR must be notified. (See 29:34.)

29:51 Should NIH/OPRR or the APHIS/AC be informed that a project has been suspended pending the completion of an IACUC investigation?

Reg. (See 29:43, 29:45.)
Opin. There is no AWAR or PHS Policy that requires APHIS/AC or NIH/OPRR notification about an activity that is under investigation. However, a gray area arises if the IACUC has properly discussed and voted to suspend an activity until an investigation is completed.

Technically, that suspension must be reported to APHIS/AC or NIH/OPRR, but because full details are not yet known or verified, the IACUC cannot properly comply with the AWAR (§2.31,d,7) or PHS Policy (IV,C,7) on reporting the suspension along with a full explanation of the problem and corrective actions. It, therefore, is suggested that an institutional policy be formulated to permit the IACUC to suspend an activity pending the results of a full investigation. The NIH/OPRR and APHIS/AC have suggested that they be informally advised of any suspension that occurs under these unique circumstances, and formally notified under the above noted regulation and policy if the IACUC verifies the complaint leading to the suspension.[5]

References

1. U.S. Department of Health and Human Services, Public Health Service, National Institutes of Health, Institutional Animal Care and Use Committee Guidebook, NIH Publ. 92-3415, 1992, Chap. D-1.
2. Silverman, J., IACUC handling of mistreatment or noncompliance, *Lab. Anim.*, 23 (8), 30, 1994.
3. OPRR Reports, 94-02, January 12, 1994. Available on the World Wide Web at: *http://www.nih.gov:80/grants/oprr/dc94-2.htm*
4. Garnett, N. and DeHaven, W.R., The view from USDA and OPRR, *Lab. Anim.*, 27 (9), 17, 1998.
5. DeHaven, W.R. and Garnett, N., personal communication, 1998.

Appendix A

Animal Welfare Act

Public Law 89-544 Act of August 24, 1966. 7 U.S.C. §2131 et. seq. (selected portions and sections of the Act are presented here.)

To authorize the Secretary of Agriculture to regulate the transportation, sale, and handling of dogs, cats, and certain other animals intended to be used for purposes of research or experimentation, and for other purposes.

Be it enacted by the Senate and House of Representatives of the United States of America in Congress assembled. That, in order to protect the owners of dogs and cats from theft of such pets, to prevent the sale or use of dogs and cats which have been stolen, and to ensure that certain animals intended for use in research facilities are provided humane care and treatment, it is essential to regulate the transportation, purchase, sale, housing, care, handling, and treatment of such animals by persons or organizations engaged in using them for research or experimental purposes in transporting, buying, or selling them for such use.

SEC. 13. The Secretary shall establish and promulgate standards to govern the humane handling, care, treatment, and transportation of animals by dealers and research facilities. Such standards shall include minimum requirements with respect to the housing, feeding, watering, sanitation, ventilation, shelter from extremes of weather and temperature, separation by species, and adequate veterinary care. The foregoing shall not be construed as authorizing the Secretary to prescribe standards for the handling, care, or treatment of animals during actual research or experimentation by research facility as determined by such research facility.

SEC. 16. The Secretary shall make such investigations or inspections as he deems necessary to determine whether any dealer or research facility has violated or is violating any provision of this Act or any regulation issued thereunder. The Secretary shall promulgate such rules and regulations as he deems necessary to permit inspectors to confiscate or destroy in a humane manner any animals found to be suffering as a result of a failure to comply with any provision of the Act or any regulation issued thereunder if (1) such animals are held by a dealer, or (2) such animals are held by a research facility and are no longer required by such research facility to carry out the research, test, or experiment for which such animals have been utilized.

SEC. 18. Nothing in this Act shall be construed as authorizing the Secretary to promulgate rules, regulations, or orders for the handling, care, treatment, or inspection of animals during actual research or experimentation by a research facility as determined by such research facility.

SEC. 20. (a) If the Secretary has reason to believe that any research facility has violated or is violating any provision of this Act or any of the rules or regulations promulgated by the Secretary hereunder and if, after notice and opportunity for hearing, he finds a violation, he may make an order that such research facility shall cease and desist from continuing such violation. Such a cease and desist order shall become effective 15 days after issuance of the order. Any research facility which knowingly fails to obey a cease and desist order made by the Secretary under this section shall be subject to a civil penalty of $500 for each offense, and each day during which such failure continues shall be deemed a separate offense.

SEC. 21. The Secretary is authorized to promulgate such rules, regulations, and orders as he may deem necessary in order to effectuate the proposes of this Act.

Appendix B

Animal Welfare Act Regulations

Introduction

9 CFR, Chapter 1, Subchapter A. Selected portions of the AWAR are presented here. For ease of reading, the appropriate Section from the AWAR is indicated; however, information within such Sections may be in abridged form. The original document should be consulted for the full text.

§1.1 Definitions

Animal means any live or dead dog, cat, nonhuman primate, guinea pig, hamster, rabbit, or any other warm-blooded animal, which is being used or is intended for use for research, teaching, testing, experimentation, or exhibition purposes or as a pet. This term excludes: birds, rats of the genus *Rattus* and mice of the genus *Mus* bred for use in research, and horses not used for research purposes and other farm animals, such as, but not limited to livestock or poultry, used or intended for use as food or fiber, or livestock or poultry used or intended for use for improving animal nutrition, breeding, management, or production efficiency, or for improving the quality of food or fiber. With respect to a dog, the term means all dogs, including those used for hunting, security, or breeding purposes.

Euthanasia means the humane destruction of an animal accomplished by a method that produces rapid unconsciousness and subsequent death without evidence of pain or distress, or a method that utilizes anesthesia produced by an agent that causes painless loss of consciousness and subsequent death.

Farm animal means any domestic species of cattle, sheep, swine, goats, llamas, or horses which are normally and have historically been kept and raised on farms in the U.S., and used or intended for use as food or fiber; for improving animal nutrition, breeding, management, or production efficiency; or for improving the quality of food or fiber. This term also includes animals such as rabbits, mink, and chinchilla, when they are used solely for purposes of meat or fur, and animals such as horses and llamas when used solely as work and pack animals.

Field study means any study conducted on free-living wild animals in their natural habitat, which does not involve an invasive procedure and which does not harm or materially alter the behavior of the animals under study.

Institutional official means the individual at a research facility who is authorized to legally commit (on behalf of the research facility) that the requirements of 9 CFR 1, 2, and 3 will be met.

Major operative procedure means any surgical intervention that penetrates and exposes a body cavity or any procedure that produces permanent impairment of physical or physiological function.

Painful procedure as applied to any animal means any procedure that would reasonably be expected to cause more than slight or momentary pain or distress in a human being to which that procedure was applied; that is, pain in excess of that caused by injections or other minor procedures.

Principal investigator means an employee of a research facility, or other person associated with a research facility, responsible for a proposal to conduct research and for the design and implementation of research involving animals.

Research facility means any school (except an elementary or secondary school), institution, organization, or person that uses or intends to use live animals in research, tests, or experiments, and that (1) purchases or transports live animals in commerce, or (2) receives funds under a grant, award, loan, or contract from a department, agency, or instrumentality of the U.S. for the purpose of carrying out research, tests, or experiments, *provided* that the Administrator may exempt, by regulation, any such school, institution, organization, or person that does not use or intend to use live dogs or cats, except those schools, institutions, organizations, or persons which use substantial numbers (as determined by the Administrator) of live animals the principal function of which schools, institutions, organizations, or persons, is biomedical research or testing, when in the judgment of the Administrator, any such exemption does not vitiate the purpose of the Act.

§2.31 Institutional Animal Care and Use Committee (IACUC)

(A) The Chief Executive Officer of the research facility shall appoint an Institutional Animal Care and Use Committee (IACUC) qualified through the experience and expertise of its members to assess the research facility's animal program, facilities, and procedures. Except as specifically authorized by law or these regulations, nothing in this part shall be deemed to permit the Committee or IACUC to prescribe methods or set standards for the design, performance, or conduct of actual research or experimentation by a research facility.

(B) IACUC membership.

1. The members of each Committee shall be appointed by the Chief Executive Officer of the research facility.
2. The Committee shall be composed of a Chairman and at least two additional members.
3. Of the members of the Committee:
    i. At least one shall be a Doctor of Veterinary Medicine with training or experience in laboratory animal science and medicine who has direct or delegated responsibility for activities involving animals at the research facility
    ii. At least one shall not be affiliated in any way other than as a member of the Committee, and shall not be a member of the immediate family of a person who is affiliated with the facility. The Secretary intends that such person will provide representation for general community interests in the proper care and treatment of animals
4. If the Committee consists of more than three members, not more than three members shall be from the same administrative unit of the facility.

(C) IACUC functions. With respect to activities involving animals, the IACUC as an agent of the research facility shall:

1. Review, at least once every 6 months, the research facility's program for humane care and use of animals, using Title 9, Chapter I, Subchapter A-Animal Welfare, as a basis for evaluation.
2. Inspect, at least once every 6 months, all of the research facility's animal facilities, including animal study areas, using Title 9, Chapter I, Subchapter A-Animal Welfare, as a basis for evaluation; *provided, however*, that animal areas containing free-living wild animals in their natural habitat need not be included in such inspection.
3. Prepare reports of its evaluations conducted as required by paragraphs (C) (1) and (2) of this section, and submit the reports to the Institutional Official of the research facility; *provided, however,* that the IACUC may determine the best means of conducting evaluations of the research facility's programs and facilities; and *provided, further,* that no Committee member wishing to participate in any evaluation conducted under this subpart may be excluded. The IACUC may use subcommitees composed of at least two Committee members and may invite ad hoc consultants to assist in conducting the evaluations; however, the IACUC remains responsible for the evaluations and reports as required by the Act and regula-

tions. The reports shall be reviewed and signed by a majority of the IACUC members and must include any minority views. The reports shall be updated at least once every 6 months upon completion of the required semiannual evaluations and shall be maintained by the research facility and made available to APHIS and to official of funding federal agencies for inspection and copying upon request. The reports must contain a description of the nature and extent of the research facility's adherence to this Subchapter, must identify specifically any departures from the provisions of Title 9, Chapter I, Subchapter A-Animal Welfare, and must state the reasons for each departure. The reports must distinguinsh significant deficiencies from minor deficiencies. A significant deficiency is one which, with reference to Subchapter A and, in the judgment of the IACUC and the Institutional Official, is or may be a threat to the health or safety of the animals. If program or facility deficiencies are noted, the reports must contain a reasonable and specific plan and schedule with dates for correcting each deficiency. Any failure to adhere to the plan and schedule that results in a significant deficiency remaining uncorrected shall be reported in writing within 15 business days by the IACUC, through the Institutional Official, to APHIS and any federal agency funding that activity.

4. Review and, if warranted, investigate concerns involving the care and use of animals at the research facility resulting from public complaints received and from reports of noncompliance received from laboratory or research facility personnel or employees.
5. Make recommendations to the Institutional Official regarding any aspect of the research facility's animal program, facilities, or personnel training.
6. Review and approve, require modifications in (to secure approval), or withhold approval of those components of proposed activities related to the care and use of animals, as specified in paragraph (D) of this section.
7. Review and approve, require modifications in (to secure approval) or withhold approval of proposed significant changes regarding the care and use of animals in ongoing acitivities.
8. Be authorized to suspend an activity involving animals in accordance with the specifications set forth in paragraph (D)(6) of this section.

(D) IACUC review of activities involving animals.

1. In order to approve proposed activities or proposed significant changes in ongoing activities, the IACUC shall conduct a review

of those components of the activities related to the care and use of animals and determine that the proposed activities are in accordance with this subchapter unless acceptable justification for a departure is presented in writing; *provided, however*, that field studies as defined in part 1 of this subchapter are exempt from this requirement. Further, the IACUC shall determined that the proposed activities or significant changes in ongoing activities meet the following requirements:

i. Procedures involving animals will avoid or minimize discomfort, distress, and pain to the animals

ii. The principal investigator has considered alternatives to procedures that may cause more than momentary or slight pain or distress to the animals and has provided a written narrative description of the methods and sources, e.g., the Animal Welfare Information Center, used to determine that alternatives were not available

iii. The principal investigator has provided written assurance that the activities do not unnecessarily duplicate previous experiments

iv. Procedures that may cause more than momentary or slight pain or distress to the animals will:

    a. Be performed with appropriate sedatives, analgesics, or anesthetics, unless withholding such agents is justified for scientific reasons, in writing, by the principal investigator and will continue for only the necessary period of time

    b. Involve, in their planning, consultation with the attending veterinarian or his or her designee

    c. Not include the use of paralytics without anesthesia

v. Animals that would otherwise experience severe or chronic pain or distress that cannot be relieved will be painlessly euthanized at the end of the procedure or, if appropriate, during the procedure

vi. The animals' living conditions will be appropriate for their species in accordance with part 3 of this subchapter, and contribute to their health and comfort. The housing, feeding, and nonmedical care of the animals will be directed by the attending veterinarian or other scientist trained and experienced in the proper care, handling, and use of the species being maintained or studied.

vii. Medical care for animals will be available and provided as necessary by a qualified veterinarian

viii. Personnel conducting procedures on the species being maintained or studied will be appropriately qualified and trained in those procedures

ix. Activities that involve surgery include appropriate provision for preoperative and postoperative care of the animals in accordance with established veterinary medical and nursing practices. All survival surgery will be performed using aseptic procedures, including surgical gloves, masks, sterile instruments, and aseptic techniques. Major operative procedures on nonrodents will be conducted only in facilities intended for that purpose which shall be operated and maintained under aseptic conditions. Nonmajor operative procedures and all surgery on rodents do not require a dedicated facility, but must be performed using aseptic procedures. Operative procedures conducted at field sites need not be performed in dedicated facilities, but must be performed using aseptic procedures.

   x. No animal will be used in more than one major operative procedure from which it is allowed to recover, unless:

     a. Justified for scientific reasons by the principal investigator in writing.

     b. Required as routine veterinary procedure or to protect the health or well-being of the animal as determined by the attending veterinarian.

     c. Or, in other special circumstances as determined by the Administrator on an individual basis. Written requests and supporting data should be sent to the Administrator, APHIS, USDA, 6505 Belcrest Road, Room 268, Hyattsville, MD 20782.

   xi. Methods of euthanasia used must be in accordance with the definition of the term set forth in 9 CFR Part 1, §1.1 of this Subchapter, unless a deviation is justified for scientific reasons, in writing, by the investigator

2. Prior to IACUC review, each member of the Committee shall be provided with a list of proposed activities to be reviewed. Written descriptions of all proposed activities that involve the care and use of animals shall be available to all IACUC members, and any member of the IACUC may obtain, upon request, full Committee review of those activities. If full Committee review is not requested, at least one member of the IACUC, designated by the chairman and qualified to conduct the review, shall review those activities and shall have the authority to approve, require modifications in (to secure approval), or request full Committee review of any of those activities. If full Committee review is requested for a proposed activity, approval of that activity may be granted only after review at a convened meeting of a quorum of the IACUC, and with the approval vote of a majority of the quorum present. No member may participate in the IACUC review or approval of an activity in which that member has a conflicting interest (e.g., personally involved in the activity), except to provide information requested by the IACUC, nor may a member who has a conflicting interest contribute to the constitution of a quorum.

3. The IACUC may invite consultants to assist in the review of complex issues arising out of its review of proposed activities. Consultants may not approve or withhold approval of an activity and may not vote with the IACUC unless they also are members of the IACUC.
4. The IACUC shall notify principal investigators and the research facility in writing of its decision to approve or withhold approval of those activities related to the care and use of animals or of modifications required to secure IACUC approval. If the IACUC decides to withhold approval of an activity, it shall include in its written notification a statement of the reasons for its decision and give the principal investigator an opportunity to respond in person or in writing. The IACUC may reconsider its decision, with documentation in Committee minutes, in light of the information provided by the principal investigator.
5. The IACUC shall conduct continuing reviews of activities covered by this subchapter at appropriate intervals as determined by the IACUC, but not less than annually.
6. The IACUC may suspend an activity that it previously approved if it determines that the activity is not being conducted in accordance with the description of that activity provided by the principal investigator and approved by the Committee. The IACUC may suspend an activity only after review of the matter at a convened meeting of a quorum of the IACUC and with the suspension vote of a majority of the quorum present.
7. If the IACUC suspends an activity involving animals, the Institutional Official, in consultation with the IACUC, shall review the reasons for suspension, take appropriate corrective action, and report that action with a full explanation to APHIS and any Federal agency funding that activity.
8. Proposed activities and proposed significant changes in ongoing activities that have been approved by the IACUC may be subject to further appropriate review and approval by officials of the research facility. However, those officials may not approve an activity involving the care and use of animals if it has not been approved by the IACUC.

(E) A proposal to conduct an activity involving animals or to make a significant change in an ongoing activity involving animals, must contain the following:

1. Identification of the species and the approximate number of animals to be used.
2. A rationale for involving animals and for the appropriateness of the species and numbers of animals to be used.

3. A complete description of the proposed use of the animals.
4. A description of procedures designed to assure that discomfort and pain to animals will be limited to that which is unavoidable for the conduct of scientifically valuable research, including provision for the use of analgesic, anesthetic, and tranquilizing drugs where indicated and appropriate to minimize discomfort and pain to animals.
5. A description of any euthanasia method to be used.

§2.32 Personnel qualifications

(A) It shall be the responsibility of the research facility to ensure that all scientists, research technicians, animal technicians, and other personnel involved in animal care, treatment, and use are qualified to perform their duties. This responsibility shall be fulfilled in part through the provision of training and instruction to those personnel.

(B) Training and instruction shall be made available and the qualifications of personnel reviewed with sufficient frequency to fulfill the research facility's responsibilities under this section and §2.31.

(C) Training and instruction of personnel must include guidance in at least the following areas:

1. Humane methods of animal maintenance and experimentation, including:
    i. The basic needs of each species of animal
    ii. Proper handling and care for the various species of animals used by the facility
    iii. Proper preprocedural and postprocedural care of animals
    iv. Aseptic surgical methods and procedures;
2. The concept, availability, and use of research or testing methods that limit the use of animals or minimize animal distress.
3. Proper use of anesthetics, analgesics, and tranquilizers for any species of animals used by the facility.
4. Methods whereby deficiencies in animal care and treatment are reported, including deficiencies in animal care and treatment reported by any employee of the facility. No facility employee, Committee member, or laboratory personnel shall be discriminated against or be subject to any reprisal for reporting violations of any regulation or standards under the Act.
5. Utilization of services (e.g., National Agricultural Library, National Library of Medicine) available to provide information:

i. On appropriate methods of animal care and use
ii. On alternatives to the use of live animals in research
iii. That could prevent unintended and unnecessary duplication of research involving animals
iv. Regarding the intent and requirements of the Act

§ 2.35 Recordkeeping requirements

(A) The research facility shall maintain the following IACUC records:

1. Minutes of IACUC meetings, including records of attendance, activities of the Committee, and Committee deliberations.
2. Records of proposed activities involving animals and proposed significant changes in activities involving animals, and whether IACUC approval was given or withheld.
3. Records of seminannual IACUC reports and recommendations (including minority views), prepared in accordance with the requirements of §2.31(c)(3) of this subpart, and forwarded to the Institutional Official.

(F) all records and reports shall be maintained for at least 3 years. Records that relate directly to proposed activities and proposed significant changes in ongoing activities reviewed and approved by the IACUC shall be maintained for the duration of the activity and for an additional 3 years after completion of the activity. All records shall be available for inspection and copying by authorized APHIS or funding federal agency representatives at reasonable times. APHIS inspectors will maintain the confidentiality of the information and will not remove the materials from the research facilities' premises unless there has been an alleged violation, they are needed to investigate a possible violation, or for other enforcement purposes. Release of such materials, including reports, summaries, and photographs that contain trade secrets or commercial or financial information that is privileged or confidential will be governed by applicable sections of the Freedom of Information Act. Whenever the Administrator notifies a research facility in writing that specified records shall be retained pending completion of an investigation or proceeding under the Act, the research facility shall hold those records until their disposition is authorized in writing by the Administrator.

Appendix C

Health Research Extension Act of 1985*

Sec. 495.

(A) The Secretary, acting through the Director of NIH, shall establish guidelines for the following:

1. The proper care of animals to be used in biomedical and behavioral research.
2. The proper treatment of animals while being used in such research. Guidelines under this paragraph shall require: (Such guidelines shall not be construed to prescribe methods of research.)
    a. The appropriate use of tranquilizers, analgesics, anesthetics, paralytics, and euthanasia for animals in such research
    b. Appropriate presurgical and postsurgical veterinary medical and nursing care for animals in such research
3. The organization and operation of animal care committees in accordance with subsection (B).

(B)

1. Guidelines of the Secretary under Subsection (A)(3) shall require animal care committees at each entity which conducts biomedical and behavioral research with funds provided under this Act (including the National Institutes of Health and the national research institutes) to assure compliance with the guidelines established under Subsection (A).
2. Each animal care committee shall be appointed by the chief executive officer of the entity for which the committee is established, shall be composed of not fewer than three members, and shall

---

* P.L. 99-158, November 20, 1985, "Animals in Research."

include at least one individual who has no association with such entity, and at least one doctor of veterinary medicine.
3. Each animal care committee of a research entity shall:
   a. Review the care and treatment of all animal study areas and facilities of the research entity at least semiannually to evaluate compliance with applicable guidelines established under Subsection (A) for appropriate animal care and treatment
   b. Keep appropriate records of reviews conducted under subparagraph (a)
   c. For each review conducted under subparagraph (a), file with the Director of NIH at least annually (1) a certification that the review has been conducted, and (2) reports of any violations of guidelines established under Subsection (A) or assurances required under paragraph (1) which were observed in such review and which have continued after notice by the committee to the research entity involved of the violations

   Reports filed under subparagraph (c) shall include any minority views filed by members of the committee.

(C) The Director of NIH shall require each applicant for a grant, contract, or cooperative agreement involving research on animals which is administered by the National Institutes of Health or any national research institute to include in its application or contract proposal, submitted after the expiration of the 12-month period beginning on the date of enactment of this section:

1. Assurances satisfactory to the Director of NIH that:
   a. The applicant meets the requirements of the guidelines established under paragraphs (1) and (2) of subsection (A) and has an animal care committee which meets the requirements of subsection (B)
   b. Scientists, animal technicians, and other personnel involved with animal care, treatment, and use by the applicant have available to them instruction or training in the humane practice of animal maintenance and experimentation, and the concept, availability, and use of research or testing methods that limit the use of animals or limit animal distress
2. A statement of the reasons for the use of animals in the research to be conducted with funds provided under such grant or contract. Notwithstanding Subsection (A)(2) of Section 553 of Title 5, U.S. Code, regulations under this subsection shall be promulgated in accordance with the notice and comment requirements of such section.

(D) If the director of NIH determines that:

1. The conditions of animal care, treatment, or use in an entity which is receiving a grant, contract, or cooperative agreement involving research on animals under this title do not meet applicable guidelines established under Subsection (A).
2. The entity has been notified by the Director of NIH of such determination and has been given a reasonable opportunity to take corrective action.
3. No action has been taken by the entity to correct such conditions; the Director of NIH shall suspend or revoke such grant or contract under such conditions as the Director determines appropriate.

(E) No guideline or regulation promulgated under Subsection (A) or (C) may require a research entity to disclose publicly trade secrets or commercial or financial information which is privileged or confidential.

Appendix D

Public Health Service Policy on Humane Care and Use of Laboratory Animals

I. Introduction

It is the Policy of the Public Health Service (PHS) to require institutions to establish and maintain proper measures to ensure the appropriate care and use of all animals involved in research, research training and biological testing activities (hereafter referred to as activities) conducted or supported by the PHS. The PHS endorses the "U.S. Government Principles for the Utilization and Care of Vertebrate Animals Use in Testing, Research, and Training" developed by the Interagency Research Animal Committee (IRAC). This Policy is intended to implement and supplement those principles.

II. Applicability and Effective Dates

This Policy is applicable to all PHS-conducted or supported activities involving animals, whether the activities are performed at a PHS agency, an awardee institution, or any other institution and conducted in the U.S., the Commonwealth of Puerto Rico, or any territory or possession of the U.S. The requirements of this Policy are effective for applications and proposals for PHS research and research training awards involving animals that are submitted for PHS consideration on or after November 1, 1986, and for all PHS-conducted or supported research and research training activities involving animals that are being conducted on or after July 1, 1987. Institutions in foreign countries receiving PHS support for activities involving animals shall comply with this Policy, or provide evidence to the PHS that acceptable standards for the humane care and use of the animals in PHS-conducted or supported activities will be met. No PHS support for an activity involving animals will be provided to an individual unless that individual is affiliated with or sponsored by an institution which can and does assume responsibility for compliance with this Policy, unless the individual makes other arrangements with the PHS. This Policy does not affect applicable state or local laws or regulations which impose more stringent standards for the care and use of laboratory animals. All institutions are required to comply, as applicable, with the Animal Welfare Act, and other federal statutes and regulations relating to animals.

III. Definitions

Animal — Any live, vertebrate animal used or intended for use in research, research training, experimentation, or biological testing or for related purposes.

Animal Facility — Any and all buildings, rooms, areas, enclosures, or vehicles, including satellite facilities, used for animal confinement, transport, maintenance, breeding, or experiments inclusive of surgical manipulation. A satellite facility is any containment outside of a core facility or centrally designated or managed area in which animals are housed for more than 24 hours.

Institution — Any public or private organization, business, or agency (including components of federal, state, and local governments).

Institutional Official — An individual who signs and has the authority to sign the institution's Assurance, making a commitment on behalf of the institution that the requirements of this Policy will be met.

IV. Implementation by Institutions

A. Animal Welfare Assurance

No activity involving animals may be conducted or supported by the PHS until the institution conducting the activity has provided a written Assurance acceptable to the PHS, setting forth compliance with this Policy. Assurances shall be submitted to the Office for Protection from Research Risks (OPRR), Office of the Director, National Institutes of Health, 9000 Rockville Pike, Building 31, Room 4B09, Bethesda, MD 20892. The Assurance shall be typed on the institution's letterhead and signed by the Institutional Official. OPRR will provide the institution with necessary instructions and an example of an acceptable Assurance. All Assurances submitted to the PHS in accordance with this Policy will be evaluated by OPRR to determine the adequacy of the institution's proposed program for the care and use of animals in PHS-conducted or supported activities. On the basis of this evaluation OPRR may approve or disapprove the Assurance, or negotiate an approvable Assurance with the institution. Approval of an Assurance will be for a specified period of time (no longer than 5 years) after which time the institution must submit a new Assurance to OPRR. OPRR may limit the period during which any particular approved Assurance shall remain effective or otherwise condition, restrict, or withdraw approval. Without an applicable PHS-approved or provisionally approved Assurance, no PHS-conducted or supported activity involving animals at the institution will be permitted to continue.

1. Institutional Program for Animal Care and Use: The Assurance shall fully describe the institution's program for the care and use of animals in PHS-conducted or supported activities. The PHS requires institutions to use the *Guide for the Care and Use of Laboratory Animals (Guide)* as a basis for developing and implementing an institutional program for activities involving animals. The program description must include the following:

    a. A list of every branch and major component of the institution, as well as a list of every branch and major component of any other institution, which is to be included under the Assurance

    b. The lines of authority and responsibility for administering the program and ensuring compliance with this Policy

    c. The qualifications, authority, and responsibility of the veterinarian(s) who will participate in the program and the percent of time each will contribute to the program

    d. The membership list of the Institutional Animal Care and Use Committee(s)*/(IACUC) established in accordance with the requirements set forth in IV.A.3 of this Policy

    e. The procedures which the IACUC will follow to fulfill the requirements set forth in this Policy

    f. The health program for personnel who work in laboratory animal facilities or have frequent contact with animals

    g. A synopsis of training or instruction in the humane practice of animal care and use, as well as training or instruction in research or testing methods that minimize the number of animals required to obtain valid results and minimize animal distress, offered to scientists, animal technicians, and other personnel involved in animal care, treatment, or use

    h. The gross square footage of each animal facility (including satellite facilities), the species housed therein, and the average daily inventory, by species, of animals in each facility

    i. Any other pertinent information requested by OPRR.

---

* The name Institutional Animal Care and Use Committee (IACUC) as used in this Policy is intended as a generic term for a committee whose function is to ensure that the care and use of animals in PHS-conducted or supported activities is appropriate and humane in accordance with this Policy. However, each institution may identify the committee by whatever name it chooses.

2. **Institutional Status:** Each institution must assure that its program and facilities are in one of the following categories:

   Category 1 — Accredited by the American Association for Accreditation of Laboratory Animal Care (AAALAC). All of the institution's programs and facilities (including satellite facilities) for activities involving animals have been evaluated and accredited by AAALAC or another accrediting body recognized by PHS.* All of the institution's programs and facilities (including satellite facilities) for activities involving animals also have been evaluated by the IACUC and will be reevaluated by the IACUC at least once every 6 months, in accordance with IV.B.1 and 2 of this Policy, and reports prepared in accordance with IV.B.3 of this Policy.

   Category 2 — Evaluated by the institution. All of the institution's programs and facilities (including satellite facilities) for activities involving animals have been evaluated by the IACUC and will be reevaluated by the IACUC at least once every 6 months, in accordance with IV.B.1 and 2 of this Policy. The initial report of the IACUC evaluation shall be submitted to OPRR with the Assurance.

3. **Institutional Animal Care and Use Committee (IACUC):**

   a. The Chief Executive Officer shall appoint an Institutional Animal Care and Use Committee (IACUC) qualified through the experience and expertise of its members to oversee the institution's animal program, facilities, and procedures.

   b. The Assurance must include the names, position titles, and credentials of the IACUC Chairperson and the members. The committee shall consist of not less than five members and shall include at least:

   - One Doctor of Veterinary Medicine with training or experience in laboratory animal science and medicine, who has direct or delegated program responsibility for activities involving animals at the institution

   - One practicing scientist experienced in research involving animals

   - One member whose primary concerns are in a nonscientific area (e.g., ethicist, lawyer, member of the clergy)

   - One individual who is not affiliated with the institution in any way other than as a member of the IACUC and is not a member of the immediate family of a person who is affiliated with the institution

---

* As of the issuance date of this Policy, the only accrediting body recognized by PHS is the American Association for Accreditation of Laboratory Animal Care (AAALAC).

c. An individual who meets the requirements of more than one of the categories detailed in IV.A.3.b of this Policy may fulfill more than one requirement; however, no committee may consist of less than five members

B. Functions of the Institutional Animal Care and Use Committee (IACUC)

As an agent of the institution, the IACUC shall with respect to PHS-conducted or supported activities:

1. Review at least once every 6 months the institution's program for humane care and use of animals, using the *Guide* as a basis for evaluation.
2. Inspect at least once every 6 months all of the institution's animal facilities (including satellite facilities) using the *Guide* as a basis for evaluation.
3. Prepare reports of the IACUC evaluations conducted as required by IV.B.1 and 2 of this Policy and submit the reports to the Institutional Official.* (Note: the reports shall be updated at least once every 6 months upon completion of the required semiannual evaluations and shall be maintained by the institution and be made available to OPRR upon request. The reports must contain a description of the nature and extent of the institution's adherence to the *Guide* and this Policy and must identify specifically any departures from the provisions of the *Guide* and this Policy and must state the reasons for each departure. The reports must distinguish significant deficiencies from minor deficiencies. A significant deficiency is one which, consistent with this Policy, and, in the judgment of the IACUC and the Institutional Official, is or may be a threat to the health or safety of the animals. If program or facility deficiencies are noted, the reports must contain a reasonable and specific plan and schedule for correcting each deficiency. If some or all of the institution's facilities are accredited by AAALAC or another accrediting body recognized by PHS, the report should identify those facilities as such.)
4. Review concerns involving the care and use of animals at the institution.
5. Make recommendations to the Institutional Official regarding any aspect of the institution's animal program, facilities, or personnel training.

---

* The Institutional Animal Care and Use Committee (IACUC) may, at its discretion, determine the best means of conducting an evaluation of the institution's programs and facilities. The IACUC may invite ad hoc consultants to assist in conducting the evaluation. However, the IACUC remains responsible for the evaluation and report.

6. Review and approve, require modifications in (to secure approval), or withhold approval of those components of PHS-conducted or supported activities related to the care and use of animals as specified in IV.C of this Policy.
7. Review and approve, require modifications in (to secure approval), or withhold approval of proposed significant changes regarding the use of animals in ongoing activities.
8. Be authorized to suspend an activity involving animals in accordance with the specifications set forth in IV.C.6 of this Policy.

C. Review of PHS-Conducted or Supported Research Projects

1. In order to approve proposed research projects or proposed significant changes in ongoing research projects, the IACUC shall conduct a review of those components related to the care and use of animals and determine that the proposed research projects are in accordance with this Policy. In making this determination, the IACUC shall confirm that the research project will be conducted in accordance with the Animal Welfare Act insofar as it applies to the research project, and that the research project is consistent with the *Guide* unless acceptable justification for a departure is presented. Further, the IACUC shall determine that the research project conforms with the institution's Assurance and meets the following requirements:

    a. Procedures with animals will avoid or minimze discomfort, distress, and pain to the animal, consistent with sound research design

    b. Procedures that may cause more than momentary or slight pain or distress to the animals will be performed with appropriate sedation, analgesia, or anesthesia, unless the procedure is justified for scientific reasons in writing by the investigator

    c. Animals that would otherwise experience severe or chronic pain or distress that cannot be relieved will be painlessly sacrificed at the end of the procedure or, if appropriate, during the procedure

    d. The living conditions of animals will be appropriate for their species and contribute to their health and comfort; the housing, feeding, and nonmedical care of the animals will be directed by a veterinarian or other scientist trained and experienced in the proper care, handling, and use of the species being maintained or studied

    e. Medical care for animals will be available and provided as necessary by a qualified veterinarian

    f. Personnel conducting procedures on the species being maintained or studied will be appropriately qualified and trained in those procedures

g. Methods of euthanasia used will be consistent with the recommendations of the American Veterinary Medical Association (AVMA) Panel on Euthanasia,* unless a deviation is justified for scientific reasons in writing by the investigator.

2. Prior to the review, each IACUC member shall be provided with a list of proposed research projects to be reviewed. Written descriptions of research projects that involve the care and use of animals shall be available to all IACUC members, and any member of the IACUC may obtain, upon request, full committee review of those research projects. If full committee review is not requested, at least one member of the IACUC, designated by the chairperson and qualified to conduct the review, shall review those research projects and have the authority to approve, require modifications in (to secure approval), or request full committee review of those research projects. If full committee review is requested, approval of those research projects may be granted only after review at a convened meeting of a quorum of the IACUC and with the approval vote of a majority of the quorum present. No member may participate in the IACUC review or approval of a research project in which the member has a conflicting interest (e.g., personally involved in the project) except to provide information requested by the IACUC, nor may a member who has a conflicting interest contribute to the constitution of a quorum.

3. The IACUC may invite consultants to assist in the review of complex issues. Consultants may not approve or withhold approval of an activity or vote with the IACUC unless they also are members of the IACUC.

4. The IACUC shall notify investigators and the institution in writing of its decision to approve or withhold approval of those activities related to the care and use of animals, or of modifications required to secure IACUC approval. If the IACUC decides to withhold approval of an activity, it shall include in its written notification a statement of the reasons for its decision and give the investigator an opportunity to respond in person or in writing.

5. The IACUC shall conduct continuing review of activities covered by this Policy at appropriate intervals as determined by the IACUC, but not less than once every 3 years.

6. The IACUC may suspend an activity that it previously approved if it determines that the activity is not being conducted in accordance with applicable provisions of the Animal Welfare Act, the *Guide*, the institution's Assurance, or IV.C.1.a through g of this Policy. The IACUC may suspend an activity only after review of

---

* *Journal of the American Veterinary Medical Association (JAVMA)*, 1986, Vol. 188, No. 3, pp. 252–268, or succeeding revised edition.

the matter at a convened meeting of a quorum of the IACUC and with the suspension vote of a majority of the quorum present.

7. If the IACUC suspends an activity involving animals, the Institutional Official in consultation with the IACUC shall review the reasons for suspension, take appropriate corrective action, and report that action with a full explanation to OPRR.

8. Applications and proposals that have been approved by the IACUC may be subject to further appropriate review and approval by officials of the institution. However, those officials may not approve an activity involving the care and use of animals if it has not been approved by the IACUC.

D. Recordkeeping Requirements

1. The awardee institution shall maintain:
   a. An Assurance which has been either approved or deemed provisionally acceptable by the PHS
   b. Minutes of IACUC meetings, including records of attendance, activities of the committee, and committee deliberations
   c. Records of applications, proposals, and proposed significant changes in the care and use of animals and whether IACUC approval was given or withheld
   d. Records of semiannual IACUC reports and recommendations (including minority views) as forwarded to the Institutional Official
   e. Records of accrediting body determinations

2. All records shall be maintained for at least 3 years; records that relate directly to applications, proposals, and proposed significant changes in ongoing activities reviewed and approved by the IACUC shall be maintained for the duration of the activity and for an additional 3 years after completion of the activity. All records shall be accessible for inspection and copying by authorized OPRR or other PHS representatives at reasonable times and in a reasonable manner.

E. Reporting Requirements

1. At least once every 12 months, the IACUC, through the Institutional Official, shall report in writing to OPRR:
   a. Any change in the institution's program of facilities which would place the institution in a different category than specified in its Assurance (see IV.A.2 of this Policy)
   b. Any change in the description of the institution's program for animal care and use as required by IV.A.1.a. though i of this Policy

c. Any changes in the IACUC membership
   d. Notice of the dates that the IACUC conducted its semiannual evaluations of the institution's program and facilities and submitted the evaluations to the Institutional Official
2. At least once every 12 months, the IACUC at an institution which has no changes to report as specified in IV.E.1.a though c of this Policy shall submit a letter, through the Institutional Official, to OPRR stating that there are no changes and informing OPRR of the dates of the required IACUC evaluations and submissions to the Institutional Official.
3. The IACUC, through the Institutional Official, shall promptly provide OPRR with a full explanation of the circumstances and actions taken with respect to:
   a. Any serious or continuing noncompliance with this Policy
   b. Any serious deviations from the provisions of the *Guide*
   c. Any suspension of an activity by the IACUC
4. Reports filed under IV.E of this Policy shall include any minority views filed by members of the IACUC.

Appendix E

U.S. Government Principles for the Utilization and Care of Vertebrate Animals Used in Testing, Research, and Training

Introduction

The development of knowledge necessary for the improvement of the health and well-being of humans as well as other animals requires *in vivo* experimentation with a wide variety of animal species. Whenever U.S. Government agencies develop requirements for testing, research, or training procedures involving the use of vertebrate animals, the following principles shall be considered, and whenever these agencies actually perform or sponsor such procedures, the responsible Institutional Official shall ensure that these principles are adhered to.

   I. The transportation, care, and use of animals should be in accordance with the Animal Welfare Act (7 U.S.C. 2131 et. seq.) and other applicable federal laws, guidelines, and policies.*

  II. Procedures involving animals should be designed and performed with due consideration of their relevance to human or animal health, the advancement of knowledge, or the good of society.

 III. The animals selected for a procedure should be of an appropriate species and quality and the minimum number required to obtain valid results. Methods such as mathematical models, computer simulation, and *in vitro* biological systems should be considered.

---

* For guidance throughout these Principles, the reader is referred to the *Guide for the Care and Use of Laboratory Animals* prepared by the Institute of Laboratory Animal Resources, National Academy of Sciences.

IV. Proper use of animals, including the avoidance of minimization of discomfort, distress, and pain when consistent with sound scientific practices, is imperative. Unless the contrary is established, investigators should consider that procedures that cause pain or distress in human beings may cause pain or distress in other animals.

V. Procedures with animals that may cause more than momentary pain or slight pain or distress should be performed with appropriate sedation, analgesia, or anesthesia. Surgical or other painful procedures should not be performed on unanesthetized animals paralyzed by chemical agents.

VI. Animals that would otherwise suffer severe or chronic pain or distress that cannot be relieved should be painlessly killed at the end of the procedure or, if appropriate, during the procedure.

VII. The living conditions of animals should be appropriate for their species and contribute to their health and comfort. Normally, the housing, feeding, and care of all animals used for biomedical purposes must be directed by a veterinarian or other scientist trained and experienced in the proper care, handling, and use of the species being maintained. In any case, veterinary care shall be provided as indicated.

VIII. Investigators and other personnel shall be appropriately qualified and experienced for conducting procedures on living animals. Adequate arrangements shall be made for their in-service training, including the proper and humane care and use of laboratory animals.

IX. Where exceptions are required in relation to the provisions of these Principles, the decisions should not rest with the investigators directly concerned but should be made, with due regard to Principle II, by an appropriate review group such as an institutional animal care and use committee. Such exceptions should not be made solely for the purposes of teaching or demonstration.

Index

(Topic, Chapter:Question Number)

A

Academic freedom, 22:15
Adoption, 14:30
Agricultural research, 8:5, 14:3, 15:12, 15:13
Alternatives
  ascites/monoclonal antibody production, 19:16
  consideration by principal investigator, 12:17
  definition, 12:15
  regulations, 12:16
  searching for, 12:18, 12:19
  to painful/distressful procedures, 16:5
Altweb, 12:18
Amendment
  procedure for, 10:4, 10:5, 10:6
  purpose, 10:1, 10:3
  regulatory requirements, 10:2
American Cancer Society, 8:6
American College of Laboratory Animal Medicine (ACLAM), 27:2
American Heart Association, 8:6
American Lung Association, 8:6
Amphibians, 14:5, 16:29, 17:17
Analgesia, 8:3, 16:4, 16:9, 16:20, 27:1
Anesthesia, 8:3, 16:4, 16:10, 16:11, 16:12, 16:20, 20:34, 27:1
  local, 16:18
  records, 16:19
  use in field studies, 16:21
Animal, definition, 12:1
Animal facility, *see* Housing facility
Animal fluids, 8:9, 8:10, 27:15
Animal numbers
  amendment, 10:7
  justification, 13:1–13:5, 13:10, 13:11
  reduction, 13:12–13:14
  tracking, 14:4, 14:5, 14:11–14:14, 14:19, 14:20, 16:24
Animal procurement, 14:1, 14:16, 27:1
  regulatory aspects of, 14:2, 14:3, 14:16
  responsibility for, 14:7–14:9
Animal shelters, 2:13
Animal tissues, 8:9, 8:10, 14:2, 14:21–14:24
  testing of cell lines, 20:22, 27:15
Animal use
  justification, 12:2, 12:5–12:8
  requirement by funding agency for, 12:8
  requirement by regulatory agency for, 12:10, 12:11
Animals, as a food source for other animals, 12:14
Antibody
  ascites, 19:16–19:18
  asepsis, 19:12
  commercial sources, 19:14, 19:15
  injection schedule, 19:13
  pain, 19:3
  production, 8:11, 19:14
Appeals, 9:51, 9:53, 29:30, 29:31
Aquarium, 2:10–2:12
Ascites, *see* Antibody production, Ascites
Aseptic technique, *see* Surgery
Assistance animals, 14:34
Attending veterinarian
  qualifications, 27:2
  requirement for, 27:3–27:5
  responsibilities, 27:6
  role on the IACUC, 5:21, 27:5, 27:6, 29:18, 29:22, 29:36

B

Biohazard, definition, 20:13, *also see* Occupational health, infectious hazard
Birds, 14:5, 14:6, 14:25
Blood collection
  exsanguination, 17:13, 17:14, 17:23
  postorbital, 16:30, 16:31
  volume, 19:19–19:21
Blood donors, 8:15
Breeding colonies, 8:13–8:15, 14:32

C

Cages, *see* Facility inspection, animal enclosures

533

Carbon dioxide, 17:2–17:6, 17:18, 17:25
Carcasses, see Dead animals
Cephalopods, see Invertebrates
Change in protocol, 11:10, also see Amendment
Chemical hazard, see Occupational health, chemical hazard
Chloralose, 16:11
Clinical trial, 14:28
Confidentiality, 8:21
Consultants, IACUC, 5:42–5:45, 9:4, 23:13
   scientific merit, 9:16–9:18
Continuing education, see Training, continuing education
Controlled substance, 8:3

D

Dead animals, 8:8, 12:5
Decapitation, 17:2, 17:7, 17:8, 17:10–17:12, 17:18, 17:20
Decerebration, pain sensation with, 16:28
Disposition of animals, 14:29–14:31
Distress, 16:1, 16:2, 16:4, 16:7, 16:8, 16:30
   alternatives 16:5
Drug Enforcement Agency (DEA), see Controlled substance
Duplication, need to assure nonduplication, 12:20, 12:21

E

Eggs, 8:12, 14:6, 16:25
Electric shock, 16:23
Electronic review, see Protocol
Embryos, see Fetuses
Endpoints, 12:11, 16:35, 16:36
Endangered species, 14:27
Enrichment
   definition, 28:1
   evaluation, 28:13, 28:14, 28:16
   exemption, 28:7
   goals, 28:1, 28:3, 28:5
   investigators, 28:11
   methods, 28:1–28:3, 28:6
   personnel, 28:8, 28:11, 28:12
   records, 28:9, 28:16
   requirement for, 28:1, 28:2, 28:9
   responsibility for, 28:8
   sources of information, 28:5, 28:15
   species to be included, 28:4
Ethical review of protocols, 12:3, 12:4
Euthanasia, 8:3, 20:39
   carbon dioxide, see Carbon dioxide
   cervical dislocation, 17:2, 17:7, 17:9–17:12, 17:20, 17:25
   decapitation, see Decapitation
   fetuses, 17:19, 17:20, 17:24
   guidelines, 17:1, 17:2
   neonates, 17:5, 17:18, 17:20, 17:21
   perfusion, 17:22
   pithing, 17:17
   stunning, 17:15
   thoracic compression, 17:16
   training, 17:11
   verification of death, 17:26
Expired pharmaceuticals, see Surgery
Exsanguination, see Blood collection

F

Facility inspection
   animal enclosures, 24:16
   animal health, 24:15, 24:20, 24:22, 26:8
   deficiencies, 23:19–23:22, 24:12, 24:13, 25:13
   environmental parameters, 24:17, 24:22
   euthanasia and carcass disposal, 24:21
   feed and bedding, 24:20, 24:22
   frequency, 23:4–23:8, 24:6, 25:3, 25:12, 26:12
   methods, 23:3, 23:11, 23:14, 23:15, 23:24, 24:3, 24:8–24:12, 24:15–24:19
   minority opinion, 23:23
   personnel issues, 24:23
   physical plant, 24:18
   purpose, 23:2, 24:2
   records and reports, 23:16, 23:17, 24:12, 24:14, 26:10, 26:11
   research laboratories, 23:9, 25:1–25:13
   review of records, 24:22
   sanitation, 24:19, 24:22, 26:8
   sites, 23:9, 24:10, 25:1, 25:2, 26:1
   team, 23:11–23:13, 24:4, 24:5, 24:8
Farms
   as a source of animals, 14:26
   standards, 15:12, 15:13, 18:14
Feed and bedding, see Facility inspection
Fetuses, 8:12, 13:11, 14:6, 14:20, 16:24–16:26, 17:19, 17:20
Field sites, 15:10, 24:10
Field study, 8:5, 12:6, 16:21, 18:12
Food deprivation, 16:22
Footpad injection, 19:11
Freedom of Information Act (FOIA), 6:21, 8:19, 22:3, 22:4
Freund's complete adjuvant, 19:1, 19:2, 19:4–19:11, 19:18, 20:23

H

Harassment of investigators, 22:16
Hazards, *see* Occupational health
Housing
    at extramural sites, 15:7, 15:8, 24:10
    facility, definition, 15:1
    facility, removal from, 15:2–15:5
    in a laboratory, 15:6
    inspection, *see* Facility inspection
    satellite facility, *see* Satellite facility
Humane societies, *see* Animal shelters
Hypothermia, 16:26, 16:27, 16:29, 16:36

I

IACUC
    activity, influence of law and authority upon, 22:1, 22:2
    breach of proprietary confidence, 22:9
    Chair, appointment of, 5:12
    Chair, qualifications, 5:13, 5:15–5:17, 5:19
    Chair, role, 5:14, 5:20, 29:5, 29:14, 29:17, 29:29, 29:35, 29:38
    compensation of members, 5:8, 5:28
    composition of, 5:2, 5:3, 5:5–5:7, 5:18
    creation of, 3:2, 3:3, 3:15
    duration of membership, 3:6
    meeting, attendance, 6:7–6:10
    meeting, electronic methods, 6:11–6:15
    meeting, frequency, 6:1, 6:2
    meeting, minutes, 6:15, 6:16, 6:18–6:20
    meeting, procedures, 6:3–6:6
    meeting, protocol review, 6:17, 6:22, *also see* Protocol review
    nonaffiliated member, 3:4, 3:5, 5:25–5:30, 6:7
    nonscientific member, 5:31–5:34, 6:7
    number of committees, 3:8–3:14
    number of members, 5:1, 5:4
    open meetings and records, 8:19, 22:10–22:14
    performance criteria, 5:9
    records, 3:16, 18:26, 23:16–23:18, 23:23, 24:14
    regulatory charges, 2:4, 3:1
    removal of members, 5:10, 5:11
    reporting lines, 4:1
    review of grants, 9:46, 9:47
    scientific member, 5:35–5:37
    specialist members
        animal care technician, 5:40
        biostatistician, 5:39
        ethicist, 5:38
    voting privileges of, 5:41
    support staff, 5:46–5:48
    Training and Learning Consortium, 3:7
    veterinary care, evaluation of, 27:9, 27:15, 27:16
    veterinary member, 5:21–5:24, 6:7
Identification of animals, 16:33
Infectious disease, 14:9
Inspection, semiannual, 3:14
Institute for Laboratory Animal Resources (ILAR), 3:7
Institutional Official, 4:1–4:7, 29:2, 29:5, 29:13, 29:28, 29:30, 29:32
Interagency Research Animal Committee, 8:4
Invertebrates, 12:13

K

Ketamine, 16:10

L

Laboratory inspection
    criteria, 25:11
    deficiencies, 25:13
    frequency, 25:3, 25:12
    necessity, 25:1, 25:2
    procedures, 25:4–25:10, 25:12
    review of records, 25:10
Lethal dose 50 test (LD_{50}), 12:11, 16:36
Livestock, for human consumption after use, 14:31

M

Mascots, 2:9
Material Safety Data Sheet (MSDS), 20:32
Minority opinions, 9:45

N

National Science Foundation (NSF), 8:4
Neuromuscular blocking agents, *see* Paralytics
Nociception 16:3, 16:29, *also see* Pain

O

Occupational health
    attending veterinarian, 20:10
    chemical hazard, 20:31–20:36, 20:38

chemical hazard committee, 20:36
Freund's complete adjuvant, 20:23
hazard oversight, 20:5–20:7
examination, 20:11
IACUC responsibility, 20:1–20:3, 20:9, 20:14, 20:16, 20:17, 20:20, 20:24, 20:27, 20:35, 20:40, 20:44, 20:46, 24:23, 26:8
 infectious hazards, 20:6–20:8, 20:13–20:18
 Institutional Biosafety Committee (IBC), 20:15–20:20, 20:24, 20:27–20:30
 nonhuman primates, 20:21, 20:24
 physical hazard, 20:37–20:40, 26:8
 personal protective equipment, 20:21
 radiation safety committee, 20:44–20:46
 radiologic hazard, 20:6, 20:41–20:46
 recapping of needles, 20:19
 recombinant DNA, 20:26–20:30
 regulatory agencies, 20:5
 respiratory hazard, 20:20
 safety officer, 20:7, 20:8
Occupational health program, 20:11, 21:11

P

Pain 16:3, 16:28–16:31, *also see* Nociception
 alternatives, 16:5, 16:9
 assessment of, 16:6, 16:8, 16:14
 fetus, 16:24, 16:25
 relief of, 16:4, 16:30, 27:1
Painful procedures, IACUC approval of, 16:7, 16:9, 16:38–16:41, 17:7, 17:8, 19:3
Paralytics, 16:4, 16:16, 16:17
Perfusion, *see* Euthanasia
Peritoneal tap, *see* Antibody production, ascites
Pet-facilitated therapy, *see* Assistance animals
Pet stores as a source of animals, 14:25
Pets as research subjects, *see* Privately owned animals, use in research
Physical hazards, *see* Occupational health
Physical plant, *see* Facility inspection
Pilot study, 13:6–13:9
Pithing, *see* Euthanasia
Poikilotherms, 16:27
Pounds, *see* Animal shelters
PREX, 12:18
Pristane, 19:18
Privately owned animals, use in research, 14:28
Procurement of animals, *see* Animal procurement
Program review
 deficiencies, 23:19–23:22

 definition, 23:1, 23:2, *also see* Facility inspection
 frequency, 23:4–23:6
 methods, 23:3, 23:10, 23:11
 minority opinion, 23:23
 records and reports, 23:16–23:18, 23:23
Proprietary information, 7:9, 22:3–22:5, 22:7–22:9, *also see* Protocol
Protocol
 content, 7:2, 7:4–7:7, 7:10, 8:3, 8:14, 12:9, 18:18, 19:13
 destruction of copies, 22:8
 distribution, 9:2
 electronic processing and review, 8:20, 8:21, 9:24
 form, 7:1, 7:3
 generic, 7:4
 identification, 7:11, 7:12
 identity of researchers, 22:6
 maintenance, 8:18, 22:8
 off-site research, 8:16, 9:48, 9:50
 protocol proprietary information 7:9, *also see* Proprietary information
 purpose of study, 7:7
 requirement for, 8:1, 8:2, 8:4–8:13, 8:15, 8:16, 14:32
 re-review, concept, 11:1, 11:2
 re-review, process, 11:7–11:9, 11:11, 11:13
 re-review, timing, 11:3–11:6
 teaching, 7:8
Protocol noncompliance
 communication of allegations, 29:4–29:6, 29:13
 definition of animals, 29:1
 documentation, 29:10, 29:12
 feedback to complainant, 29:20, 29:21
 IACUC Chair, 29:5, 29:14, 29:17, 29:29, 29:35, 29:38
 identification of complainant, 29:7
 investigation, 29:3, 29:11, 29:12, 29:15, 29:17
 investigative subcommittee, 29:17–29:19
 procedures, 29:16, 29:17, 29:20, 29:23, 29:24, 29:27–29:29, 29:34–29:40
 protection of complainant, 29:8
 reporting of allegations and sanctions, 29:43–29:51
 responsibility, 29:2, 29:3, 29:14
 sanctions, 29:25–29:30, 29:34–29:36
 sanctions, appeal of, 29:30–29:32
 suspension of project, 29:26, 29:36–29:42, 29:45, 29:48–29:51
 veterinarian, 29:5, 29:18, 29:22, 29:36, 29:37
Protocol review
 committee actions, 9:35–9:39
 committee vote, 9:43, 9:44

consistency, 3:13
ethical review, 12:3, 12:4
expedited/designated review, 9:19–9:24, 10:5, 10:6
extramural protocols, 9:48, 9:49
full committee review, 9:25–9:29, 10:5, 10:6
influences, 9:32, 9:33
internally funded research, 9:3
investigator response, 9:40–9:42, 9:51, 9:53, 9:54
length of approval, 11:3, 11:4, 11:12
prereview, 9:5–9:13
regulatory principles, 9:1
resubmission, 9:30, 9:31, 9:52
scientific merit, 9:14–9:18, 12:7

Q

Quarantine, 14:9, 27:1

R

Radiologic hazards, *see* Occupational health
Recombinant DNA, *see* Occupational health
Regulatory documents, description, 2:2, 2:3
Reptiles, 14:5, 14:25
Research laboratories, *see* Facility inspection
Restraint, chronic, 16:37

S

Safety, *see* Occupational health
Sanctions, *see* Protocol noncompliance
Sanitation, *see* Facility inspection
Satellite facility, 15:2, 24:10
Schools
    elementary, 2:6–2:8
    secondary, 2:6–2:8
Scientific merit, *see* Protocol review
Sedation, 16:4, 16:30, 17:10, 27:1
Seeing eye dogs, *see* Assistance animals
Sentinel animals, 8:15, 27:15
Separation of species, 15:14, 27:1
Slaughterhouse, samples from, 14:21, 14:22
Statistical analysis, *see* Animal numbers
Stress, 16:1, 27:1
Stunning, *see* Euthanasia
Suffering, 16:2, 27:1
Surgery
    aseptic technique, 18:8, 18:9, 18:13, 26:8, 26:20, 26:21
    aseptic technique, farm animals, 18:14
    aseptic technique, field studies, 18:12
    aseptic technique, nonsurvival surgery, 18:11
    aseptic technique, rodents, 18:10
    expired pharmaceuticals, 18:19, 26:8
    inspection, deficiency, 26:16
    inspection, frequency, 26:2
    inspection, issues, 26:8, 26:9
    inspection, review of records, 26:10–26:15
    inspection, sites, 26:1
    facilities, 26:3, 26:4, 26:6, 26:7
    laparoscopy, 18:4
    major vs. minor, 18:3, 18:4, 26:5
    nonsurvival, protocol, 18:2
    perioperative care, 18:16–18:18, 18:22–18:25, 26:8, 27:1
    personnel and training, 18:15, 21:10, 27:1
    records, 18:26, 26:10–26:15, 27:1, 27:10
    survival for subsequent slaughter, 18:7
    survival, multiple, 18:5, 26:17
    survival, performed by vendor, 18:6
    survival, protocol, 18:1
    training, 26:19
    wound closure, 18:20, 18:21, 27:1
Suspension, *see* Protocol noncompliance

T

Tail clipping, 16:32
Teaching, use of animals in, 12:21
Thoracic compression, *see* Euthanasia
Tissues, *see* Animal tissues
Toe clipping, 16:34
Tracking, animal numbers, *see* Animal numbers
Training
    animal care personnel, 21:5
    continuing education, 21:20, 21:21, 27:1
    IACUC members, 3:7, 23:24, 24:5, 26:9
    information sources, 21:8–21:11
    investigators, 7:5, 17:11, 18:15, 21:22
    methods, 4:8, 21:6, 21:7
    program, 21:4, 21:6
    requirement for, 21:1–21:3, 21:19
    students, 21:16–21:19
Training and qualifications, evaluation, 21:12–21:15, 21:23, 26:19, 27:1
Transfer of animals, 14:17, 14:18
Transportation, 15:4, 15:5
Trapping, 15:9
Treadmill, 16:23

Tumor size, 16:35

U

Urethane, 16:12
United States Fish and Wildlife Service (USFW), 8:5

V

Vendor assessment, 14:8–14:10, 27:1
Veterinary care
   availability, 27:4, 27:5, 27:8, 27:13
   definition of adequacy, 27:1
   preventive medicine, 27:1, 27:15
   records, 27:10–27:12, 27:14
   role of investigator, 27:17
   support staff, 27:5, 27:7, 27:8

W

Water deprivation, 16:22
Weight loss, 16:14, 16:15
Well-being of animals, 24:15, *also see* Enrichment

X

Xylazine, 16:10

Z

Zoos, 2:10–2:12